TELECOMMUNICATIONS
An Introduction to Electronic Media

THIRD EDITION

TELECOMMUNICATIONS
An Introduction to Electronic Media

LYNNE SCHAFER GROSS
California State University-Fullerton

wcb
Wm. C. Brown Publishers
Dubuque, Iowa

Book Team

Editor *Stan Stoga*
Developmental Editor *Michael Lange*
Production Editor *Carla Aspelmeier*
Photo Research Editor *Mary Roussel*
Visuals Processor *Vickie Werner*

wcb

Chairman of the Board *Wm. C. Brown*
President and Chief Executive Officer *Mark C. Falb*

wcb group

Wm. C. Brown Publishers, College Division

President *G. Franklin Lewis*
Vice President, Editor-in-Chief *George Wm. Bergquist*
Vice President, Director of Production *Beverly Kolz*
Vice President, National Sales Manager *Bob McLaughlin*
Director of Marketing *Thomas E. Doran*
Marketing Communications Manager *Edward Bartell*
Marketing Information Systems Manager *Craig S. Marty*
Marketing Manager *Kathy Law Laube*
Production Editorial Manager *Colleen A. Yonda*
Production Editorial Manager *Julie A. Kennedy*
Publishing Services Manager *Karen J. Slaght*
Manager of Visuals and Design *Faye M. Schilling*

Cover © 1988 Digital Art/Click/Chicago

Cover and interior design by Tara L. Bazata

Part One—United States Catholic Conference—CHD; **Part Two**—
© NASA; **Part Three**—© James L. Shaffer; **Part Four**—© Richard
Good; **Part Five**—© James L. Ballard; **Part Six**—© James L.
Shaffer.

Library of Congress Catalog Card Number: 88–71148

ISBN 0–697–03037–7

Printed in the United States of America by Wm. C. Brown Publishers
2460 Kerper Boulevard, Dubuque, IA 52001

10 9 8 7 6 5 4 3

TO MY HUSBAND

CONTENTS

Contents ix

Purpose

Telecommunications is one of the most potent forces in the world today. It influences society as a whole, and it influences every one of us as an individual. As each year passes, telecommunications grows in scope. The early pioneers of radio would never recognize today's vast array of electronic media—broadcast television, cable TV, teletext, and videocassettes, just to name a few. Neither would they recognize the structure that has evolved in such areas as regulation, advertising, and audience measurement. They would marvel that their early concepts of equipment have led to such developments as audio tape recorders, cameras, video tape recorders, switchers, editors, computer graphics, and satellites. If they could see the quantity and variety of programming available today, they might not recognize that it all began with amateurs listening for radio signals on their "primitive" crystal sets.

All indications are that telecommunications will continue to change and develop at a rapid pace. As it does, it will further affect society, and individuals will find themselves interacting with electronic media to an even greater extent. All people, whether they be individuals working in the telecommunications field or individual members of society, have a right to become involved with media and have an obligation to understand why people need to interact with the media. Some knowledge of the background and structure of the industry is an essential basis for this understanding.

A major goal of this book is to provide just that kind of knowledge so that intelligent decisions about the role of telecommunications can be made both by those who are practitioners in the field and those who are members of the general society.

Organization of the Book

This is the third edition of this book, the first appearing in 1983 and the second in 1986. A number of major changes have occurred in telecommunications since 1986, so corresponding changes needed to be made in this book. The importance of electronic media in the international arena and the continuing growth of corporate telecommunications led to the addition of two chapters on those subjects.

The multitude of changes in the newer media led to the total revision of Chapter 6, "Other Electronic Media." Improvements in equipment also led to major revisions in that chapter. Other major structural alterations, such as management changes at ABC, CBS, and NBC, the introduction of peoplemeters, the continuing aspects of deregulation, and variances in audience viewership, led to a great deal of updating in all areas.

PREFACE

To enhance organization, the text is divided into six parts. Part I, the introduction, deals with the social implications of telecommunications and sets the scene for the importance of the electronic media industry. Although the entire first chapter is devoted to social impact, the subject is not dropped at this point. The "social implications of telecommunications" theme is carried throughout the book and is one that should be considered in conjunction with all of the chapters.

Part II deals with communication systems—commercial radio, commercial TV, public broadcasting, cable TV, and other media. Each chapter in this part covers the historic development of a media form.

The technical aspects of production and distribution of all the previously discussed forms of telecommunications are covered in Part III.

Part IV deals with a subject familiar to most readers—programming. Here, again, both the older and newer media are discussed.

Part V covers regulation and business, ending with a chapter on telecommunications personnel that should be particularly helpful to those contemplating a career in the field.

The final section, Part VI, consists of two new chapters dealing with corporate, government, and personal uses of telecommunications and with international broadcasting.

All the chapters should lead the reader to assess the strengths and weaknesses of the particular subject being discussed. Information relating to future directions telecommunications may possibly take is also woven throughout the appropriate chapters.

Special Features

Each part of the book begins with an overall statement that relates the chapters to one another. Each chapter begins with an overview of the major topics covered in that chapter. This alerts the reader to important points and also serves as a review of chapter material once the reader has completed that chapter.

Each chapter conclusion summarizes major points but does so in an organizational manner slightly different from that given within the chapter. For example, if the chapter is ordered chronologically, the conclusion may be organized in a topical manner. This should help the reader form a gestalt of the material presented.

Further aids in understanding the material are the thought questions at the end of each chapter. These questions do not have "correct" answers, but rather are intended to lead the reader to form his or her own judgments. Discussions centering around these questions will indicate that varying opinions surround telecommunications issues.

Chapters are broken down into major divisions and marginal notes appear within each division. Each marginal note highlights the main subject being discussed in the adjacent paragraph or paragraphs. Taken together, these notes serve as review points for the reader. Important concepts are in boldface, which can also help the reader review the chapter.

The chapters may be read in any sequence; however, some of the terms that are defined early in the book may be unfamiliar to people who read later chapters first. The glossary can help overcome this problem. It includes important technical terms that the reader may want to review from time to time, as well as terms that are not necessary to an understanding of the text but that may be of interest to the reader. The glossary also includes abbreviations used frequently in the telecommunications field.

Chapter notes, which appear at the end of the book, are extensive and provide many sources for further study of particular subjects. In addition, a selected bibliography of some of the comprehensive books and articles dealing with telecommunications is provided.

The photographs that appear throughout the book supplement the textual information. Similarly, the quotations that appear at the beginning of the chapters provide additional food for thought.

Supplementary Materials

Instructor's Manual   The instructor's manual available with *Telecommunications* offers two sample course outlines that can be adapted to semesters or quarters, as well as learning objectives, audiovisual sources, suggested lecture topics/activities, and a bibliography for each text chapter.

TestPak   wcb TestPak is a computerized system that enables you to make up customized exams quickly and easily. Test questions can be found in the Test Item File, which is printed in your instructor's manual or as a separate packet. For each exam you may select up to 250 questions from the file and either print the test yourself or have wcb print it.

Printing the exam yourself requires access to a personal computer—an IBM that uses 5.25- or 3.5-inch diskettes, an Apple IIe or IIc, or a Macintosh. TestPak requires two disk drives and will work with any printer. Diskettes are available through your local wcb sales representative or by phoning Educational Services at 319–588–1451. The package you receive will contain complete instructions for making up an exam.

If you don't have access to a suitable computer, you may use wcb's call-in/mail-in service. First determine the chapter and question numbers and any specific heading you want on the exam. Then call Pat Powers at 800–351–7671 (in Iowa, 319–589–2953) or mail information to: Pat Powers, Wm. C. Brown Publishers, 2460 Kerper Blvd., Dubuque, IA 52001. Within two working days, wcb will send you via first-class mail a test master, student answer sheet, and an answer key.

Acknowledgments

This book represents the combined efforts of many people, including the following reviewers who offered excellent suggestions.

Charles F. Aust, West Georgia College
Don Flournoy, Ohio University
H. Bruce Fowler, University of Arizona
Craig Klein, Jacksonville University
Linda Krug, University of Minnesota-Duluth
Stanley Lichtenstein, Chabot College
Dennis Pack, Winona State University
C. Joseph Pusateri, University of San Diego
Edward Roehling, Indiana Central University
Vincent Spadafora, Ondonga Community College
Greg Swanson, University of Minnesota-Duluth

In addition, my colleagues and students at California State University, Fullerton were very supportive and helpful while I was writing this book.

Finally, I would like to thank my husband and three sons for their advice and encouragement while I was working on the text.

Lynne Schafer Gross

TELECOMMUNICATIONS
An Introduction to Electronic Media

Introduction

Telecommunications is a powerful force in society. Radio and television permeate our lives, yet many generations of people existed without electronic media. The pervasive influence has occurred in a short space of time, but its intensity compensates for its youth.

PART

I

Social Implications of the Electronic Media

1

Television is less a means of communication (the imparting or interchange of thoughts, opinions, and information by speech, writing, or signs) than it is a form of communion (act of sharing or holding in common; participation, association; fellowship).

Richard Schickel
The Urban Review

Overview

The influence that telecommunications exerts upon our society is obviously extensive. The mere ability to communicate instantaneously affects the process of communication. Beyond this, the permeation of opinions, emotions, and even fads can often be attributed to various elements of the media. The pervasiveness of radio and television, whether applauded or condemned, cannot be denied. The influence extends from the individual through the social structure, the economy, technology, and politics.

This chapter emphasizes the influence of electronic media by presenting the following information:

Reasons for the study of telecommunications

Points of criticism and praise for radio and TV

Problems of terminology in the field

Statistics detailing the extent to which radios and TVs are owned and used

Opinions about the media

Communications models

Examples of scientific research undertaken in such areas as television teaching and children's television

Historical and biographical research

Effects of telecommunications on individuals in terms of positive and negative education, passivity, relaxation, role models, and stereotypes

The issue of giving people what they want or what they need

Effects of radio and TV on such societal elements as family, school, culture, sports, fads, and crime

Telecommunications as a reflector and improver of society

Spiraling interrelationships of technology and electronic media relating to obsolescence, artistic quality, and communication

Direct and indirect effects of telecommunications on the economy

Profit versus public responsibility

Effects of telecommunications on politicians and the political process

A Rationale for Study

Everyone has an opinion about radio or television fare, and everyone can exhibit a certain amount of expertise about a force that is seen and heard on a daily basis.

Then why study this field? Some of the answers to this question are obvious. Anyone who is aiming toward a career in this area will profit from an intimate knowledge of the history and inner workings of the industry. Radio and television are highly competitive fields, and those armed with knowledge have a greater chance for career survival than those who are naive about the inner workings and interrelationships of networks, stations, cable TV facilities, advertisers, unions, program suppliers, telephone companies, the government, and a whole host of other organizations that affect the actions and programming of the industry.

career information

On a broader scope, individuals owe it to themselves to understand the messages, tools, and communication facilities that belong to our society because they are so crucial in shaping our lives. Rare is the individual who has not been emotionally touched or repulsed by a scene on TV. Rare, too, is the individual who has never formed, reinforced, or changed an opinion on the basis of a presentation seen or heard on one of the electronic media. A knowledge of the communications industry and its related areas can lead to a greater understanding of how this force can influence and affect both individual lives and the structure of society as a whole.

understanding of influence

The practitioners and the critics of the media need to understand each other so that barriers of ignorance do not deter communication. Those who generalize that employees of networks, stations, and other related companies are profit-hungry, glory-hungry deceivers will not gain a sympathetic ear from the very people they wish to influence. Similarly, those within the telecommunications establishment who look upon their critics as cause-happy, glory-seeking profit hamperers will not be willing to listen to reason even when it does prevail. A television executive must be able to understand the feelings of a mother who is watching her child turn into a TV junkie, and the mother must be able to realize that, in our capitalistic society, the media as a whole cannot exist if it does not obtain reasonable profits.

critic-practitioner understanding

Whether a person is interested in a radio or TV career, in being an informed media consumer, or in making social changes within the present structure, that person must study and understand the electronic media.

Criticism and Controversy

Radio and television have been blamed, at least in part, for a vast array of society's ills. Many of these comments may sound familiar.

Johnny can't read because he spends too much time watching television.
The lyrics played on radio encourage drug usage.

critical comments

Commercials inspire poor eating and health habits.

A rising crime rate is caused by violence on TV.

Interpersonal communications suffer because people watch TV instead of talking to each other.

Prejudice against people is a result of the stereotypical roles given to minorities on TV shows.

The continual depiction of foul language and nudity causes a rise in sexual promiscuity.

TV causes obesity in children.

Advertising encourages unabashed materialism.

The abundance of sports programming overemphasizes winning at all costs.

The mere act of watching television is a waste of time that prevents people from engaging in more worthwhile activities.

Because of television, politicians are now attractive and personable rather than intelligent and statesmanlike.

praise

On the other hand, there are times when radio and television receive praise for programming and services. Immediacy and quick thinking on the part of broadcasters have saved many lives during times of floods, hurricanes, and other disasters. Documentaries on such subjects as breast cancer and heart disease have been responsible for early detection and cure. During periods of both national pain and national pride, broadcasting has led the nation to a sense of communion and participative association. Fine cultural programming usually receives praise, albeit not from a large audience, and inspired entertainment also rates high marks.

Of course, some of the praise handed to the electronic media is due, at least in part, to inherent features such as the ability for immediacy; likewise, some of the criticism is leveled because the media cannot be all things to all people. For example, traditional radio and television, with their ability to transmit only to the audience and not vice versa, cannot be expected to be socializing agents, and yet they are criticized for their lack of social interaction.

The Broad Context

broadcasting

Studying this field used to be fairly simple. There were two media—radio and television, and together they were called broadcasting. As time progressed, broadcasting was divided into two categories—commercial and public (originally called educational). These two coexisted fairly harmoniously because public broadcasting was small and not really a threat to its commercial kin. In fact, it often relieved commercial broadcasting of its more onerous public service requirements, because the commercial broadcasters could point out that public broadcasting served that interest.

Then in the mid 1970s a number of other media came to the fore to challenge radio and TV, and students had to begin to learn an alphabet soup that included CATV, VCR, DBS, STV, MMDS, SMATV, and LPTV. The word *broadcasting* no longer seemed to apply because that word implied a wide dissemination of information through the airwaves. Many of these newer media were sending information through wires, and cable TV was even going around touting itself as a "narrowcaster" because its programs were intended for specific audience members. For a while these new forms were referred to as new media or new technologies, so people studied broadcasting and the new media. When they weren't so new anymore, they began being referred to as developing technologies, but some of them didn't develop very well. In fact, a number of them just plain died. Generally, the term *electronic media* was used to describe broadcasting and the newer competitive forces, but sometimes the word *telecommunications* was used to label the entire group.

newer electronic media

Now in the 1980s the field of study may be broadening even more. The telephone and computer have teamed up and may be on the verge of creating a new round in the information evolution. In a way, this brings broadcasting full circle. Radio can be seen to have its antecedents in the telephone because, at one point, the telephone was seen as a mass medium and the radio as an individual, private medium. In 1877, a song called "The Wondrous Telephone" contained the following lyrics: "You stay at home and listen to the lecture in the hall, Or hear the strains of music from a fashionable ball!"[1] The original idea for the telephone was that it would deliver words and music to large groups of people. When radio was first developed, many people tried to invent ways to make the signals private so that two people could have their own confidential conversation.

the telephone

Of course, over the years the two media switched roles—telephones being the private medium and radio becoming the mass medium. The two also went their separate ways academically and socially. Rarely were they studied in the same curriculum and rarely did people trained for broadcasting obtain jobs in the telephone industry. The social, economic, and political issues affecting each were quite dissimilar.

Then along came the computer and a device called a modem. When this modem was attached between a computer and a telephone, data generated by the computer could be sent over phone wires to another computer with its keyboard, disc drives, and screen. Suddenly information sent through the telephone was appearing on what looked like (or were) television screens. Some of the information being transmitted over this computer-telephone system was not private, but was intended for anyone in the population who wanted it or was willing to pay for it. It included news, weather, sports, and other information traditionally provided by radio and TV, as well as newspapers and magazines.

the computer

The word *telecommunications* was somewhat taken over by the telephone industry to encompass both the old telephone services and all the new data transmission and other fancy services the computer enabled the telephone to undertake.

telecommunications

Now the telephone, computer, radio and TV broadcasting, cablecasting, and the alphabet soup of newer technologies seem to be merging into some form of information supplier as yet undetermined. The most common word used to encompass all of this is *telecommunications,* but the word or even the concept could change drastically in the near future.

The words *telecommunications, electronic media,* and *broadcasting* are used somewhat interchangeably in this book because, at the time of writing, the general use of all three words is somewhat ambiguous. The book will concentrate, however, on commercial radio, commercial television, public radio, public television, cable television, and the various newer forms that came to the fore in the 1970s. It will also deal with the telephone, but mainly as it relates to and is becoming merged with the media taught in the traditional radio and TV curriculum.

Statistics of Pervasiveness

senders

This force by whatever name—telecommunications, electronic media, or broadcasting—is very pervasive in our society. On the transmitting end, there are over 11,000 radio and TV stations in the country, approximately 10,000 for radio and 1200 for TV. Of these, about 1200 are public radio stations and 300 are for public TV. In addition, there are about 7800 operating cable TV systems,[2] 30 million videocassette recorders, 2 million backyard satellite dishes,[3] and 450 low-power TV stations in operation.[4]

receivers

On the receiving end, 99 percent of families have radios, 98 percent have TVs, and 96 percent have telephones. In fact, the average household has 5.4 radio sets, and 57 percent of households own two or more TV sets.[5] Almost 50 percent of the national households subscribe to cable TV,[6] and the same number own a videocassette recorder.[7]

use

More important, people do not just own their TV and radio sets; they use them. The average household TV is on about seven hours a day with the average person watching about four hours.[8] Viewing is amazingly similar among different viewing groups. For example, the rich and the poor and the educated and uneducated watch TV just about the same amount of time. However, people in the center of the country watch considerably more TV than those on the coasts, and women over fifty-five watch the most TV while teenage girls watch the least.[9]

The average person also tunes in radio about three hours a day. During the course of a week, radio reaches 96 percent of all people and nearly all teenagers. Car radios reach three out of four adults a week.[10]

opinions

People also have definite opinions about their media—both positive and negative. One survey conducted for *U.S. News and World Report* found that only 51 percent of respondents were satisfied with TV's entertainment programming.[11] Another poll conducted by an advertising agency asked people their sources of greatest pleasure and 68 percent mentioned watching TV.[12]

FIGURE 1.1

Roper statistics.

	Years														
	1959	61	63	64	67	68	71	72	74	76	78	80	82	84	86
	%	%	%	%	%	%	%	%	%	%	%	%	%	%	%
Most believable:															
Television	29	39	36	41	41	44	49	48	51	51	47	51	53	53	55
Newspapers	32	24	24	23	24	21	20	21	20	22	23	22	22	24	21
Radio	12	12	12	8	7	8	10	8	8	7	9	8	6	8	7
Magazines	10	10	10	10	8	11	9	10	8	9	9	9	8	7	6
Don't know/ No answer	17	17	18	18	20	16	12	13	13	11	12	10	11	9	12

The most famous of the pulse takers is the Roper Organization, which has periodically been asking the public questions about television and other mass media since 1959. One of its typical questions involves the relative credibility of the media.[13]

As can be seen, TV wins hands down as the most credible medium. Similarly, TV rates first as the source of most news and as the source that is most likely to acquaint people with candidates running for political office. Other questions asked by the Roper interviewers include how many hours people watch TV, how people get information about TV programs, the types of programs people watch, and public attitudes toward the portrayal of women and minorities.[14]

All of these quantitative statistics point to the fact that telecommunications is assuredly an important element of today's society.

Communication Models

Various models have been designed to explain the communication function. One model that can apply to both personal communications and telecommunications is presented by Bert E. Bradley.[15] This model shows communication as an ongoing process that includes a source, a message, a channel, a receiver, barriers, and feedback.

For telecommunications, the **source** is composed of the total number of people needed to communicate a message. In radio or television, this is a large number because no one person can play all of the required roles. For example, a person on camera as talent cannot also act as camera operator, producer, audio engineer, video tape recorder engineer, station manager, and all the other positions necessary for programming to be successful.

source

Of course, some people are more important sources than are other people. For example, the people who decide which news stories will be broadcast, which will receive the most time, and which will be presented at the beginning or end of the newscast are more important in the communication process than are the camera operators who focus on the anchorpeople.

Some communication models refer to the people who make important decisions as **gatekeepers.** They do, in effect, control what information passes through.

FIGURE 1.2

Communication model.

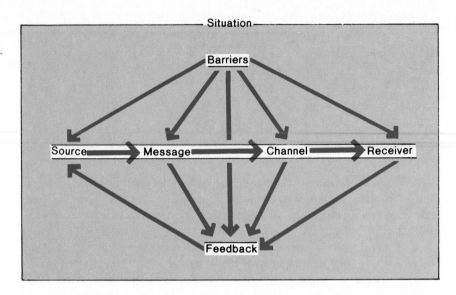

The people who act as the source are concerned with sending a **message**. The message will have a purpose—some messages are intended to inform, some are designed to entertain, and some are planned to persuade. Transmission of a message is the primary reason for communication although, obviously, both the message and its process of transmission are often far from perfect.

Messages are transmitted through **channels**. For telecommunications, some of the most common channels are radio, broadcast television, and cable TV. Various devices and processes can influence the effectiveness of a channel. For example, instant replay, which is a particular quality of the TV medium, enables sportscasting to communicate to the fans in ways that other channels, such as radio and newspapers, cannot.

The **receiver** of the message is the audience. For telecommunications, the audience is a composite of individuals that is generally large, heterogeneous, and fairly anonymous. The people representing the source send their message through the channel to the receiver.

However, many **barriers** can obstruct the communication process or reduce its effectiveness. The source can have built-in physical or psychological barriers. For example, if all the people representing the source live and work in California, they may not have a proper frame of reference to design a message that will communicate to people in Massachusetts. If the majority of people working in a newsroom are Democrats, they may not communicate in a way that is considered appropriate by Republicans. Individual biases and experiences can affect how the source sends its messages.

Messages also have barriers. A poorly written joke may not entertain. An informational piece delivered in Spanish will not communicate to someone who does not understand that language. A campaign speech intended to persuade might only inform.

message

channel

receiver

barriers

Static is an example of a physical barrier that can affect the communication ability of a channel. The government, through its regulations, can also create barriers, whether advantageous or disadvantageous. If a TV station feels it must give equal time to opposing viewpoints on a controversial issue, this might inhibit the station from discussing the issue at all. Consumer groups that threaten boycotts can affect the behavior of the channel. On the other hand, a message can be distorted simply because information presented on radio or TV takes on added importance. For example, sometimes criminal acts that are fairly minor get blown out of proportion because they receive extensive coverage.

The receiver, too, is affected by many barriers. Some of these are in the receiver's background or environment. For example, a person who has lived in France might perceive a program about French politics in an entirely different way than a person who has never even visited France. A person listening to a radio station in a hot room might react differently to the music than someone listening in a cold room. Interruptions or distractions such as a phone ringing or dinner being placed on the table can also be barriers for the receiver.

The message, channel, and receiver all serve as sources of **feedback** to the source. If no one laughs at a joke during rehearsal, this can be a signal to the creator that it is not funny. Managers of affiliated radio and TV stations often express their opinions to the people who provide the network programs. But the receivers are the major source of feedback for those who constitute the source. Ratings are the primary form of feedback, but fan letters and complaints also count. These can alter content, as is evidenced by the cancellation of shows with poor ratings.

feedback

There are also barriers to feedback. People can laugh out of politeness, not bother to complain, or fill out a rating form incorrectly.

Despite all the barriers and the inefficiencies of feedback, the telecommunications process does work, and most messages do proceed from the source to the receiver with some degree of clarity.

Media Research

A great deal of qualitative research about the media has also been undertaken, particularly by college professors. Rather than dealing exclusively with numbers and percentages of people who own sets or state their opinions about programming, academic studies attempt to uncover causes and effects. These studies often involve scientific research using experimental and control groups, random samples, and statistical methodology. These studies are particularly helpful because they enable media practitioners, regulators, advertisers, individuals, and social agents to understand the effects of telecommunications on society and, hence, to provide better service to the public.

importance

Qualitative studies are often difficult to administer, mainly because radio and television are such pervasive forces in our world today. If one is going to have an experimental group that watches television, how can a group of people who never watch television be found to act as a control for the experiment?

difficulties

How can effects of radio and television be isolated when people are constantly under the influence of so many other social factors such as family, school, and church? How can effects of the media be determined when individuals vary so greatly among themselves in terms of reaction?

teaching ability of TV

Despite the difficulties inherent in such questions, a great deal has been learned about media and society from academic research. In the early days of television, its effectiveness as a teaching medium was frequently tested. During the 1950s and early 1960s, many researchers conducted experiments in the area of teaching and television. The very early experiments tried to determine whether television could teach at all. For example, one researcher found that people who watched televised French lessons knew more French words after seeing the program than they knew before,[16] and another researcher found that dentists exposed to television courses scored significantly higher on tests than did dentists not exposed to the material at all.[17] This research seems very naive today because it is obvious that people can learn from TV, but the research was conducted when TV was very new and its effects unknown.

Later research compared TV-delivered instruction with "conventional teaching" and generally found "no significant difference." Students learned about the same amount from TV as they did from other means of instruction.[18]

Many studies varied aspects of TV teaching to determine optimum teaching effectiveness. For example, one study that considered close-ups versus long shots, dynamic versus static editing, and distracting versus nondistracting background found that students learned more from dynamic editing and nondistracting background than from static editing and distracting background.[19] But most of these types of studies found no significant difference when format, production values, graphics, or other elements were varied.

Much of this educational research was criticized as being research for the sake of research or research for the sake of grant money, which was then relatively plentiful. But the results, taken as a whole, could be summarized and made available to educators as a guide for preparing instructional television lessons.[20] Thus, they did have merit.

children and TV

One area of research that became very popular during the 1960s and 1970s was the effects of television on children. Much of this research dealt with the effects of television violence on children and generally was not conclusive. Some studies found that violence did affect children adversely while other studies found that it did not.

In one study, forty-four third and fourth graders were randomly divided into two groups, both of which were shown a new trailer on the school grounds and told that it was used for kindergarten children. The experimenter pointed out a TV camera on the wall and said that it would take pictures of all that took place inside the trailer. One group of children was then taken into a room and shown a violent western, while the control group was taken into another room and not shown any film. The experimenter then said he had to go to see the principal and asked each group of third and fourth graders to watch the kindergarten children for him. He turned on the TV monitor that showed a still empty trailer and asked both groups to watch the room and come to him in the principal's office if anything went wrong. Both groups were shown a videotape of two little children coming into the trailer, starting to play, and then getting into a fight and pushing each other until they apparently broke the camera. The group of children who had seen the violent western took significantly longer to seek adult help than did the children who had not seen the film. The conclusion of this study was that exposure to TV violence taught children to accept aggression as a way of life.[21]

In another study, children assessed to be emotionally vulnerable were shown, at various times, television programs containing minimal, moderate, and maximal violence. Afterward, their behavior was observed, they were led in group discussions concerning the programs, and they were given psychological tests to measure changes in aggressive feeling and fantasy. The results showed that exposure to aggressive programs produced more aggressive fantasies but did not lead to heightened aggressive behavior.[22] Other studies, however, have contradicted this.[23]

These subjects of academic research—learning through TV and effects of TV on children—are illustrative but by no means the only types of research done concerning telecommunications. Other common research subjects are radio age groups, effects of sex and violence on TV, broadcast economics, advertising, minorities and women, news coverage, and political influence. Research journals such as *Journal of Broadcasting* and *Journal of Communication* are excellent sources for keeping up to date on the latest research in the field.

other areas

The fact that many bodies of research about media and society are inconclusive should not be surprising. Our complex society would make a statement such as "TV violence causes 75 percent of murders" virtually impossible to substantiate. So many factors are at play that research can only alert us to danger zones and show probable cause. But this, in itself, serves a very useful function in that research can aid society in tackling likely problems before they become conclusive.

Not all academic research is of a scientific, experimental nature. Historical and biographical research is also undertaken so that a clear understanding of the past and the people who created this past can help pave the way for the future. Researchers scour primary materials, such as letters written by individuals instrumental in broadcasting's history and documents printed throughout the years, in order to uncover new facts and interpretations.

historical and biographical

Recently, radio and television archives have been established in various places around the country. In these archives both early and recent programs are collected so that they can be viewed by scholars and interested members of the public. In some instances this viewing is purely for information, entertainment, or nostalgia, but more and more these programs are being utilized for sociological research. For example, situation comedies can depict the changing complexion of families through the last several decades and public affairs programs can document the changing views of liberals and conservatives.[24]

archives

As the newer technologies such as cable TV, low-power TV, and videocassettes become even more pervasive, a whole new body of academic research is being undertaken.[25] How many channels does a viewer use? What are the best ways to make people aware of a product or service when a vast array of products and services are available? Will entertainment still be the predominant force in the future? If more sexually explicit material is available, how will this affect society?

All of these questions and many more can be researched and then researched again as both society and the media change in our fast-paced world.[26] Research will continue to provide a base level of knowledge and direction for the future.

Individual Effects

Individual preoccupation with telecommunications is enormous as evidenced by the statistics on viewership. But what effect does all of this viewing and listening have on the individual? This is not an easy question to answer. For one thing, as previously noted, difficulty exists in trying to isolate the effects of radio and TV on an individual from the effects of other elements of society, such as family, school, and church. What is more, different individuals are affected in different ways. What follows is a discussion summarizing some of the effects, both pro and con, that have been espoused with varying degrees of commitment by consumers, critics, and practitioners of the media.

Television can be an educational force for many individuals, not only through purposeful educational programs, but also through documentaries and entertainment programs. A movie that is set in a foreign country can unobtrusively teach a viewer about the customs and physical features of that country. News programs can make everyone more knowledgeable about current events than was the case in pretelevision days. Science fiction can convey minilessons in astronomy or physics. Commercials can inform people of new products that may meet their specific needs. Some people even claim that game shows are educational because of the content of their questions.

Yet, if what is presented is educational, then much of it is negatively educational. Television, with its happy, well-scrubbed families who solve all problems in thirty minutes, does not represent reality. A person from a deprived background can be frustrated and discontent if television is considered the only window to the "real world." By fictionalizing for the sake of drama, television practitioners can "teach" that life is far more exciting and rewarding than the average individual's humdrum existence. Individuals who react to this with aggressive self-pity can be dangerous to themselves and to society.

The mere act of watching TV, regardless of program content, can be harmful to individuals. Watching TV is a passive, sedentary antiactivity that often requires little or nothing of the viewer so that neither the mind nor the body receives exercise. The time spent watching aimless hours of TV subtracts from the time an individual has for more meaningful activities, such as study and human interaction. In some cases watching TV becomes a major factor in procrastination or even psychic paralysis.

Particularly dangerous is watching TV as opposed to watching programs. Many individuals turn on the TV set and watch whatever is on regardless of its quality or interest to them. Rather than planning to watch programs that they particularly want to see, they flip the dial until they hit upon the least objectionable fare—a program that may not really appeal to them, but one that is more desirable than any other programs on at that time. Would they be better off using the time for something more constructive?

Yet television viewing makes some people's time worthwhile. The old and infirm can use TV as something to live for, something to pass the time, something to keep them in touch with the world. Others can use radio or TV to unwind from a hectic day. All individuals need time for pure relaxation— a time to escape vicariously and temporarily from their trials and tribulations. Why not use least objectionable fare to fill these times?

Radio and television provide role models, especially for the young. A role models youth who watches a medical show and decides to become a doctor and a child who joins a sports team after listening to an interview with a sports star both profit. Minority children and adults can develop increased self-esteem by watching minority individuals in significant roles. In a time when many children are from broken homes or homes in which they see little of their working parents, positive role models are needed.

But what if the role models are negative? What if a role model is a loose-living amoral person or an aggressive fugitive? Certainly, with the drama for drama's sake fare that appears, role models can be both oversimplified and dangerously stereotyped. Many programs, particularly the older reruns, portray mothers as passive, fathers as buffoons, and children as insensible patsies. Worse yet are portrayals of servile blacks and lazy Mexicans, all of which can affect the impressions of children, or adults for that matter, concerning personality traits and social positions of races, sexes, or creeds of people.

Individuals do tend to select and retain those elements of a radio or TV program that reinforce previously held beliefs, attitudes, and values. A person who feels that women are and should be passive will notice those programs or parts of programs that demonstrate women's passivity and tend to ignore program material to the contrary, even though this may be the crux of the message. In this way TV, even though it may not be trying to do so, can magnify stereotypes.

For decades arguments have abounded regarding whether the electronic media should give individuals what they want or what they need. Judging by want versus need what is watched, most people want the more frivolous, light, unthinking type of program material. Those who favor the "what they need" theory feel the media should present classical, informative, and educational materials. As telecommunications advances into the 1990s, with a potential for many more sources of program material, perhaps the two schools can be reconciled. The consumer will have a broader "least objectionable fare" choice at any given time, and perhaps more of the fare will be food for the mind. However, the choice will most assuredly remain with the individual, who can always choose to turn off the TV or radio.

Sociological Effects

Television can have a personal and intimate effect upon a person, almost as if he or she were in a vacuum. But in reality people do not live in vacuums; they live in a society where they must interact with one another through various organized methods. Television affects these sociological organizations and, of course, the individuals within them.

family

One of the backbones of our social structure is the family. In the days of early radio, several generations of a family would live in close proximity, forming a secure social structure for one another. However, as transportation and communication improved, family ties weakened and the young often left home to seek their fortune in other parts of the country or world. The once "ideal" family of father-provider, mother-homemaker, son, and daughter, which was prevalent even when television began, is now a distinct minority. Women are entering the work force and the divorce rate has climbed. Through all of the transformations of the family during the past sixty years, radio and then television have been depicting family life through both comedy and drama. Has this depiction been realistic? Has it been beneficial?

Comedy, because it is constantly in pursuit of a laugh, and drama, because it aims for suspense, cannot be truly realistic without being boring. No families lead lives that are a laugh a minute or a trauma a moment, but TV programs are produced this way in order to sustain interest. On the other hand, a perusal of family radio shows of the 1930s compared with family TV shows of the 1980s shows a vast difference in family values and structures that are reflected, at least to some degree, in the changing society. Some blame TV for the rising divorce rate by pointing out that the increasing number of shows centering around divorced parents seem to be condoning and even glamorizing divorce. Others point to the "old-fashioned" programs that emphasize family closeness and integrity and say that these are role models for present families.

The mere presence of a television set within a family also causes controversy. Is this "box" preventing family members from talking and interacting with one another, or is it giving them subjects of common interest about which to talk? When a family gathers before a TV set are they engaging in togetherness, or is each member of the family in an individual shell relating only to the electronic set and not to the other humans in the room? If family members were not watching television, would they be engaged in other family activities, or would they be spread out among other, perhaps destructive, activities? Does the music blaring from a radio increase the generation gap, or does it serve as a mild form of teenage independence that prevents other more dangerous forms of rebellion against parental domination?

Arguments can be made for all points of view on these questions, and certainly no one answer is correct. Families, like individuals, differ in and among themselves, but each family, regardless of its structure, does have the obligation to control television rather than let television control the family.

schools

Schools are also sociological institutions that all people encounter for at least part of their lives. The quality of education and the quality of school graduates has frequently been a heated subject that has drawn radio and television into its midst. As college aptitude scores declined during the 1960s, many critics blamed television, which, they said, had lured children away from books and into the insane world of endless cartoons that made no demands upon reading or spelling abilities. Others, mitigating the effects of television, blamed the schools themselves for relaxing disciplinary strictness and for devising courses of study that placed greater emphasis on social well-being than on academic basics.

There were also those who felt that television actually stimulated an increased intellectual awakening in young people. They pointed to the fact that when a famous story was dramatized on television, the libraries were inundated with requests from people who wanted to read the book. Teachers sometimes assigned television programs as homework and used them as a basis for an intellectual lesson. After "Sesame Street" hit the airwaves, studies showed that kindergarten children who had watched it were further advanced in cognitive skills than were previous kindergarten children.

Countering this, however, were the kindergarten teachers who noted that children did not play together as well as they had in pretelevision days. They blamed this on all the hours the preschoolers spent in front of the TV that deprived them of creative play with one another. And on the homefront many parents experienced the "stop-watching-TV-and-do-your-homework" battle just as many of their parents had had to restrict radio listening in order to see that the schoolwork was completed.

culture

The cultural aspects of society generally receive stepchild treatment on radio and TV, although public television and some aspects of cable TV are giving recognition to their enhancement. Historically the mass media of radio and television have, as their name implies, catered to the masses and culture does not sell on a mass basis. A sort of social aloofness also seduced those who were culturally inclined into thinking that culture was somehow debased if it appeared on TV, even though this allowed many more people to view it than could ever view it in person. When classic plays, ballets, or concerts do appear on radio or TV, they often have a social impact far beyond what they would have otherwise.

sports

Sports, on the other hand, receive wide play on radio and television, to the point where the sheer amount of sports programming becomes controversial. Once again the spectrum of questions is raised regarding whether sports programming inhibits or encourages social interaction. The "sports widow" phenomenon is bandied about by cartoonists and comedians, but some women become as addicted to watching sports as their husbands do. Cable TV enables the true addict to watch sports twenty-four hours a day. But because sports is such a widespread topic of conversation, watching sports can lead to sociable and sometimes heated conversations. Watching often encourages people to participate in sports, certainly not a sedentary activity. The vicarious "thrill of victory and agony of defeat" can provide people with the excitement they need in their lives and, perhaps in this way, stifle more socially unacceptable aggressive behavior.

fads

Society's fads are frequently created or at least fanned by the media. Everything from Little Orphan Annie rings to Madonna look-alikes have been the result of media exposure. Many "Mickey Mouse Club" fans grew into "Star Trek" appreciators called Trekkies, and the sociological implications were the same—adoration and belonging.

Sometimes the same phenomenon that creates a unity among Mouske-teers and Trekkies can create a common belief of a more serious nature. Many people feel that the uprisings against the Vietnam War were mainly media induced and that once radio and television attached themselves to the antiwar sentiment, it spread uncontrollably through the country.

violence

When the sociological phenomenon of violence comes into discussion, the opinions about television's role become very heated. Over and over again citizens' groups, church groups, and individuals have equated rising violence and crime in the streets with violence on TV. The dramatic nature of television programming emphasizes conflict and, more often than not, this conflict is shown in the form of violence. The critics of such programming worry that it is creating a society in which violence is taken as commonplace and its rising tide is accepted as a natural phenomenon. Worse yet, they feel that the pro-grams are emotionally damaging and lead people toward aggressive acts that they might not contemplate if they did not see them on TV. Even more alarming, they feel that the violent programs are instructional and that those watching, including prison inmates, learn better ways of committing crimes by observing techniques and mannerisms.

To counter these arguments, network executives point out that Homer's *Iliad* and Shakespeare's plays have a higher degree of violence than do modern teleplays and that violence is used as an interest holder and not an instrument intended for social destruction. People have watched violence for centuries and have not suffered apparent ill effects. But, counter the critics, several people watching one violent program once in awhile, as happened prior to TV, is not the same as millions of people watching violent programs every evening. The cumulative effect of the television watching, they feel, is what is harmful to society.

reflect or improve

In the area of violence, as well as in other sociological areas, media prac-titioners are constantly confronted with the dilemma of whether they should try to reflect society or try to improve it. If they merely reflect what is going on around them, they will be accused of a lack of social conscience. If they try to change society, they will be accused of using their power, which is con-siderable, to create change in directions that they feel are desirable. Undoubt-edly, neither of these directions will be unanimously approved by all elements of society. Some people are bothered by the fact that Hollywood appears to them to be attempting to impose its own mores and morals on the rest of the nation when, in fact, the country might be a better place if the opposite were true and Hollywood more accurately portrayed the rest of the country. So the dilemma is circular and, for all practical purposes, unresolvable. Those in telecommunications must certainly take stock of the fact that their product does have a significant impact upon society.

Technological Effects

The electronic media and technology exhibit an interesting spiraling relationship. Improved technology leads to improved communications that in turn leads to a desire for improved technology. In early times the discovery of radio waves led to the transmission of Morse code without the use of wires, and the ability to transmit Morse code led to the desire to transmit voice through this wireless means.

      spiraling relationship

    As each technological step is taken, new uses are found that push the technology even further. When special effects were first introduced into television in the form of corner inserts and split screens, the "gee whiz" approach of directors and technical directors led to a desire for more special effects. This led to switchers with provisions for soft wipes, reversed images, swirling digital images, and many other effects to challenge the creativity of the operators, who then thought of new effects with which to challenge the engineers. When the technology was developed to enable cable to carry twenty channels, many wondered what type of programming could possibly fill all those channels. Within a brief period of time twenty were easy to fill and the technology was developed for over thirty channels, which, in turn, were filled. And on and on goes the saga. Humans' curiosity and ingenuity chase each other around the technological maypole with the end result of either constant improvement or constant turmoil, depending on one's point of view.

    Certainly, the constant technological development keeps everyone in the media business on their toes. Any production method or delivery system that tries to coast on its laurels will find itself standing in the dust as the newest of the new technologies rushes by.

    Many entertainment and information forms have been severely injured, permanently or temporarily, by this constant rush of technological change. Vaudeville barely exists anymore, its demise due largely to radio and films. Radio experienced a skid when television came to the fore and had to readjust drastically in order to revitalize itself. Radio and television have led newspapers to cast but a shadow of their former power, and movie theaters are now under threat from videocassettes. The conventional broadcasting system looks over its shoulder at cable TV while cable TV glances furtively at direct satellite broadcast. Survival of the fittest often translates into survival of the newest and the shiniest.

    media damage

    Technical developments have a definite bearing on the artistic quality of programs. The audio tape recorder brought significant changes to radio, and

    art

the video tape recorder did likewise for TV. Programs could be produced at times other than when they were aired, and they could even be produced in small sequences out of order and then assembled by means of editing. As equipment became miniaturized, production possibilities were maximized. Instant replay, still-frame, and computerized graphics all added a new dimension to creative capability.

communication

The public at large is invariably brought into the technological spiral. Events that were not known about for months in the 1700s are now known about in moments. When handheld cameras were developed, the public was able to see news events as it had never been able to see them before. Consumer-quality cameras and video tape recorders make possible a whole range of activities, many of which have not even been conceived as yet. Home video components, such as discs and cassettes, can bring new dimensions to individuals and society.

In the short space of one hundred years technological changes have been enormous. In the 1880s great mystery surrounded electronic possibilities. In the 1980s great uncertainty surrounds the direction that all of the technological development will take. During that century, radio and television became important sociological tools. The future direction of telecommunications will be determined by the degree to which individuals and society feel that the services that can be offered will be of value and interest to them.

future

The technology exists to create a society in which people sit poised before the TV set for most of their waking hours, using it not only as a source of entertainment and information, but as a lifeline and pipeline to accomplish the routines of their business and personal lives. But the existence of the technology does not mean that it should function in this manner—the benefits and/or harms to society must be evaluated. The technological possibilities are enormous, but the social consequences are yet to be determined.

Economic Effects

The exact interrelationship of broadcasting and the economy is nebulous. The actual number of dollars passing through radio, television, and cable TV, primarily in the form of advertising money, represents only about .3 percent of the nation's gross national product.[27] More money per year is spent on buying and repairing TV sets than is spent on advertising; the amount is absolutely dwarfed when compared with major expenditures such as national defense, welfare, food, and transportation.[28] Needless to say, the social and political impact of the electronic media far outweighs the economic impact.

small monetary amount

The indirect economic effect is larger but much more difficult to measure. Advertising stimulates buying, but relating particular purchases to particular advertisements ranges somewhere between difficult and impossible. Rarely will a person see or hear an ad and run right out to buy the product. Advertising builds an awareness over time so that an eventual purchase could be the result of several radio or TV ads, a billboard, a magazine ad, the yellow

advertising

pages, or perhaps the recommendation of a friend. An accurate figure cannot be derived for how much telecommunications contributes to the nation's actual purchases.

Taken a step further, most purchases would be made even if there was no radio and TV advertising. This advertising serves more to develop brand awareness than product awareness. People would buy bread even if it were not advertised, but advertising can lead them to buy brand A instead of brand B. The extent to which advertising develops false needs and leads people to buy products that are frivolous to their existence is highly criticized. Advertising is accused of creating a materialistic society that overspends on the needless and the useless, often at the expense of the worthwhile and needed. Cereal ads are an example. Highly advertised sugary cereals low in nutritional value are sold and used as replacements for more healthful breakfast foods.

Advertising also influences programming. Some feel that programs are merely a lure to draw an audience for commercials and that the real purpose of radio and television is advertising, not entertainment or information. Programming material is selected not on its merit but on the number of people it will deliver. In some instances the desired people are of a particular age, sex, and/or income level as they are more likely to buy the products advertised.

profit

Within our capitalistic society, most forms of telecommunications are profit oriented. Some forms, such as public broadcasting, are nonprofit in nature, but commercial radio and TV stations, cable TV systems, networks, advertising agencies, syndicators, and a host of other media entities are working to make a profit and a return for their shareholders. Videocassette machines as well as other electronic equipment are sold for profit.

In the area of profit, most radio and TV entities do well, although there are notable exceptions. Radio stations did poorly after TV was introduced, and conventional TV has been having financial difficulties since the introduction of the newer media. Overall, though, most stations, networks, and cable systems end up in the black.

station sales

Because of the enormous social impact of the electronic media, the price at which radio and television stations and cable systems sell is far greater than their tangible assets, such as equipment and buildings. The first TV station sold in 1949 went for only $300,000, while the highest price registered as of the late 1980s was $510 million.[29] The reason for this apparently inflated price is that what is actually being bought is a right to use the airwaves and "goodwill," a catchall industry term that underscores the fact that broadcasting is related more to sociopolitical power than to economic gain.

profit versus public responsibility

Closely allied to the economic health of the industry is the issue of profit versus public responsibility. In order to pay salaries and bills, companies must have income, and in order to reward stockholders, many of whom are members of the general public, the income must exceed the general expenses. But broadcasters, unlike furniture manufacturers or automobile dealers, are mandated to serve in the public interest. Defining public interest and profit causes a great deal of consternation among media people and members of the public.

To what extent should a station forego profit to air local-issue material that brings in little or no revenue either because it will not deliver an audience or because it is too controversial for advertisers? How much money should a network pour into a money-losing documentary regardless of its subject matter? How much money should a cable TV system spend on access equipment and personnel so that members of the community can create their own programs, which other members of the community may or may not watch? Drawing limits on issues of this nature is never easy and necessitates constant give-and-take between the telecommunications industry and the public served.

The economic future of telecommunications is not assured. In the final analysis the public, by accepting or rejecting what is offered, decides the health of both the industry as a whole and the particular elements within that industry, such as individual radio stations and individual networks. A constant idea exchange between society and media practitioners is necessary to ensure continued economic well-being.

Political Effects

Would George Washington or Abraham Lincoln have been elected president if radio and television had been invented two hundred years ago? No one knows, of course, but many theorize that the low-keyed George and gawky Abe would not have succeeded in an electronic age. Like it or not, candidates today must be TV personalities. A politician who shuns the TV camera shuns a major source of exposure to his or her constituency.

candidates

This applies not only to political campaigning but also to daily activities. News programs report the actions of senators, governors, and other public officials on a regular basis, and politicians who make buffoons of themselves on the evening news are not long for the world of elected officialdom.

Nevertheless, controversy surrounds the media making of a president and other political figures. Are the articulate and attractive really the best people to run the country? Is the choice of a public relations firm more important to a candidate than the choice of a foreign affairs policy? Should elections go to the candidates who can afford to spend the most on political advertising?

objectivity

The intertwining of politics and the media creates some strange bedfellows. Reporters are supposed to ferret out news objectively, but when they themselves harbor political ambitions or become friends with those they are covering, their judgment may not remain objective. People who cover politics effectively must, by nature, be independent and inquisitive. Perhaps too many of the same types of people wind up covering the political scene and, thus, inadvertently bias it in terms of their own dispositions.

Radio and television are regulated by the very people they cover. For example, senators pass laws related to communications, city council members award cable TV franchises, and judges rule on First Amendment rights. In order to win favor with these government people, the media practitioners might be benign in their reporting. On the other hand, the media can be so powerful in creating a politician's image that the politician might be inclined to treat the media kindly. In both instances the public could be the loser.

There are many positive points to make about the media-political interrelationship, however. Politicians are much more visible now than they were before the age of telecommunications. Constituents can now see their representatives in more revealing ways than ever before and can make voting decisions based on firsthand visual observation rather than on secondhand written information.

The political process is scrutinized by the media. Innumerable changes occurred in political nominating conventions when the watchful eye of the camera made its debut. Major speeches and events were scheduled or orchestrated to coincide with prime-time hours. Delegates paid attention and were present for votes. Cameras in the halls of Congress have resulted in a bit more decorum than was present in years gone by as lawmakers realize that someone out there may be watching.

public scrutiny

The entire media process aids in our country's dedication to peaceful change. Views can be expressed through radio and television, debates can be held, issues can be discussed, and in the end, changes are made without resorting to revolution as is the case in so many countries of the world.

peaceful change

The media serves as another balance of power by acting as a watchdog over the government and by reporting government corruption or wrongdoing. Sometimes, however, the media is accused of overreporting on wrongdoing and, therefore, inflicting more harm than good. Sometimes in order to obtain a story or a scoop, radio and TV reporters resort to sensationalism and blow minor infractions way out of proportion. At other times they prolong reporting on wrongdoing to the extent that it impedes the normal business of government. Some have concluded that many people qualified to run for public office do not in order to spare themselves and their families from the often inhumane scrutiny of the media.

balance of power

The nation can be informed instantly of what is occurring in the political process. Sometimes this immediacy breeds danger, especially if all the facts have not been gathered, but the overall result is an informed public.

immediacy

Because our country is strongly committed to freedom of the press, politics and the media will undoubtedly continue to interrelate. The balance between the two is delicate, but the forces of our society can act as a rational mediating force.

A Forward Look

Society and the electronic media enjoy a marriage that, like any marriage, must be constantly fine-tuned. Just as a husband and wife must adjust to growth and changes in each other as the years go by, so must society realize that technological and philosophical developments will change the media, and so must the media realize that they will have to adapt to the inevitable changes in social patterns.

possible changes

The criticisms of the media that exist today, such as the one-way nature of radio and TV, may become totally outdated as society and the media change. Indeed, some of the primary strengths of telecommunications, such as immediacy, may no longer be needed. Controversies that exist today, both national and international, may seem trite and inconsequential in years to come.

Even if people continue to use their radios and TVs to the extent that they presently do, the type of use they put them to may differ drastically from one of pure receiver to one of sender or creator. This will alter communication models by changing the function of the gatekeepers, strengthening the feedback, and opening up vast new areas for research.

The individual will still be the key in telecommunications usage even though content, role models, and attitudes may change. Individuals will also be the key in determining the manner in which telecommunications will affect society and the world in terms of the family, schools, culture, sports, fads, and violence.

Technological advances are bound to determine a great deal of the future direction of telecommunications as the upward spiral of technology and the media continues. This fact may affect the economic pattern of our advertising-based media system, as well as the general economic health of the present radio and TV structure.

Telecommunications has become ingrained in the political process to such an extent that future changes in the media will be both challenging and frustrating to politicians as well as to those people who deal with the field of telecommunications.

purpose of book

One of the purposes of this book is to help those who will be dealing with the field of telecommunications to understand its present idiosyncrasies and future potential. With this knowledge, they should be able to make contributions that will create a healthy relationship between society and telecommunications.

Conclusion

Whether this field be called telecommunications, electronic media, or broadcasting, it is an area that deserves careful study. For the individual who each day spends four hours with TV and three hours with radio, it is important to know their effects in terms of education, use of time, and acceptance of role models. TV can teach, and teach effectively, but the instructional messages are not always beneficial to the learners. Individual emotional responses to TV are crucial and should be understood by those who experience them. The positive effects of the media on individuals during emergencies must also be noticed.

From a sociological point of view, telecommunications is often a unifying phenomenon that engenders a sense of communion in times of both trial and triumph. The media permeate the home and affect the family structure and particularly children in ways that are still not totally understood by either formal research or observation. Schools, culture, sports, fads, and outlooks toward violence are all affected by the electronic media. The basic structure of communication, with its source, message, channel, receiver, barriers, and feedback, affect society and the media in an interactive way. The fact that TV is considered a primary source of news and information as well as entertainment makes its relationship to society all the more crucial.

From an economic point of view, advertising seems to be the price that is paid for our telecommunications, and yet this advertising represents a small portion of the country's economic base, especially in relationship to the sociological importance of telecommunications. This importance determines the value of stations and other media entities. The American capitalistic system, based on profit, must be carefully considered when one studies both the positive and negative effects of media.

Telecommunications and technology interrelate in a spiraling fashion. New developments in one cause new developments in the other. The various media also interrelate, and one medium has the potential to destroy or significantly alter another. Technology affects both artistic quality and communication ability of all telecommunications.

In the realm of politics, the electronic media are the primary method by which the public is now acquainted with candidates. They are also the primary scrutiny device for overseeing political actions. They must be given credit for the ability to bring about peaceful change, but they also must be carefully watched for unavoidable and avoidable bias in reporting politics.

In studying telecommunications, one should be very careful to assess criticism and understand reality. Nothing is perfect, but all elements of the media can be improved. The issues of giving people what they want versus what they need, reflecting or improving society, and balancing profit with public responsibility must be kept uppermost in the mind.

Thought Questions

1. Why are you interested in studying telecommunications?
2. In your opinion, what are the most valid of all of the criticisms leveled against radio and TV?
3. Suggest some worthwhile research studies that ought to be conducted concerning electronic media.
4. In what specific ways and to what extent has television affected your life and the life of your family?

Communication Systems

A wide variety of electronic media exists for the dissemination of entertainment and information. The oldest, of course, is radio, which came to the fore in the 1920s. Within less than the average lifetime, these media have proliferated into over a dozen forms. While some wonder what the market can bear, others marvel at what the market does bear. The various media both complement and compete with each other, experiencing both the slings and security of the free enterprise system. As these media develop, change is inevitable, brought about both by external and internal forces. Although all the media forms are young, they are already rich in history and adaptation.

Commercial Radio

2

Overview

Radio, the oldest of the telecommunications media, formed most of the models of entertainment and information that are common to the media today. Its beginnings are veiled in dispute. The early inventors lived in various countries and, in some instances, devised virtually the same inventions. Ironically, this was partly due to the fact that no communication system was available for people to learn what others were inventing. This led to innumerable rivalries, claims, counterclaims, and patent suits.

The earliest inventions crucial to the field were not even intended to be used for radio broadcasting. When radio waves were first discovered, there was consternation over the fact that they were so public. Many experimenters were involved in devising methods to make the airwaves private so that messages could be sent confidentially. Only a few visionaries foresaw the use of radio broadcasting as we know it today, and none of them could have predicted the trials and tribulations, successes and failures that radio has undergone.

This chapter chronicles those trials and tribulations, successes and failures by discussing the following material:

Early radio developments by such pioneers as Maxwell, Hertz, Marconi, Fleming, Fessenden, and De Forest

Early radio's relationship to the sinking of the *Titanic*

Patent problems before, during, and after World War I

The AT&T, GE, and Westinghouse formation of RCA

The contributions to radio networking by David Sarnoff and William Paley

Early radio station pioneers such as Terry, Scripps, and Conrad

The beginnings of KDKA and other early stations

A brief history of the NBC, CBS, ABC, and Mutual radio networks

The Radio Act of 1927 and the Communications Act of 1934

Equipment and programming advances of the "golden era of radio," including the contribution of Correll and Gosden

The conflict between newspapers and radio during the 1930s

Radio during World War II

The economic and programming ups and downs of postwar radio

The rise of FM radio

The status of radio stations and networks today, including the issues related to deregulation and technological changes

It is inconceivable that we should allow so great a possibility for public service as broadcasting to be drowned in advertising chatter.

*Herbert Hoover
while serving as secretary
of commerce*

(a)

(b)

FIGURE 2.1

(a) James Clerk Maxwell. (b) Heinrich Hertz. (Smithsonian Institution, Photo Nos. 56859 and 66606)

Early Inventions

Many people believe that radio originated in 1873 when James Clerk **Maxwell,** a physics professor at Cambridge University, England, published his theory of electromagnetism. His treatise predicted the existence of **radio waves** and how they should behave based on his observations of how light waves behave.[1]

Maxwell

Experiments to prove Maxwell's theory were undertaken by the German physics professor Heinrich **Hertz** during the 1880s. Hertz actually generated at one end of his laboratory and transmitted to the other end the radio energy that Maxwell had theorized. He thus proved that variations in electrical current could be projected into space as radio waves similar to light waves. In 1888 he published a paper that served as a basis for the theory of modern radio transmission. Originally radio waves were called "Hertzian waves," and today Hertz's name is used as a frequency measurement meaning cycles per second.[2]

Hertz

A battle was waged in the 1890s between General Electric (GE) and Westinghouse over whose patent would be adopted for nationwide electrical use. GE favored **direct current** (DC) and Westinghouse favored **alternating current** (AC). In 1896 Westinghouse won the contest and AC became the national standard. Long-distance radio wave radiation is dependent on AC generation, so it is fortunate for radio that this was the adopted standard.

DC versus AC

Guglielmo **Marconi,** often referred to as the "Father of Radio," expanded upon radio principles. Marconi, the son of a wealthy Italian father and an Irish mother, was scientifically inclined from an early age. Fortunately, he had the leisure and wealth to pursue his interests. Shortly after he heard of Hertz's ideas, he began working fanatically in his workshop, finally reaching a point where he could actually ring a bell with radio waves.

Marconi

FIGURE 2.2

Guglielmo Marconi, shown here with wireless apparatus about 1902. *(Smithsonian Institution, Photo No. 52202)*

Marconi then incorporated the Morse key into his system with the goal of transmitting Morse code by radio waves. Until this time, the transmission of Morse code had required the laying or stringing of wires from one reception point to another. To set his radio waves in motion, Marconi used Hertz's method, which was to generate a spark that leaped across a gap. To receive the signal, he placed metal filings in a glass tube. When the radio wave contacted the metal filings, they cohered and the glass tube then had to be tapped to loosen the filings to receive the next impulse. Marconi's first crude but effective system thus consisted of a Morse key, a spark, a coherer, and a tapper.

After he tested his invention outside his workshop by successfully transmitting throughout his estate and beyond, Marconi wrote to the Italian government in an attempt to interest them in his project. They replied in the negative. His determined Irish mother decided that he should take his invention to England. There, in 1897, he received a patent and the financial backing to set up the Marconi Wireless Telegraph Company, Ltd. Under the auspices of this company Marconi continued to improve on wireless and began to supply equipment to ships. In 1899 he formed a subsidiary company in the United States, the Marconi Wireless Company of America.[3]

Although Marconi maintained a dominant international position in wireless communication, many other people were experimenting and securing patents in Russia, Germany, France, and the United States. Until this time the primary use of wireless had been as a means of Morse code communication by ships at sea. Now some people were becoming intrigued with the idea of voice transmission.

Fleming

A significant step in this direction was taken by John **Fleming** of Britain in 1904. He developed the vacuum tube, which led the way to voice transmission. It was later developed further by others, particularly Reginald Aubrey Fessenden and Lee De Forest.[4]

FIGURE 2.3

Lee De Forest, shown here with wireless apparatus about 1920. (Smithsonian Institution, Photo No. 52216)

Fessenden, a Canadian-born professor who worked at the University of Pittsburgh, proposed that radio waves not be sent out in bursts—which accommodated the dots and dashes of Morse code—but rather as a continuous wave on which voice could be superimposed. He succeeded in obtaining financial backing from two Pittsburgh financiers and on Christmas Eve of 1906 broadcast to ships at sea his own violin solo, a few verses from the Book of Luke in the Bible, and a phonograph recording of Handel's "Largo."[5]

Fessenden

Lee **De Forest** is known primarily for the invention of the audion, an improvement on John Fleming's vacuum tube. It contained three electrodes instead of two and was capable of amplifying sound to a much greater degree than was previously possible. This tube was the most crucial key to voice transmission.

DeForest

Like Marconi, De Forest was fascinated with electronics at an early age and later secured financial backing to form his own company. However, De Forest experienced management and financial problems that frequently rendered him penniless and led him eventually to sell his patent rights to the Marconi Company and to American Telephone and Telegraph (AT&T).

De Forest was farsighted in his views of radio wave utilization, and he strongly advocated voice transmission for entertainment purposes. In 1910 he broadcast the singing of Enrico Caruso from the New York Metropolitan Opera House. Several years later he started a radio station of sorts in the Columbia Gramophone Building, playing Columbia records in hopes of increasing their sales. He was also hopeful of increasing the sale of wireless sets so that more of his audion tubes would be sold.[6]

Therefore, by 1910 radio waves had been theorized by James Maxwell, proven to exist by Heinrich Hertz, put to use with Morse code by Guglielmo Marconi, and developed for voice transmission by John Fleming, Reginald Fessenden, and Lee De Forest.

FIGURE 2.4

David Sarnoff working at his radio station position atop the Wanamaker store in New York. It was here he heard the *Titanic's* distress call, causing him to stay at his post for seventy-two hours to report the disaster. *(Courtesy of RCA)*

Early Control

During these early stages radio grew virtually without government control. The first congressional act to mention radio was the Wireless Ship Act of 1910, which required all ships holding more than fifty passengers to carry radios for safety purposes.

Wireless Ship Act of 1910

However, the rules concerning this safety requirement were not very stringent, as was proven with the 1912 sinking of the *Titanic*. As the "unsinkable" *Titanic* sped through the night on its maiden voyage, radio operators on other ships warned it of icebergs in the area, but the *Titanic's* radio operator, concerned with transmitting the messages of the many famous passengers to friends in other parts of the world, did not heed the warnings. When the *Titanic* did strike the fatal iceberg at about midnight of April 14, the wireless operator frantically transmitted SOS signals, but none of the nearby ships, which could have helped to save some of the one thousand passengers who died, heard the distress calls because their wireless operators had signed off for the night. The fact that seven hundred people were saved can be attributed to the fact that distant ships heard the SOS calls and steamed many miles to the rescue.

sinking of the *Titanic*

The wireless operators who had been receiving the passengers' messages also heard the distress calls, so for the first time in history people knew of a distant tragedy as it was happening. The wireless operator decoding the messages in New York was David **Sarnoff,** who later became president of RCA. He stayed at his post for seventy-two hours, relaying information about the rescue efforts to anxious friends and relatives and to the newspapers. This brought wireless communication to the attention of the general public for the first time.[7]

Radio Act of 1912

Congress then passed the **Radio Act of 1912,** which emphasized safety and required everyone who transmitted on radio waves to obtain a license from the secretary of commerce. The secretary could not refuse a license, but could assign particular wavelengths to particular transmitters. Thus, ship transmissions were kept separate from amateur transmissions, which were, in turn, separate from government transmissions. All this was done without any thought of broadcasting as we know it today.

World War I

At the beginning of World War I the government took over all radio operation. Ship-to-shore stations were operated by the Navy, and many ham radio operators were sent overseas to operate radio equipment. Perhaps even more important, patent disputes were set aside for the good of the country. Marconi's company, still the leader in wireless, had aroused the concern of American Telephone and Telegraph (AT&T) by suggesting the possibility of entering the wireless phone business. AT&T, in an effort to maintain its supremacy in the telephone business, had acquired some wireless patents, primarily those of Lee De Forest. The stalemate that grew out of the refusal of the Marconi Company, AT&T, and several smaller companies to allow one another to interchange patents had virtually stifled the technical growth of radio communications.

patent problems

With the onset of the war these disputes were set aside so that the government could develop the transmitters and receivers needed. World War I was also responsible for ushering into the radio field two other large companies, General Electric and Westinghouse. Both were concerned with electrical energy and were established manufacturers of light bulbs. Because both light bulbs and radio tubes require a vacuum, GE and Westinghouse assumed responsibilities for manufacturing tubes. GE had also been involved in the development of Ernst F. W. Alexanderson's construction of the alternator to improve long-distance wireless. During the war this alternator was perfected.

the alternator

After the war the patent problem returned and, as a result, GE began negotiating with the Marconi Company to sell the rights to its Alexanderson alternator. The Navy, which had controlled radio during the war, feared that this sale would enable Marconi, a primarily British company, to achieve a monopoly on radio communication. The Navy intervened and convinced GE president Owen Young to renege on the Marconi deal. Rumors circulated that President Woodrow Wilson himself encouraged the Navy action because he felt communication was a key to world power and he did not want Britain to obtain this power. This cancellation left GE sitting with an expensive patent from which it could not profit because it did not control other patents necessary for its utilization. But the patent placed GE in an excellent negotiating position because of its value for long-distance transmission.

government intervention

The Founding of RCA

Marconi ousted

What ensued from this situation was a series of discussions among American **Marconi, AT&T, GE,** and **Westinghouse** that culminated in the formation of **Radio Corporation of America (RCA)** in 1919. American Marconi, realizing with reluctance that it would not receive Navy contracts as long as it was associated with British Marconi, transferred its assets to RCA. The U.S. government was not directly involved in the negotiations carried on by Owen Young, but it was obvious to the Marconi people that their British ownership was an incurable liability.

GE, Westinghouse, AT&T, and RCA divisions

Individual stockholders received RCA shares for Marconi shares, and GE purchased the shares of American Marconi held by British Marconi. AT&T, GE, and Westinghouse also bought blocks of RCA stock and agreed to make patents available to one another, thus averting the patent problem and allowing radio to grow.

In the original agreement, GE and Westinghouse had exclusive rights to manufacture receiving sets; RCA had exclusive rights to sell them; and AT&T had sole rights to make, lease, and sell broadcast transmitters. Again, this was not undertaken with entertainment broadcasting in mind, but with emphasis on ship-to-shore transmission. It was accomplished in a postwar era when the government had very little direct control over radio and when the primary companies involved—British Marconi, American Marconi, AT&T, GE, Westinghouse, and RCA—were leery of one another and insecure about their own futures.[8]

Owen **Young** had put together this complicated international agreement, but it remained for David Sarnoff to convert it into a successful company operation. Sarnoff, a Russian immigrant who had supported his family from a very young age, was among the RCA employees who came from American Marconi. At the age of fifteen he had become an employee of Marconi and as such, at age twenty-one, received the tragic distress messages from the *Titanic*.

Sarnoff

Additional laws were passed that required radio equipment on ships, and these helped American Marconi's business to boom and Sarnoff's career to escalate. In 1915, at the age of twenty-four, he wrote a memo to Marconi management suggesting entertainment radio. It read in part as follows:

> I have in mind a plan of development which would make radio a "household utility" in the same sense as the piano or phonograph. The idea is to bring music into the home by wireless. . . . The problem of transmitting music has already been solved in principle and therefore all the receivers attuned to the transmitting wavelength should be capable of receiving such music. The receiver can be designed in the form of a simple "Radio Music Box" and arranged for several different wavelengths, which should be changeable with the throwing of a single switch or pressing of a single button.
>
> The "Radio Music Box" can be supplied with amplifying tubes and a loudspeaker telephone, all of which can be neatly mounted in one box. The box can be placed on a table in the parlor or living room, the switch set accordingly and the music received. . . .

The same principle can be extended to numerous other fields as, for example, receiving lectures at home which can be made perfectly audible; also, events of national importance can be simultaneously announced and received. Baseball scores can be transmitted in the air by the use of one set installed at the Polo Grounds. The same would be true of other cities. This proposition would be especially interesting to farmers and others in outlying districts removed from cities. By purchase of a "Radio Music Box," they could enjoy concerts, lectures, music recitals, etc., which may be going on in the nearest city within their radius. . . .

It is not possible to estimate the total amount of business obtainable with this plan until it has been developed and actually tried out; but there are about 15 million families in the United States alone, and if only one million or 7 percent of the total families thought well of the idea, it would, at the figure mentioned, ($75 per outfit) mean a gross business of about $75 million, which would yield considerable revenue.[9]

This idea was filed away as a harebrained notion, and Sarnoff had to wait for a more propitious time.

Early Radio Stations

Meanwhile, with restrictions lifted, many of the amateur radio enthusiasts began to experiment again. One of these was Professor Earle M. **Terry** of the University of Wisconsin, who broadcast weather reports and occasional music to farm areas near Madison over what is now WHA.[10]

The publisher of *The Detroit News,* William E. **Scripps,** fostered interest in radio by using his newspaper to advertise his radio broadcasts on WWJ, which in turn advertised his newspaper. In 1920 the Harding-Cox presidential returns were broadcast. An office boy, bearing news of the latest development, rushed from the newspaper editorial office to the conference room where the radio "studio" was in operation. It was estimated that in the neighborhood of five hundred amateurs heard these results.[11]

In Pittsburgh, Frank **Conrad,** a physicist and an employee of Westinghouse, resumed his amateur activities in his garage, programming music and talk during his spare time. A local department store began selling wireless reception sets and placed an ad for these in a local newspaper, mentioning that these sets could receive Conrad's concerts. One of Conrad's superiors at Westinghouse saw the ad and envisioned a market. Up until this time both radio transmission and reception had been for the technical-minded who could assemble their own sets. But it was obvious that sets could be preassembled for everyone who wished to listen to what was being transmitted.

Conrad was asked to build a stronger transmitter at the Westinghouse plant, one capable of broadcasting on a regular schedule so that people purchasing receivers would be assured listening fare. Thus, in 1920, Westinghouse became the first company to apply to the Department of Commerce for a special type of license to begin a broadcasting service. The station was given the call letters **KDKA** and was authorized to use a frequency away from amateur interference. KDKA launched its programming schedule with the

FIGURE 2.5

The *Detroit News* radio station first known as 8MK and then as WWJ. This picture from 1920 shows the method by which music was broadcast from a phonograph. Much of the transmitting equipment was developed by Lee De Forest. *(Courtesy of the Automotive History Collection, Detroit Public Library)*

Harding-Cox election returns, interspersed with music, and then continued with regular broadcasting hours. Public reaction could be measured by the length of the lines at department stores where radio receivers were sold.[12]

KDKA's success spurred others to enter broadcasting. Foremost among these was David Sarnoff, who could now dust off his old memo and receive more acceptance for his idea. He convinced RCA management to invest $2,000 to cover the Jack Dempsey-Georges Carpentier fight on July 2, 1921, and a temporary transmitter was set up in New Jersey for the fight. Fortunately, Dempsey knocked Carpentier out in the fourth round, for shortly after that the overheated transmitter became a molten mass. This fight, however, helped to popularize radio, and both radio stations and sets multiplied rapidly.

Dempsey-Carpentier fight

By 1923 radio licenses had been issued to over six hundred stations, and receiving sets were in nearly one million homes.[13] The stations had low power (usually 10 to 15 watts) and were owned and operated primarily by those who wanted to sell sets (Westinghouse, GE, RCA), and retail department stores, as well as radio repair shops, newspapers wanting to publicize themselves, and college physics departments wishing to experiment. One licensing aspect that would lead to later problems was that all stations were on the same frequency—**360 meters.** Stations in the same reception area worked out voluntary arrangements whereby they could share the frequency by broadcasting at different times of the day.

1923 status

FIGURE 2.6

Frank Conrad, who
worked with
experimental
equipment and
supervised the
construction of KDKA.
(Courtesy of KDKA,
Pittsburgh)

Early Programming

Programming was no problem in the early days. People were mainly interested in the novelty of picking up any signal on their battery-operated crystal headphone receivers. Programs consisted primarily of phonograph record music, call letter announcements, and performances by endless free talent who wandered in the door eager to display their virtuosity on this new medium. For a time all stations in some cities would go off the air one night a week so that listeners could pick up signals from distant places. This was usually referred to as "black night."

It is only natural, then, that early radio would produce some humorous anecdotes. For example, one early station, WJZ, operated from a shack on the roof of a building in Newark, New Jersey. It was accessible only by an iron ladder and hatchway and programmed mainly time signals, weather, and selections played on the Edison phonograph. On one occasion a woman was invited up to read a story, but after climbing the ladder and being pushed through the hatchway, she fainted.[14]

early anecdotes

A woman who was a strong, speech-making advocate of birth control asked to speak on radio. The people at the station were nervous about what she might say, but when she assured them that she only wanted to recite some nursery rhymes, they allowed her into the studio. She then broadcast, "There was an old woman who lived in a shoe/She had so many children because she didn't know what to do." She was not invited back.[15]

FIGURE 2.7

A tent atop a Westinghouse building in East Pittsburgh served as KDKA's first studio. It was the cause of some of early radio's unusual moments—such as the whistle of a passing freight train heard nightly at 8:30, and a tenor's aria abruptly concluded when an insect flew into his mouth. *(Courtesy of KDKA, Pittsburgh)*

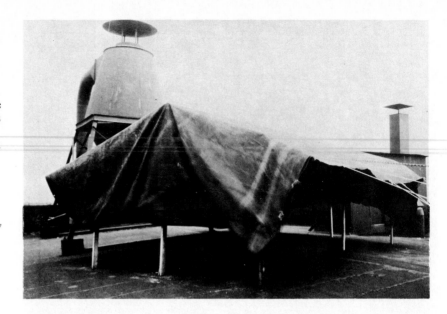

A young man in New Jersey wanted to let his mother know how he sounded over the air, so he dropped in at WOR, which had just opened a studio near the music department of a store. The singer the studio was expecting had not arrived yet, so this young man was put on the air before he even had time to notify his mother. He sang to piano accompaniment for over an hour as a messenger rushed sheet music from the music counter to the studio.[16]

A Chicago man wanted to discuss Americanism over a Chicago station and even submitted a script ahead of time. When he appeared at the station, he was with a group of bodyguards who covered the station premises to make sure that no buttons were pushed to take him off the air. It turned out that he was a potentate of the Ku Klux Klan, and, digressing from the script, he extolled the virtues of white supremacy.[17]

music

The primary programming of the era was was dubbed "potted palm" music—the kind played at teatime by hotel orchestras. Sometimes a vocalist was featured and sometimes a pianist or small instrumental group played. Sopranos outnumbered all other "potted palm" performers.

drama

Drama was also attempted even though engineers at first insisted that men and women needed to use separate microphones placed some distance from each other. Performers found it difficult to play love scenes this way. Finally it was "discovered" that men and women could share a mike.

public affairs

From time to time radio excelled in the public affairs area. Political conventions and presidential speeches were broadcast, as well as the funeral service for Woodrow Wilson. When the six-year-old son of Ernst F. W. Alexanderson, the builder of the alternator, was kidnapped, a radio report of the child's description was responsible for his recovery.

FIGURE 2.8

Los Angeles's first
station, KFI, began
broadcasting in 1922.
Its studio shows that
although only the
audio was received by
the audience, special
attention was given to
the decor, including the
potted palms. (Courtesy
of KFI, Los Angeles)

The Rise of Advertising

As the novelty of radio wore off, performers were less eager to appear and some means of financing programming had to be found. Many different ideas were proposed, including donations from citizens, tax levies on radio sets, and manufacturer and distributor payment for operating stations. Commercials came about largely by accident.

AT&T was involved mainly in the telephone business and was unwilling to see radio grow because the demand for wired services might be diminished. Therefore, one of its broadcasting entries was closely akin to phone philosophy. It established station **WEAF** in New York as what it termed a **"toll"** station. AT&T stated that it would provide no programming, but anyone who wished to broadcast a message could pay a "toll" to AT&T and then air the message publicly in much the same way as private messages were communicated by dropping money in pay telephones. In fact, the original studio was about the size of a phone booth. The idea did not take hold. People willing to pay to broadcast messages to the world did not materialize.

WEAF "toll" experiment

AT&T realized that before people were going to pay to be heard, they wanted to be sure that someone out there was listening. As a result, the non-programming idea was abandoned, and WEAF began broadcasting entertainment material, drawing mainly on amateur talent found among the employees. Still there were no long lines of people willing to pay to have messages broadcast.

Finally, on August 22, 1922, WEAF aired its first income-producing program—a ten-minute message from the **Queensboro Corporation,** a Long Island real estate company, which paid $50 for the time. The commercial was

Queensboro ad

just a simple courtesy announcement because AT&T ruled out direct advertising as poor taste and an invasion of privacy. (There was also strong sentiment against advertising toothpaste because it was an intimate product.)[18] Many people of the era said that advertising on radio would never sell products. In fact, every dollar of income that WEAF obtained was a painful struggle.

WEAF frequency change

What eventually made the station succeed was the fact that AT&T was able to convince the Department of Commerce that WEAF should have a different frequency. The argument was that other broadcasters were using their stations for their own purposes while WEAF was for everyone and therefore should have special standing and not be made to broadcast on 360 meters like everyone else. As a result, WEAF and a few other stations were assigned to the 400-meter wavelength. This meant less interference and more broadcast time. The phone booth was abandoned, a new studio was erected, and showmanship took hold.[19]

The Formation of Networks

Because AT&T was still predominantly in the phone business, it began using phone lines for remote broadcasts. It aired descriptions of football games, which came over long-distance lines, from Chicago and Harvard. It also established "toll" stations in other cities and interconnected them by phone lines—in effect, establishing a network.

AT&T's network

During this time AT&T did not allow other radio stations to use phone lines and also claimed sole right to sell radio "toll" time. At first, other stations were not bothered because they were not considering selling ads. In fact, there was an antiadvertising sentiment in the early 1920s. Herbert Hoover, then secretary of commerce, stated that it was inconceivable that a service with so much potential for news, entertainment, and education should be drowned in advertising chatter.

However, as the AT&T toll network emerged and began to prosper, other stations became discontent with a second-class status. The fires of this flame were further fanned by patent disputes, by AT&T's apparent attempt to enter the receiver-manufacturing business, and by a Federal Trade Commission inquiry that accused AT&T, RCA, GE, and Westinghouse of creating a monopoly in the radio business.

formation of NBC

A series of "behind-closed-doors" hearings was held by the major radio companies. The result of these complicated negotiations was the formation in 1926 of the **National Broadcasting Company (NBC)**—owned by RCA, GE, and Westinghouse—that was to handle broadcasting activities for the radio group. AT&T agreed to withdraw from the programming area and from the radio group in exchange for a long-term contract assuring that NBC would lease AT&T wires. This agreement was to bring the phone company millions of dollars per year. NBC also purchased WEAF from AT&T for $1 million, thus embracing the concepts of both "toll" broadcasting and networking. This was frequently referred to as chain broadcasting in those days because a chain of stations was interconnected.

In November of 1926 the NBC Red Network, which consisted of WEAF and a twenty-two-station national hookup, was launched in a spectacular debut that aired the New York Symphony Orchestra from New York, Mary Garden singing "Annie Laurie" from Chicago, Will Rogers mimicking President Coolidge from Kansas City, and dance bands from various cities throughout the nation. A year later NBC's Blue Network was officially launched. This consisted of different stations that had been part of a loosely knit RCA network. Both networks rapidly added stations to their chains, but the Red Network was able to outdistance the Blue.

At first, advertising on both networks was brief and low-key. Many advertisers simply associated their names with the programs—"Eveready Hour," "Ipana Troubadours," and "Maxwell House Hour."

In 1932 GE and Westinghouse withdrew from RCA, largely because of a U.S. attorney general's order that the group should be dispersed and partly because David Sarnoff, now president of RCA, felt his company should be an entity. Again a series of closed-door meetings resulted in a divorce settlement. RCA retained the radio manufacturing business, but GE and Westinghouse could now compete. RCA became the sole owner of NBC, and GE and Westinghouse received RCA debentures and some of the RCA real estate. In retrospect it appears that RCA walked off with the lion's share of value. But all this happened during the depression, and GE and Westinghouse were not overly eager to hold onto what they thought might be an expensive broadcasting liability. NBC, in spite of the depression, moved to new mid-Manhattan headquarters, dubbed Radio City.[20]

GE and Westinghouse withdraw from NBC

As can be seen, both RCA and NBC have an interesting parentage. NBC was originally owned by RCA, GE, and Westinghouse, who ousted AT&T in the process of NBC's formation. RCA was formed by GE, Westinghouse, and AT&T, who ousted Marconi during the process of RCA's formation. The exact details of all these corporate maneuvers will probably never be known.

What eventually became the **Columbia Broadcasting System (CBS)** began in 1927. Arthur **Judson,** disgruntled because Sarnoff had not accepted his offer to supply talent for the NBC network, established, with several associates, the United Independent Broadcasters. The original intent of the company was to supply talent to stations not in the network.

However, in the search for capital, the company joined with the Columbia Phonograph Company to form the Columbia Phonograph Broadcasting System. Judson's group was to supply talent and programs, and the Columbia group was to sell the programming to sponsors. Unfortunately, sponsors were not eager to oblige, and the project failed, with Columbia Phonograph Record Company pulling out.

Several investors tried to keep the project afloat and changed the name to Columbia Broadcasting System. Finally, in 1928, the family of William S. **Paley** supplied the needed capital, Paley became president, and the CBS network began to develop. It progressed steadily with Paley at the helm. He developed a workable relationship between the network and its affiliate stations and during the 1930s managed to lure much of the top radio talent from NBC to CBS.[21]

Paley and CBS

FIGURE 2.9

chain regulations and
ABC

The **American Broadcasting Company** (ABC) came to the fore because of **chain regulations** issued by the Federal Communications Commission (FCC) in the early 1940s. By 1940 the networks had established a power base that the FCC felt could be detrimental to local broadcasting. The contract between the network and its affiliated stations generally favored the network because each station in the chain needed the network more than the network needed any particular station. As a result, the NBC and CBS contracts with affiliates made it difficult to reject network programs. The stations had to prove that the rejected program was not in the public interest. In its chain regulations of 1941 the FCC stated that an affiliated station should have the right to reject a network program simply because it wanted to broadcast something else.

Another aspect of the chain regulations that resulted in the formation of ABC was the **duopoly rule,** which prohibited one company from owning and operating more than one national radio network. The FCC issued this ruling because it felt such concentration of control was unhealthy. This created a situation whereby NBC had to sell one of its networks. NBC at first formed a separate company to operate its Blue Network, then in 1943 sold that company to a group headed by Edward J. Noble, the Lifesavers millionaire. In 1945 this group changed the name of the Blue Network to the American Broadcasting Company.[22]

A fourth radio network, **Mutual,** was formed in 1934 when four stations decided to work jointly to obtain advertising. Unlike the other networks, Mutual owned no stations. It kept enough money from the ads it sold to recover its costs and to buy programs. Then it paid the rest to the stations in its network so that they would carry the programs and the network ads. It also allowed stations to sell their own ads.

Mutual

Mutual, too, was involved with the chain regulations. One of the stipulations of the NBC and CBS affiliate contracts was that the local stations could not carry programs from a different network. In 1938 Mutual gained exclusive rights to broadcast the world series, but the NBC and CBS contracts would not allow their affiliated stations to carry these games even in cities where there were no Mutual stations. The people wanted the games, the stations wanted to carry them, and advertisers wanted to pay for the coverage. Nevertheless, many Americans never did hear the 1938 world series. The FCC determined that this type of program thwarting was not in the public interest. As part of its chain regulations it stated that no station could have an arrangement with a network that hindered that station from broadcasting programs of another network.

Mutual gained radio affiliates more slowly than NBC, CBS, and ABC and never did enter television. In fact, it experienced unusually hard times when TV emerged and changed owners frequently.[23]

A fifth network, the **Liberty Broadcasting System,** was organized in 1946 and at one time served about four hundred stations. However, this network did not survive the rapid expansion of television and so suspended operations in 1951.[24]

Liberty

Chaos and Government Action

The problem of broadcast frequency overcrowding continued to grow during the 1920s. Secretary of Commerce Herbert **Hoover** was besieged with requests that the broadcast frequencies be expanded and that stations be allowed to leave the 360-meter quagmire, the frequency band on which all of them were broadcasting. He made various attempts to improve the situation by altering frequencies, powers, and broadcast times, and he called four national radio conferences to discuss with broadcasters problems and solutions to the radio situation. But he was unable to deal with the problem in any systematic manner because he could not convince Congress to give him the power to do so.

national radio conferences

By 1925 the situation had so deteriorated that the only remedy would have been to reassign frequencies being used for other purposes. However, under the existing law, the secretary of commerce was powerless to act in this regard. Hoover threw up his hands and told radio station operators to regulate themselves as best they could.[25]

In 1926–1927 there were some two hundred new stations, most of them using any frequency or power they wished and changing at whim. The airwaves were complete chaos. To help remedy this situation, Congress passed the **Radio Act of 1927.** The act proclaimed that radio waves belonged to the people and could be used by individuals only if they had a license and were broadcasting in the public "interest, convenience, and necessity."

All previous licenses were revoked and applicants were allowed sixty days to apply for new licenses from the newly created **Federal Radio Commission (FRC).** The commission gave temporary licenses while it worked out the jigsaw puzzle of which frequencies should be used for what purposes. In the end it granted 620 licenses in what is now the AM band. The FRC also designated the power at which each station could broadcast. Ten stations were authorized to operate at 50,000 watts; seventeen were to use between 10,000 and 50,000 watts; most were to use between 100 and 10,000 watts; but over 150 were left in the quicksand of less than 100 watts.[26]

Several years after the Radio Act of 1927, Congress passed the **Communications Act of 1934,** which created the **Federal Communications Commission** (FCC) and made permanent most of the provisions of the 1927 act. This act still governs broadcasting today.

The Golden Era of Radio

With the chaotic frequency situation under control, radio was now ready to enter the era of truly significant programming development—a heyday that lasted some twenty years. Improvements in radio equipment helped. **Earphones** had already been replaced by **loudspeakers** so that the whole family could listen simultaneously. The early **carbon microphones** were replaced by ribbon microphones, which had greater fidelity, and **single-dial tuning** replaced the **three-dial system** required on earlier receivers. For portability and use in automobiles, battery sets were introduced. (However, the first portables were cumbersome because of the size of early dry batteries.)

Radio became the primary entertainment medium during the depression. In 1930, 12 million homes were equipped with radio receivers, but by 1940 this number had jumped to 30 million. During the same period, advertising revenue rose from $40 million to $155 million. In 1930 NBC Red, NBC Blue, and CBS offered approximately sixty combined hours of sponsored programs a week. By 1940 the four networks (Mutual had been added) carried 156 hours.[27]

The first program to generate nationwide enthusiasm was **"Amos 'n' Andy."** It was created by Freeman Fisher **Gosden** and Charles J. **Correll,** who met while working for a company that staged local vaudeville-type shows throughout the country. Gosden and Correll, who were white, worked up a blackface act for the company and later tried this on WGN radio in Chicago as "Sam 'n' Henry." When WGN did not renew their contract, they took the show to WMAQ in Chicago and changed the name to "Amos 'n' Andy" because WGN owned the title "Sam 'n' Henry."

(a)

(b)

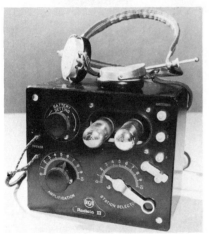

(c)

(d)

(e)

FIGURE 2.10

(*a*) A carbon microphone that was the best quality available during the formative years of radio. (*b*) An early station setup that included a carbon microphone, a multitubed audio board, and Westinghouse receivers. (*c*) A battery-operated radio receiver from about 1923. (*d*) Home radio receiver with speaker from about 1924. (*e*) Early backpack equipment for remote radio broadcasting. *(a, b, e courtesy of KFI, Los Angeles; c, d courtesy of RCA)*

FIGURE 2.11

Amos 'n' Andy as they
appeared when
broadcasting from
Studio B in NBC's
Hollywood Radio City.
Freeman Fisher
Gosden is on the left
side of the table with
Madaline Lee, Charles
Correll is at the right,
and sitting in the left
foreground is the
"Here th' are" man,
announcer Bill Hay.
*(Courtesy of KFI, Los
Angeles)*

Correll and Gosden wrote all the material themselves and played most of the characters by changing the pitch, volume, and tone of their voices. Gosden always played Amos, a simple, hardworking fellow, and Correll played Andy, a clever, conniving, somewhat lazy individual who usually took credit for Amos's ideas. According to the scripts, Amos and Andy had come from Atlanta to Chicago to seek their fortune, but all they had amassed was a broken-down automobile, also known as the Fresh-Air Taxicab Company of America. Much of the show's humor revolved around a fraternity-type organization called the Mystic Knights of the Sea headed by a character called Kingfish, who was played by Gosden.

WMAQ allowed Correll and Gosden to syndicate the show on other stations. Its success caught the attention of the NBC Blue Network, which hired the two in 1929 at $100,000 a year. Their program, which aired from 7:00 to 7:15 P.M. eastern time, became such a nationwide hit that it affected dinner hours, plant closing times, and even, on one notable occasion, the speaking schedule of the president of the United States.[28]

comedy

Many other comedians followed in the wake of the success of Correll and Gosden—Jack Benny, Lum and Abner, Ed Wynn, George Burns and Gracie Allen, Edgar Bergen and Charlie McCarthy, Bob Hope, Fibber McGee and Molly, Arthur Godfrey, and the Aldrich Family.

music

Music, especially classical music, was also frequently heard. There were broadcasts of the New York Philharmonic concerts and performances from

FIGURE 2.12

Jack Benny and his wife and costar, Mary Livingston, in 1933. The "Jack Benny Show," sponsored for many years by Jell-O and Lucky Strike, featured such sure-fire laugh provokers as an ancient Maxwell automobile that coughed and sputtered, Benny's perennial age of thirty-nine years, a constant feud with Fred Allen, and Benny's horrible violin playing. *(Courtesy of NBC)*

(a)

(b)

FIGURE 2.13

(*a*) Ed Wynn, also known as "the perfect fool." He received this label originally because he performed in the first Broadway stage show broadcast on radio, "The Perfect Fool." (*b*) Lum and Abner, played by Chester Lauck (*left*) and Norris Goff. This comedy took place in the Jot 'Em Down grocery store in the supposedly fictional town of Pine Ridge, Arkansas. In 1936 the town of Waters, Arkansas, changed its name to Pine Ridge in honor of Lum and Abner. *(a courtesy of NBC; b courtesy of KFI, Los Angeles)*

the Metropolitan Opera House. As a pet project of David Sarnoff, NBC established its own orchestra led by Arturo Toscanini. For lighter music, "Your Hit Parade," which featured the top-selling songs of the week, was introduced in 1935 and people who later became well-known singers, such as Kate Smith and Bing Crosby, took to the air. As the big bands developed, they too went out over the airwaves.

One program innovation was to involve the audience. Among many amateur hours, perhaps the most famous was the one hosted by Major Bowes. Quiz shows, such as "Professor Quiz," rewarded people for responding with little-known facts. Stunt shows, such as "Truth or Consequences," on which people had to undertake silly assignments if they did not answer questions correctly, were attracting large and faithful audiences.

audience participation

FIGURE 2.14

George Burns and Gracie Allen. Many of the jokes of this program were plays on words based on Gracie's supposed empty-headedness. At one point Gracie started searching for her "lost brother" by suddenly appearing on other shows to inquire about him. *(Courtesy of NBC)*

children's shows

Many programs were developed for children including "Let's Pretend," a multisegment program that emphasized creative fantasy; "The Lone Ranger," a western; "Uncle Don's Quiz Kids," on which a panel of precocious children answered questions; and "Little Orphan Annie," a drama about a child's trials and tribulations.

soap operas

During the day there were continuing dramas called "soap operas" because soap manufacturers were frequent sponsors. The segments always ended with an unresolved situation in order to entice the listener to tune in tomorrow. Most did. The scripts for a major portion of the soap operas were developed by a husband-wife team, Frank and Ann Hummert. They defined the basic idea for each series and wrote synopses of programs, then farmed the actual script-writing to a bevy of writers around the country, some of whom never even met the Hummerts.

drama

In the field of drama, the networks first tried to rebroadcast the sound of Broadway plays, but discovered that this was akin to sitting in a theater blindfolded. As a result, the networks hired writers such as Norman Corwin, True Boardman, Maxwell Anderson, and Stephen Vincent Benet to script original dramas for radio. These dramas usually employed many sound effects and were sponsored by one company that often incorporated its name into the program, such as "Lux Radio Theater" or "Collier's Hour." In 1938 Orson Welles produced **"War of the Worlds,"** a fantasy about a Martian invasion in New Jersey. Upon hearing the broadcast, an estimated 1.2 million people succumbed to hysteria. They panicked in the streets, fled to the country, and seized arms to prepare to fight—despite the fact that the "Mercury Theater" program included interruptions to inform the listener that the presentation was only a drama. Some people felt that radio had become overly realistic.[29]

commercials

It was mainly the depression that brought about the growth of commercials. During the 1920s, advertisements were brief, tasteful, and did not mention price. However, as radio stations and all the facets of the American

economy began digging for money at any price, the commercial standards dissolved. Some advertisers felt commercials should irritate, and broadcasters, anxious for the buck, acquiesced. The commercials became long, loud, dramatic, hard-driving, and cutthroat.

Most of the radio programs were produced not by the networks but by **advertising agencies.** They found they could combine advertising effectiveness with human misery. Thus, a large number of personal help programs developed. Listeners would send letters to radio human relations "experts" detailing traumas, crimes, and transgressions and to ask for help. Often this help had commercial tie-ins. Box tops accompanying the letter qualified it for an answer; or suggested help might involve the sponsor's drug product; or the contentment to be derived from puffing on the sponsor's brand of cigarette

FIGURE 2.17

Katherine Naiht,
House Jamison, Ann
Lincoln, and Ezra
Stone of "The Aldrich
Family." This
program's familiar
opening was "Henry,
Henry Aldrich," to
which Ezra Stone
replied, "Coming
Mother." The program
was based on the
Broadway play *What a
Life. (Courtesy of NBC)*

FIGURE 2.18

Arturo Toscanini and
the NBC orchestra.
Toscanini was coaxed
out of retirement in
Italy by David Sarnoff,
head of NBC and
classical music lover. A
special studio, 8H, was
built for the orchestra
and was referred to as
the world's only
floating studio because
of its unique
construction. *(Courtesy of
NBC)*

might be recommended. By 1932 more air time was spent on commercials than on news, education, lectures, and religion combined. The commercials did bring in profits for NBC, CBS, and some individual radio stations. They also brought profits to the advertising agencies that were intimately involved in most details of programming, including selecting program ideas, overseeing scripts, selling and producing advertisements for the shows, and placing the programs on the network schedule.[30]

stunt broadcasts

    There were also many events that could be termed **stunt broadcasts,** such as those from heights, depths, widely separated points, gliders, and underwater locations. A four-way conversation involved participants in Chicago, New York, Washington, and a balloon. One music program featured a singer in New York accompanied by an orchestra in Buenos Aires.

public events

    These stunt broadcasts paved the way for the broadcast of legitimate public events from distant points. In 1931 nineteen separate locations around the world participated in a program dedicated to Marconi. People were able to hear the farewell address of King Edward VIII when he abdicated the British

throne and the trial of the man who kidnapped Charles Lindbergh's baby.[31]

Radio also figured in politics of the day. President Franklin Delano **Roosevelt** effectively used radio for his **"fireside chats"** to reassure the nation during the depression. Louisiana's firebrand Governor Huey Long was often heard on the airwaves, and Father Charles E. Coughlin, a Detroit priest, tried to build a political movement through radio.[32]

politics

Commercial Radio   51

FIGURE 2.21

Rudy Vallee (*right*) with Charles Butterworth (*left*), Helen Vinson, and Fred Perry. Some of the many stars who got their start on this variety format show were Bob Burns, Bob Hope, Eddie Cantor, Alice Faye, Milton Berle, and Ezra Stone. *(Courtesy of KFI, Los Angeles)*

FIGURE 2.22

Dave Garroway (*right*) in Pittsburgh. The future host of the "Today" show was KDKA announcer from 1938 to 1940. *(Courtesy of KFI, Los Angeles)*

The Press-Radio War

newspaper objections

News was destined to become one of radio's strongest services, but not without a struggle. At first announcers merely read newspaper headlines over the air, but gradually networks began purchasing news from the wire services. In 1932 Associated Press sold presidential election bulletins to the networks and programs were interrupted with news flashes. Newspapers objected to all of this on the grounds that news on radio would diminish the sale of papers and, from 1933 to 1935, a press-radio war ensued.

FIGURE 2.23

Major Edward Bowes of "The Original Amateur Hour." Some of the winners from this amateur competition formed a touring Major Bowes company that provided talent employment during the depression. *(Courtesy of KFI, Los Angeles)*

A meeting of newspaper publishers, network executives, and wire service representatives, held at the Biltmore Hotel in New York in 1933, established the **Biltmore Agreement.** It stipulated that networks could air two five-minute newscasts a day consisting of material that they received from the established wire services. These newscasts had to be aired in the morning after 9:30 A.M. and in the evening after 9:00 P.M. so that they would not compete with the primary hours of newspaper sales. No"hot-off-the-wire" news was to be broadcast, and newscasts were not to have advertising support because this might detract from newspaper advertising. Newspapers were able to see that these provisions were placed in the Biltmore Agreement because they were the most numerous, most powerful, and wealthiest of the participants at the meeting.

Biltmore Agreement

But the ink on this agreement was barely dry when its intent began to be subverted. For one thing, a new company, Trans Radio Press, began selling news items to the radio world and the networks produced sponsored newscasts with this information. The newspapers countered by forcing Trans Radio Press out of business and, in some instances, newspapers boycotted offending radio networks by refusing to carry radio program schedules in their papers.

Trans Radio Press

Another way around the news restrictions was to use commentators. The publishers had agreed that radio and networks could have commentators, but often these commentators became thinly disguised news reporters.

commentators

Also, NBC and CBS began their own news-gathering activities. In the case of NBC, one man, Abe Schechter, gathered news simply by making telephone calls. Sometimes he scooped newspaper reporters because just about anyone would answer a call from NBC. In addition, Schechter could reward news sources with tickets to Rudy Vallee's show, which were highly prized. Most of the material Schechter collected was broadcast by NBC's prime

network news gathering

▲ THE INCOMPARABLE AMOS 'N' ANDY, returning to air via NBC Friday Oct. 8.

NBC PROUDLY

Presents:

T RAMP, TRAMP, TRAMP.
It's NBC's parade of stars marching along to open the fall and winter season of happy listening.

For dialers, the biggest news of all is the return of Amos 'n' Andy. The two old favorites introduce a brand new show on Friday, October 8, complete with guest stars, music and the kind of laughter which made Freeman Gosden and Charles Correll famous.

TRAMP, TRAMP, TRAMP.

"The Great Gildersleeve" started the parade by huffing and puffing his way back to his fans late in August. This is Hal Peary's third season on the air with his own program, and from the way the polls were going when he went off in June, it looks like his biggest.

TRAMP, TRAMP, TRAMP.

Fanny Brice, with more antics of

▲ THE INCORRIGIBLE BABY SNOOKS and Frank Morgan on Maxwell House show Thursday, 8:30 p. m.

▼ THE INVENTIVE ARKANSAS TRAVELER, Bob Burns, back on air Thursday night.

▼ THE INGENIOUS JACK BENNY, airing with all the gang at 4 p. m., Sunday.

▼ THE INFALLIBLE H. V. KALTENBORN, commentator, heard four afternoons weekly.

Page Four

FIGURE 2.24

A wartime plug for NBC's programs. *(Courtesy of KFI, Los Angeles)*

A Star-Bedazzled Parade Of Fast-Stepping Radio Entertainers, on March To Storm Your Listening

her inimitable Baby Snooks, and Frank Morgan, with a new batch of tall stories, followed "Gildersleeve" to NBC microphones on the first Thursday in September.

TRAMP, TRAMP, TRAMP.

Edgar Bergen and Charlie McCarthy marched back from Newfoundland, where they entertained the troops stationed there. This season they are presenting Victor Moore and William Gaxton, in addition to Ray Noble's orchestra and the songs of Dale Evans.

TRAMP, TRAMP, TRAMP.

That bad little boy, Red Skelton, was next in line, and with him were the popular members of his cast—Harriet Hilliard and Ozzie Nelson and his band.

TRAMP, TRAMP, TRAMP.

An account of Bob Hope's travels while away from his radio show for the summer sounds like a review of the war headlines—England, Bizerte, Tunis, Algiers, Sicily. Back on the air with him for the new season comes another grand trouper, Frances Langford, who also went into the battle areas with Bob. And, of course, Jerry Colonna and Vera Vague will be on hand.

TRAMP, TRAMP, TRAMP.

And so they come. Those two top

comedians, Jim and Marian Jordan, who have more delightful sessions with "Fibber McGee and Molly" ready for their listeners.

Eddie Cantor with another season of Wednesday night laughfests.

Jack Benny, another of radio's globe-trotters, only recently returned from the European and North African battlefronts.

And, of course, there are all the favorites who have been on NBC this summer and who will continue to make radio listening America's Number One pastime—"One Man's Family;" Bing Crosby; the Standard Symphony Hour; H. V. Kaltenborn and the other commentators who bring the world into our homes; Kay Kyser's "College of Musical Knowledge;" Ginny Simms; the Joan Davis-Jack Haley show; and the Sunday morning Westinghouse program.

Happy listening? Yes, indeed!

▲ THE IMPERTINENT CHARLIE McCARTHY and Bergen for Chase & Sanborn, Sunday, 5 p. m.

▲ THE INGRATIATING FATHER AND MOTHER BARBOUR on "One Man's Family," Sunday, 5:30 p. m.

▼ THE INIMITABLE FIBBER McGEE AND MOLLY, back with their friends Tuesday, 6 p. m.

▼ THE INEXHAUSTIBLE LAUGH CREATOR, Bob Hope, back in his regular Tuesday spot, 7 p. m.

▼ THE IRREPRESSIBLE LITTLE KID, as played by Red Skelton, Tuesday, 7:30 p. m.

Page Five

The NBC radio mobile unit making contact with an airplane. This 1929 experimentation led to future possibilities for news coverage. *(Courtesy of NBC)*

FIGURE 2.26

President Franklin Delano Roosevelt delivering a "fireside chat." *(Courtesy of NBC)*

newscaster, Lowell Thomas, but an item or two usually wound up on Walter Winchell's Sunday night gossip program. CBS set up a larger news force that included stringers—reporters paid only for material actually used. That network's top news commentator was H. V. Kaltenborn.

As world tensions grew, the public became increasingly aware of news. Advertisers became interested in sponsoring news radio programs because of the growing potential listener market. At one point, two services agreed to make their news available to advertisers who would then broadcast it over radio, but they would not make it available to radio stations directly. This arrangement led to a total breakdown of broadcast news blackouts, and radio began to develop as an important news disseminator. Americans heard actual sounds of the Spanish Civil War and of Germany's march into Austria, and they heard the voices of Hitler, Chamberlain, and Mussolini.[33]

FIGURE 2.27

Lowell Thomas began broadcasting in 1929 with one of the earliest programs, called "Headline Hunters," and remained on the radio regularly until after his eightieth birthday. For many years he preceded "Amos 'n' Andy" with the news, prompting him to say of himself, "Here is the bird that everyone heard while waiting to hear 'Amos 'n' Andy.'" *(Courtesy of Lowell Thomas)*

World War II

The government did not take over broadcasting during World War II as it had during World War I. However, it did solicit the cooperation of radio for morale and public service announcements, bond purchase appeals, conservation campaigns, and civil defense instructions. Among the most famous of these solicitations were singer Kate Smith's marathon broadcasts for war bonds. Her appeals sold over $100 million worth of bonds. Many of the plays and soap operas produced during the period dealt with the war effort, and some even tried to deal with the segregation problem, which was coming to a head because of segregation in the armed forces. Several soap operas presented Negroes (the preferred term at that time) in esteemed professional roles.

war efforts

(a) Walter Winchell with the signal key he used to accent his rapid-fire speaking style. He always worked with his hat on in the studio and always began his programs, "Good evening Mr. and Mrs. North and South America and all the ships at sea. Let's go to press." (b) H.V. Kaltenborn, the dean of radio commentators, received his greatest recognition during the 1938 Munich crisis, when he didn't leave the CBS studios for eighteen days and went on the air eighty-five times to analyze news from Europe. *(a courtesy of NBC; b courtesy of KFI, Los Angeles)*

(a)

(b)

The news function greatly increased as up-to-date material was broadcast at least every hour. Some of the best-known voices heard from overseas were H. V. Kaltenborn, Eric Sevareid, Charles Collingwood, and probably most important of all, Edward R. Murrow with his "This Is London" broadcasts, which detailed what was happening to the English during the war.

news

David Sarnoff, president of NBC, joined the Signal Corps and supervised the communication coverage for D day. William Paley, president of CBS, became a colonel who dealt in psychological warfare. Special events, such as President Roosevelt's war message to Congress and the signing of the surrender papers, were heard over radio.[34]

audio tape recorders

One result of the war was the perfection of audio tape recorders. Events could now be recorded and played back whenever desired. Prior to the war, NBC and CBS had policies forbidding the use of recorded material for anything other than sound effects, and even most of those were executed live. This policy was abetted by the musicians' union, which insisted that all broadcast music utilize musicians rather than phonograph records. The Mutual Broadcasting Company permitted some use of recorded speech, but was considered

(a)

(b)

(c)

FIGURE 2.29

(*a*) Correspondent Edward R. Murrow at his typewriter in wartime London. (*b*) David Sarnoff was promoted to brigadier general during World War II. (*c*) William S. Paley serving as colonel in the U.S. Army during World War II. *(a from United Press International; b courtesy of RCA; c courtesy of CBS)*

second-rate for doing so. As a result, the live programs usually had to be performed twice—once for the East and Midwest, and once again three hours later for the West Coast.

The recording technique used before the **audio tape recorder** usually employed phonograph discs, for the only magnetic recording known in America prior to World War II was **wire recording.** In order to edit or splice, a knot had to be tied in the wire and then fused with heat, making it a cumbersome and essentially unusable technique. During the war American troops entering German radio stations found them operating without any people. The broadcasting was handled by a machine that used plastic tape of higher fidelity than

Americans had ever heard from wire. This plastic tape could be cut with scissors and spliced with adhesive. The recorders were confiscated, sent to America, improved, and eventually revolutionized programming procedures.[35]

economic prosperity

Radio stations enjoyed great economic prosperity during the war. There were about 950 stations on the air when the war began. No more were licensed during the war, so these 950 received all the advertisements. A newsprint shortage reduced ad space in newspapers, and some of that advertising money was channeled into broadcasting. Institutional advertising became common because of high wartime taxes; companies preferred to pay for advertising rather than turn money over to the government. And with few consumer products to sell because industry was geared to the war effort, companies were happy to sponsor such prestige programs as symphony orchestra concerts. Thus radio station revenue increased from $155 million in 1940 to $310 million in 1945.[36]

Postwar Economics

Postwar radio was, above all, prosperous. Advertisers were standing in line, and the main problem was finding a way to squeeze in the commercials. To the networks, especially NBC, this boon provided the necessary capital to support the then unprofitable television development. In order to invest even more in the new baby, nonsponsored public affairs radio programs dropped by the wayside, as did some expensive entertainment. Radio fed the mouth that bit it.

station expansion

On the local level this prosperity created a demand for new radio station licenses as both entrepreneurs and large companies scrambled to cash in on the boom. The 950 wartime stations expanded in a rabbitlike fashion to well over two thousand by 1950.[37]

Part of this increase was the result of actions by the FCC and the courts. In the 1930s the FCC required that companies or individuals applying for a new station license prove that this station would cause little or no interference to existing local or distant radio stations. Shortly after the war, potential licensees had only to demonstrate lack of interference to local stations. This opened the door for many more stations in the frequency band.

Also in the 1930s the potential licensee had to prove that establishment of another station would not cause economic injury to other stations in the area. In other words, an applicant had to prove to the FCC's satisfaction that there was enough potential advertising revenue to go around. This thwarted many new stations, especially during the depression years. Shortly before the war, however, the U.S. Supreme Court ruled in *FCC v. Sanders Brothers Radio* that the effect on the public should be considered in designating new radio stations. It stated that the FCC was to rule in the public interest and should not be concerned with the economic well-being of particular stations. After the war, when the rash of new station applications came in, the FCC did not require that the issue of economic injury be dealt with.[38]

For several years it appeared that there was no such thing as economic injury to radio. Advertising revenues increased from $310 million in 1945 to $454 million in 1950.[39] But the bubble burst as advertisers deserted radio to try the medium that featured both sound and sight. This left radio networks as hollow shells. The two thousand local stations found that the advertising dollars remaining in radio did not stretch to keep them all in the black. In 1961 almost 40 percent of radio stations lost money.[40]

The period of the TV takeover tried the souls of those in radio. Some attempted to hang onto yesteryear and maintain traditional programming, but this was grasping at a straw in the wind. Others succumbed to economic pressures and left the business. Still others groped for solutions to the problem and eventually settled for disc jockey domination. After a long, hard pull, the disc jockeys won the favor of the public—and, hence, the advertisers—and once again radio assumed economic and social status.

This revival was due in part to group station owner Todd Storz. According to radio lore, Storz was in a bar one night trying to drown his sorrows over the sinking income of his radio stations and noticed that the same tunes seemed to be played over and over on the juke box. After almost everyone else had left, one of the waitresses went over to the jukebox. Rather than playing something that had not been heard all evening, she inserted her nickel and played the same song that had been played most often. Storz observed this and decided to try playing the same songs over and over on his radio stations. Thus was born top-40 radio, one of the main ingredients in radio's economic comeback.[41]

TV takeover (margin note)

top-40 radio (margin note)

Postwar Format

After the war, radio networks appeared for a time to return to prewar programming—comedy, drama, soap operas, children's programs, news, and public service. But the new phenomenon was beginning to appear on the scene— the **disc jockey.** Several conditions precipitated this emergence.

rise of disc jockeys (margin note)

For one, the FCC altered a ruling about identification of recorded material. Previously it had been necessary to identify all recordings as they were broadcast. Such frequent announcing would have stigmatized a DJ show. But in the 1940s the FCC ruled that such announcements could be made only each half hour.

Also, a court decision in 1940 ruled that if broadcasters purchased a record, they could then play it without further financial obligation. This ended the practice of stamping records "not licensed for radio broadcast" and added legal stature to disc jockey programs.

Then in the mid-1940s, the musician's union, which had voted to halt recording, was appeased with a musicians' welfare fund to which record companies would contribute. This opened the door to mass record production.

Radios became more portable at the same time that Americans were becoming more mobile. The public appreciated the disc jockey shows, which could be enjoyed while listeners were engaged in other activities.

Commercial Radio   61

As a final important reason, station management appreciated the lower overhead, fewer headaches, and higher profits associated with disc jockey programming. A DJ did not need a writer, a bevy of actors, a sound effects person, an audience, or even a studio. All that was needed were records, and these were readily available from companies who would eagerly court disc jockeys in the hope that they would plug certain tunes, thus assuring sales of the records.

payola

This courtship slightly tarnished the disc jockeys' image in the late 1950s when it was discovered that a number of disc jockeys had been engaged in **payola,** the practice of accepting money or gifts in exchange for favoring certain records. To remedy the situation, Congress amended the Communications Act so that if station employees received money from individuals other than their employers for airing records or other material, they had to disclose that fact prior to broadcast time under penalty of fine or imprisonment. This requirement and personal conscience effectively wiped out payola.[42]

decline of networks

As the stations began their romance with the disc jockeys and top talent left radio for TV, the stations had less and less need for the networks. The increasing number of stations also meant that more stations were in existence in each city, so more of them were programming independently of networks. Therefore, although the actual number of network-affiliated stations did not decrease, the percentage decreased dramatically because the new stations were independently programmed. The overall result was a slow but steady erosion of network programming that began in the 1950s. Eventually, radio stations essentially ceased to exist as conventional entertainment sources and became primarily news and music sources.

Frequency Modulation—FM

Armstrong

In the early 1930s David Sarnoff mentioned to Edwin H. **Armstrong** that someone should invent a black box to eliminate static. Armstrong did not invent just a black box, but a whole new system—**frequency modulation (FM).** He wanted RCA to back its development and promotion, but Sarnoff had committed RCA funds to television and was not interested in underwriting an entirely new radio structure despite its obviously superior clarity and fidelity.

Armstrong continued his interest in FM, built an experimental 50,000-watt FM station in New Jersey, and solicited the support and enthusiasm of GE for his project. In the late 1930s and early 1940s an FM bandwagon was rolling, and some 150 applications for FM stations were submitted to the FCC. As a result, the FCC altered channel 1 on the TV band and awarded spectrum space to FM. It also ruled that TV sound should be frequency modulated. Armstrong's triumphant boom seemed just around the corner, but the war intervened and commercial FM had to wait. However, FM went to war installed on virtually every American tank and jeep.

FIGURE 2.30

Edwin H. Armstrong.
*(Smithsonian Institution,
Photo No. 43614)*

After the war the FCC reviewed spectrum space and decided to move FM to another part of the broadcast spectrum, ostensibly because it felt that sunspots might interfere with FM. This move was violently protested by Armstrong and other FM proponents because it rendered all prewar FM sets worthless and saddled the FM business with heavy conversion costs.

Armstrong was further infuriated by the fact that although FM sound was to be used for TV, RCA had never paid him royalties for the sets it manufactured. In 1948 he brought suit against RCA. The suit proceeded for over a year and the harassment and illness it caused led him to leap from the window of his thirteenth-floor apartment to his death.[43]

FM continued to develop slowly. With television on the horizon, there was little interest in a new radio system. Many of the major AM stations acquired FM licenses as insurance in case FM replaced AM, as its proponents were predicting. They simply duplicated their AM programming on FM, which naturally did not increase the public's incentive to purchase FM sets.

However, as general interest in high-fidelity music grew, FM's interference-free signal became a greater asset. In 1961 the FCC authorized stereophonic sound transmission for FM, which led to increased awareness of the medium by hi-fi fans. At first classical and semiclassical music dominated the FM airwaves. But as hi-fi equipment became less expensive and the baby-boom generation reached its teens, rock music became prominent FM fare. This led to an increased number of listeners followed by an increased number of advertisers.

A further aid to FM's success was a 1965 FCC ruling stating that in cities of over 100,000 population, AM and FM stations with the same ownership had to have separate programming at least 50 percent of the time. This helped FM gain a foothold to the extent that it eventually overtook AM and the rule was dropped.[44]

moving FM

slow start

FM success

Deregulation

The Reagan era at the beginning of the 1980s ushered in an overall national philosophy toward **deregulation** that greatly affected the broadcast industry and led to the FCC eliminating many of the regulations that had governed radio for many years.

changes in station rules

Radio station licenses were granted for seven years rather than three. Radio stations were no longer limited in the amount of commercial minutes they could program and were no longer required to program news and public affairs programs. The stations did not need to keep logs and they did not need to **ascertain** the needs of the community in order to have their licenses renewed.[45] In addition, the number of radio stations any one company could own was raised from seven AM and seven FM to twelve of each. Companies did not need to keep stations for at least three years before they could sell them as had been previously required.[46]

Although many radio stations did not initiate any perceivable changes after the deregulation policies were initiated, many did greatly reduce or eliminate public affairs programs and some totally eliminated news. Quite a few companies did acquire additional stations and the pace of station trading picked up considerably from 424 stations changing hands in 1980 to 1558 in 1985.[47]

Reemergence of Networks

Today radio audiences listen to stations rather than programs. This is because most stations have a sound that usually translates into a particular format— country and western, top-40, easy listening, religious, jazz, all news, or talk. However, within this station-oriented structure, there has been a reemergence of networks. These do not serve the same function as the old radio networks that brought common programming to the entire nation. Instead, they tend to supply bits and pieces of programming or a particular format of music to a variety of stations.

ABC

ABC was the first to redevelop its network. In 1968 its shell of a network was providing primarily news to a limited number of affiliates. It received a waiver from the FCC's duopoly rule so that it could, in essence, operate four networks—American Contemporary, American Information, American Entertainment, and American FM. Each network provided some combination of news, sports, and features for a different target audience (e.g., teens, young adults, adults).[48]

CBS

Shortly thereafter CBS restructured its news network to add features and old-fashioned radio drama. The drama did not draw the audiences hoped for and was phased out.[49]

NBC

NBC experimented with News and Information Service, the intent of which was to make it economically possible for stations in small markets to afford an all-news format. This was discontinued in 1977 because of insufficient affiliates, but NBC then formed a network called The Source that was comprised of news and features for young adults. It also formed Talknet, a weekend talk radio service.[50]

Mutual also made some minor changes in its offerings and several other companies started radio networks, quite a few of which folded. Two companies that did survive, however, were Satellite Music Network and Transtar Radio Network. Both of these offered twenty-four hour music, news, and features in a variety of different formats (country-western, big band, rock) from which stations could choose.[51]

other networks

Others that have survived are networks aimed at blacks or Spanish-speaking audiences and networks that serve single states or geographic areas supplying various types of fare, such as farm reports, sports, and local news.[52]

The upstart in the network business is **Westwood One,** which purchased Mutual in 1985 and then in 1987 pulled an even bigger coup by buying the three NBC networks—NBC Radio Network, The Source, and Talknet—for $50 million. That meant the original radio network was no longer in the business although Westwood One continued to use the term NBC News.[53]

Westwood One

Going into the 1990s, ABC has the largest share of the network market (40 percent) with five networks, Westwood One has 30 percent of the market, and CBS and other networks have the rest.

Healthy Growth

During recent times, radio has been fairly healthy economically, leaving room for innovation and an increased number of networks and stations.

AM, in an effort to combat the success of high fidelity FM, initiated **AM stereo.** Several different companies proposed stereo transmission systems that, unfortunately, were not compatible with each other. The FCC, in its deregulatory mood, refused to rule on one common standard, letting the marketplace decide instead. All but Kahn and Motorola have pulled out of the race, and Motorola seems destined to become the standard because most of the stations broadcasting stereo are using that method.[54]

AM stereo

Engineers were able to carve additional stations out of the FM band and the FCC, under what it referred to as "Docket 80–90," began accepting applications for what could be as many as two thousand additional FM stations nationwide.[55]

Docket 80–90

Many AM stations received good news when Mexico and the United States signed an agreement that allowed over two thousand AMs to operate more hours per day than they had previously been allowed.[56]

longer hours

Although radio has undergone some cataclysmic changes, it has rebounded nicely, thank you, and today it is a healthy medium.

Commercial Radio Chronology

1873   Maxwell publishes theory of electromagnetism
1888   Hertz publishes theory of radio transmission
1896   AC becomes standard

1901	Marconi transmits letter "s" across the Atlantic
1904	Fleming develops the vacuum tube
1906	Fessenden broadcasts to ships at sea
1910	Wireless Ship Act of 1910
1910	De Forest broadcasts Caruso
1912	Sinking of the *Titanic*
1912	Radio Act of 1912
1915	Sarnoff writes radio memo
1917	Government takes over radio operations for war
1919	RCA is formed
1920	Harding-Cox presidential returns broadcast
1921	Dempsey-Carpentier bout aired
1922	Queensboro Corporation message aired
1926	NBC is formed
1927	Radio Act of 1927
1928	Paley becomes president of CBS
1929	Correll and Gosden on NBC as "Amos 'n' Andy"
1930–	
1950	Golden era of radio programming
1932	Westinghouse and GE withdraw from RCA
1933	Armstrong invents FM
1933–	
1935	Press-radio war
1934	Mutual Network formed
1934	Communications Act of 1934
1940–	
1945	World War II broadcasts
1941	FCC chain regulations
1945	Blue Network becomes ABC
1946	Liberty Broadcasting System organized
1950s	TV creates hard times for radio
1950s	Rise of the disc jockey
1958	Payola scandals
1961	FM stereo authorized
1965	FCC ruling that commonly-owned AM and FM stations must have separate programming
1968	ABC forms four networks
1979	RKO begins a radio network
1981	Radio deregulated
1982	FCC refuses to set AM stereo standard
1984	Radio ownership rules changed
1985	RKO is purchased by United Stations
1985	Mutual is purchased by Westwood One
1987	Westwood One buys NBC

Conclusion

Radio has survived through periods of experimentation, glory, and trauma. Early inventors, such as Maxwell, Hertz, Marconi, Fleming, Fessenden, and De Forest, would not recognize radio in its present form. Indeed, many of those who knew and loved radio during the 1930s and 1940s do not truly recognize it today. But radio has endured and along the way chalked up an impressive list of great moments: picking up the *Titanic's* frantic distress calls, broadcasting the Harding-Cox presidential election returns, the World War II newscasts of Edward R. Murrow, and its mere survival after the television takeover.

Along the way the government has interacted with radio in significant ways that illustrate the medium's growth as a broadcasting entity. The Wireless Ship Act of 1910 and the Radio Act of 1912 dealt with radio primarily as a safety medium. The fact that government took over radio during World War I but did not do so during World War II indicates that radio had grown from a private communication medium to a very public one that most Americans relied upon for information. The need for the government to step in to solve the problem of overcrowded airwaves in the late 1920s proved the popularity and prestige of radio. The ensuing Communications Act of 1934 and the various chain regulations helped solidify the government's role in broadcasting. The post-World War II rulings by the FCC led to a great increase in the number of radio stations on the air, including the late-blooming FM stations. The deregulation of radio in recent times is further evidence that radio has indeed survived.

Companies from the private enterprise sector have also played significant roles in the history of radio, starting with the Marconi Company and progressing through the founding of RCA by the still-powerful AT&T, GE, and Westinghouse. These early companies contributed a great deal in terms of technology, programming, and finance. As the networks were formed, each in its own peculiar way, they set the scene for both healthy competition, and elements of unhealthy intrigue. Intrigue also characterized the free enterprise rivalry between newspapers and radio in the prewar days, and certainly it was free enterprise in its purest sense that altered the format of radio when television stole its listeners. The recent formation of new networks is further proof that radio has survived.

Radio programming is indebted to early pioneers who filled the airwaves with boxing matches, "potted palm" music, and call letters, and to "Amos 'n' Andy," Jack Benny, and other contemporaries who are remembered for creating the golden era of radio. Today it is countless disc jockeys and newscasters who let us know that radio has survived.

Thought Questions

1. What are some methods of financing radio, other than advertising, that might have been successful? What would have been the advantages and disadvantages of each?
2. How might patents have been handled so that they did not impede the development of radio technology?
3. How do you think radio will change in the next ten years?

Commercial Television

Overview

To some people it may sound presumptuous to suggest that anything as young as television even has a history. To other people it
may seem as if there were no history before television. Some say
television is still in its infancy, and others argue that it is past its
prime. Television has been dominant only since 1952, but the years
have been a blur of technological change and programming turn-
over.[1]

The following elements of TV history are considered in this
chapter:

Early mechanical and electronic scanning developments

Sarnoff's decision to develop TV and display it at the 1939 world's fair

Experimental programming of the early 1940s

The adoption of the 525–line black-and-white system for channels 2
through 13

World War II stifling of TV

The postwar emergence of TV as a popular medium

The explosion of TV during the "freeze"—1948 to 1952

"Texaco Star Theater" and other early programs

The allocation table worked out for the lifting of the "freeze" and the
ensuing rush to obtain stations

Blacklisting and its demise during the 1950s

The golden age of television

The innovations of Pat Weaver

The controversies surrounding color TV and its eventual development

The invention of videotape and the influence of film

The quiz scandals of the late 1950s

Problems of UHF stations

The journalistic coming of age of TV during the 1960s

Cycles of violent and nonviolent programming

The interrelationships of TV and government during the 1970s and
1980s

The takeovers of many broadcasting entities

The formation of Fox Broadcasting Company

The varied programming of recent years

3

*Until a few years ago, every
American assumed he possessed
an equal and God-given expertise
on three things—politics, religion,
and weather. Now a fourth has
been added—television.*

*Eric Sevareid
longtime broadcast
newsperson and commentator*

FIGURE 3.1

Early mechanical
scanning equipment.
Through a peephole,
J.R. Hefele observes
the image recreated
through the rotating
disc. The scanning disc
at the other end of the
shaft intervenes
between an illuminated
transparency and the
photoelectric cell. This
cell is in the box that is
visible just beyond the
driving shaft. *(Courtesy
of AT&T Co. Phone Center)*

Early Experiments

The first experiments with television employed a **mechanical scanning** process originally invented by German Paul **Nipkow** in 1884. This process was dependent on a wheel that contained tiny holes positioned spirally. Behind the wheel was placed a small picture. As the wheel turned, each hole scanned one line of the picture.

Even though this device could scan only very small pictures, attempts were made to promote it commercially. C. F. Jenkins, an American, developed a workable system and formed a company in 1930 to exploit the idea. John Baird of Britain obtained a television license in 1926 and convinced the British Broadcasting Corporation to begin experimental broadcasting with a mechanical system.

Nipkow

Jenkins and Baird

At GE's plant in Schenectady, New York, Ernst F. W. **Alexanderson** began experimental programming in the 1920s using a revolving scanning wheel and a three-by-four-inch image. One of his "programs," a science fiction thriller of a missile attack on New York, scanned an aerial photograph of New York that moved closer and closer and then disappeared to the sound of an explosion.[2]

Alexanderson

While mechanical scanning was being promoted, other people were developing **electronic scanning,** the system that has since been adopted. One was Allen B. **Dumont,** who developed the oscilloscope, or cathode-ray tube, a basic electronic research tool that is similar to the TV receiver tube. Dumont was able to capitalize on this invention when the TV receiver market took hold in the 1940s.[3]

Dumont

Another early electronic inventor was Philo T. **Farnsworth,** who in 1922 astounded his Idaho high school teacher with diagrams for an electronic TV system. He convinced a backer to provide him with equipment and in 1927 transmitted still pictures and bits of film. He applied for a patent and found himself battling the giant of electronic TV development, RCA. In 1930 Farnsworth, at the age of twenty-four, won his patent and later received royalties from RCA.[4]

Farnsworth

FIGURE 3.2

Early television
experimenters. Allen
Dumont (*right*) giving
a personalized tour of
the Dumont
Laboratories to Lee De
Forest. *(Smithsonian
Institution, Photo No.
73–11097)*

FIGURE 3.3

Philo T. Farnsworth in
his laboratory about
1934. *(Smithsonian
Institution, Photo No. 69082)*

The RCA development was headed by Vladimir **Zworykin,** a Russian
immigrant and onetime Westinghouse employee who had patented an electronic pickup tube called the **iconoscope.** In 1930 GE, RCA, and Westinghouse merged TV research programs at RCA's lab in Camden, New Jersey,
and incorporated Zworykin, Alexanderson, and other engineers into a team
to further develop television.

Zworykin

This group systematically attacked and solved such problems as increased lines of scanning, definition, brightness, image size, and frequency of
scanning. They started with a system that scanned sixty lines using a model
of Felix the Cat as the star. Gradually they improved this scanning to 441
lines.

In 1932 experimental broadcasts were transmitted from the Empire State
Building. Three years later, in the midst of the depression, David Sarnoff, now
president of RCA, announced that the company would invest millions in the
further development of television. Experimental broadcasts continued with

experimental broadcasts

FIGURE 3.4

The television receiving screen used during the first intercity TV broadcast on April 7, 1927. Dr. Herbert E. Ives, former director of electro-optical research at Bell Telephone Laboratories, stands beside this screen, which consisted of fifty neon-filled tubes. Each tube was divided into small segments, creating a pattern of light and dark areas to form a picture. Dr. Ives is holding a photoelectric cell from the transmitter. Early scanning equipment is seen at the left. *(Courtesy of AT&T Co. Phone Center)*

programs emanating from a converted radio studio, 3H, in Radio City. Because the iconoscope was not a very sensitive pickup tube, actors had to wear green makeup and purple lipstick and simmer under intensive lights in order for a discernible black-and-white picture to be produced.[5] However, quality at this point was better than it had been under the mechanical scanning or the sixty-line "Felix the Cat" system.

The "Coming Out" Party

Sarnoff decided to have television displayed at the 1939 New York world's fair. President Roosevelt appeared on camera and was seen on sets with five- or seven-inch tubes.[6]

"mobile" productions

In 1939 RCA's program schedule usually included one program a day from 3H, one from a mobile unit traveling the streets of New York, and several assorted films. The studio productions included plays, bits of operas, singers, comedians, puppets, and household tips. The mobile unit consisted of two huge buses, one jammed with equipment to be set up in the field and one containing the transmitter that broadcast back to the Empire State Building. It covered such events as baseball games, wrestling, ice skating, airport interviews with dignitaries, fashion shows, and the premiere of *Gone With the Wind*. The films were usually cartoons, travelogues, or government documentaries.

early color TV controversy

Other companies established experimental stations and broadcast in New York. Sets, mostly manufactured by RCA and Dumont, had increased tube sizes to twelve inches and sold for $200 to $600. CBS began experimentation with color television, utilizing a mechanical color wheel of red, blue, and green that transferred color to the images. This color system was not compatible with the RCA-promoted system. In other words, the sets being manufactured could not receive either color or black-and-white pictures from the CBS mechanical system, and proposed CBS receivers would not be able to pick up existing black-and-white pictures.

(a)

(b)

In 1940 a group led mainly by RCA personnel tried to convince the FCC to allow the operation of the **441-line system.** However, the FCC was not certain that this system had adequate technical quality so it established an industrywide committee of engineers, the **National Television System Committee (NTSC),** to recommend standards. This committee rejected the 441-line system and recommended the **525-line system,** which the United States presently uses. CBS approached the committee with the idea of color television, but the committee did not think the system was of sufficient quality.

In May of 1941 the FCC authorized the full operation of 525-line black-and-white television. Originally there were to be thirteen very high frequency (VHF) channels, but channel 1 was eliminated to allow spectrum space for FM radio. Twenty-three stations went on the air, ten thousand sets were sold, and commercials were sought. The first commercial was bought by Bulova for $900 and consisted of a shot of a Bulova clock with an announcer intoning the time. *beginning authorization*

All of this, however, was called to a halt in 1942 because of World War II. During the war only six stations remained on the air, and most sets became inoperable because spare parts were not being manufactured for civilian use. The NBC studio was used to broadcast air raid warden-training programs; volunteers traveled to their local police station to watch this instruction. So, in a very limited way, television went to war.[7] *World War II*

The Emergence of Television

TV activities did not resume immediately after the war. The delay was in part due to a shortage of materials. Also, it was expensive to build and operate a TV station—the initial investment ranged from $75,000 to $1.5 million. And then the owners had to assume that they would operate at a loss until there were enough receivers in the area to make the station attractive to advertisers.

David Sarnoff dedicating the RCA pavilion at the 1939 New York World's Fair. This dedication marked the first time a news event was covered by television. Sarnoff's speech, entitled "Birth of an Industry," predicted that television one day would become an important entertainment medium. *(Courtesy of RCA)*

UHF possibilities

color

enormous 1948 growth

To add to the risk, there were rumors that all television stations might be moved to the **ultrahigh frequency** band (UHF), which had been explored during the war. CBS once again raised the question of **color,** stating that its color system was so well developed that TV station allocations should not be made until the color question was resolved. RCA represented a contrary position on the color issue and promised black-and-white sets on the market by mid-1946 and a color system shortly thereafter. The system RCA proposed was electronic and compatible with current sets as opposed to the CBS system, which was mechanical and incompatible. In the fall of 1946 RCA did demonstrate a color system that was compatible, albeit unstable and unreliable.

In 1947 the FCC declared that CBS's color system would be a hardship on set owners because they would have to buy new sets. It therefore stated that television should continue as black and white and in the VHF range it had been using, channels 2 to 13.

In the next year, 1948, television emerged as a mass medium. Stations, sets, and audience all grew over 4,000 percent within that one year. Advertisers became aware of the medium, and networks began more systematic programming.[8]

TV networks had already existed before 1948 as offshoots of the radio networks. As early as 1945–46 television networks had been organized by NBC, CBS, and ABC. A fourth network, Dumont, was organized by Allen B. Dumont. However, most cities only had one or two TV stations, and NBC and CBS usually recruited them as affiliates, making it difficult for ABC and Dumont to compete. ABC survived because it merged in 1953 with United Paramount Theaters and thus gained an increase in operating funds. Dumont, however, went out of business in 1955.[9]

The Freeze

Television grew so uncontrollably in 1948 that in the fall of that year the FCC imposed a **freeze** on television station authorizations because stations were beginning to interfere with each other. Mindful of the days of radio chaos, the FCC wanted to nip the problem in the bud. Additional reasons were that some

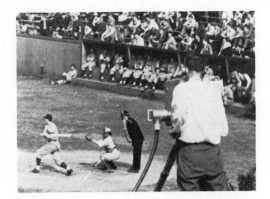

of the VHF characteristics had not been predicted, there were not enough channels to meet the demand, and CBS once again raised the question of color. The freeze, which was originally predicted to last about six months, lasted until July 1, 1952. During this period 108 stations were on the air and no more were authorized to begin operation.

What occurred between 1948 and 1952 could be termed an explosive lull. Many cities, including Austin, Denver, and Portland, had no television stations. Others, including Pittsburgh, St. Louis, and Milwaukee, had only one. Twenty-four cities boasted between two and six stations. New York and Los Angeles were the only cities with seven.

Although TV networking still could not be considered truly national, the number of sets, audience size, advertising, and programming continued to grow. By 1952 there were sets in 15 million homes. The largest ones had twenty-inch tubes and sold for about $350. TV advertising revenues reached $324 million. In TV cities, movie attendance, radio listening, sports event attendance, and restaurant dining were all down—especially on Tuesday night, which was Milton **Berle** night.

People with TV sets stayed home and often invited their friends over (or allowed their friends to invite themselves over) to watch this ex-vaudevillian's show, which included outrageous costumes, slapstick comedy, lavish productions, and a host of guest stars. "Uncle Miltie," on his "Texaco Star Theater," became a national phenomenon and the reason many people bought their first television set.[10]

continued growth

Milton Berle

Early Programming

During 1948 and 1949, 30 percent of sponsored evening programs were sports—basketball, boxing, bowling, wrestling, roller skating. Among the wrestlers, both men and women, there was competition to outdo one another in costumes, hairdos, and mannerisms. The emphasis on sports was due, at least in part, to the fact that a large number of the first sets were in bars and taverns. But during the 1949–50 season sports comprised less than 5 percent of evening programming and children's programming was tops, indicating that TV had moved to the home.

sports

Commercial Television   75

FIGURE 3.8

(*a*) Milton Berle in one of his outlandish costumes. His "Texaco Star Theater" was TV's first big hit. (*b*) "Your Show of Shows," starring Imogene Coca (*right*) and Sid Caesar. The regular cast included Carl Reiner and Howard Morris. Some of the writers for the show were Mel Brooks, Woody Allen, and Neil Simon. Much of the comedy stemmed from Sid's masterful use of verbal and facial expressions. *(Courtesy of NBC)*

(a)　　　　　　　　　　　(b)

comedy and variety

Other broadcasting fare of the freeze era included comedy team Sid Caesar and Imogene Coca in "Your Show of Shows," Ed Sullivan hosting a variety of acts for "Toast of the Town," and Groucho Marx's quiz show "You Bet Your Life."

"Amos 'n' Andy" was brought to TV and featured black actors trained by Gosden and Correll. The program was condemned as an insult by the National Association for the Advancement of Colored People at its 1951 convention, much to the astonishment of the lily-white broadcasting fraternity.[11]

children's programming

For children there was Buffalo Bob and his puppet Howdy Doody, and for children of all ages, Burr Tillstrom's puppets starred with Fran Allison in "Kukla, Fran, and Ollie." Talent shows abounded, led by Arthur Godfrey's "Talent Scouts" and Ted Mack's "The Original Amateur Hour."

drama

Drama was initiated with "Kraft Television Theater," "Philco Playhouse," and "Studio One." The first soap opera was "A Woman to Remember," which was forgotten after its first year, but others followed successfully in its wake. For example, "Love of Life," which began in 1951, lasted until the late 1970s with one of the original members still in the cast.

"I Love Lucy"

In 1951 **"I Love Lucy"** began as a maverick of the TV world because it was filmed ahead of time, while other shows were aired live. There was a stigma against film at the time, partly because it added to the cost of the show and partly because the TV networks inherited the live tradition from radio and assumed that all shows should be produced live. But the film aspect was particularly useful and dramatic when Lucille Ball became pregnant and, hence, the story line dealt with Lucy's pregnancy. The episode involving the birth of Lucy's baby was filmed ahead of time, and Lucille's real baby was born the same day the filmed episode aired—to an audience that comprised 68.8 percent of the American public. Eventually, the filming more than paid for itself because the program became the first international hit. Copies of the film were made, dubbed into numerous languages, and sold overseas.

TV newscasts of the 1950s developed slowly. Networks found it easy to obtain news and voices, but pictures were another matter. At first they contracted with the companies who supplied the newsreels then shown in theaters. This did not exactly fill television's bill because much of it was shot for in-depth stories, not news of the day. However, this system could not begin to cover all the news. Networks set up their own film crews but limited budgets and bulky film equipment meant that camera operators could attend only planned events such as press conferences, coronations, political conventions, and ribbon-cutting grand openings. The fifteen-minute newscasts tended to be reports on events that had been filmed earlier, which limited their timeliness.

Interview-type news shows were a further development. "Meet the Press" began its long run of probing interviews with prominent people. "See It Now" started in 1951 as a news documentary series featuring Edward R. **Murrow**

news

public affairs

FIGURE 3.11

Lucy tries to make a
hit playing the
saxophone. The "I
Love Lucy" series
starring Lucille Ball
captured America's
heart and can still be
seen in reruns. *(Photo/
Viacom, Hlwd.)*

and was produced by Fred **Friendly.** A historical feature of their first program
was showing for the first time both the Atlantic and Pacific Oceans live on
TV.

politics

The Senate crime hearings were also shown in 1951. At one point witness Frank Costello objected to having his face televised, so the cameras focused on his hands as the audience tried to gain clues from his gestures.

The importance of TV to the political process was already becoming
evident as the 1952 political nominating conventions were covered by the TV
networks. NBC's coverage was sponsored by Westinghouse and brought fame
to Betty Furness as she demonstrated refrigerators over and over in live commercials, including one never-to-be-forgotten spot when the "easy-to-open"
refrigerator door she had just described refused to open at all.[12]

At this point television could be proud of its achievements. Sets had increased in size and quality, TV programming had increased in hours and variety, the public was fascinated with the new medium, and advertisers were
providing increasingly strong financial backing.

Lifting the Freeze

In April of 1952 the FCC issued a report that ended the freeze on new stations
that had begun four years earlier. This report included a station allocation
that was worked out by FCC engineers on the basis of various parameters.
For AM radio, the FCC had allocated stations on a first-come, first-served

basis with the burden resting on the applicant to prove that the proposed station would not cause interference with other established stations. What the FCC discovered with this process was that first-come stations were generally sought in the more populous and hence more economically viable areas, and that by the time people were interested in establishing stations in small towns or rural areas, appropriately cleared frequencies were difficult to get. This meant residents of nonpopulous areas were not as well served as were residents of populous areas.

allocation tables

The FCC decided this injustice should not be repeated with TV, and as a result it allocated specific stations to particular geographic areas. These stations were to be reserved for such areas until someone was interested in establishing the stations.

Another fact apparent to the FCC as it undertook its work between 1948 and 1952 was that the twelve VHF channels originally planned would not be sufficient to meet demand; additional channels needed to be found. The engineers determined that there was not adequate room left in the VHF band, and they opted for adding seventy stations in the UHF band, for a total of eighty-two channels. UHF was at a much higher frequency than VHF, and very little was known about the technical characteristics of UHF at that time. However, the FCC engineers felt that by increasing the power and tower height of UHF stations, they would be equal in coverage to VHF stations.

Another item that had to be provided for in the allocation table was reservation for educational channels. Hearings on this subject were held during the freeze period, and the FCC decided to set aside 242 station assignments for noncommercial educational purposes. The allocation table that the FCC published listed 2,053 station allocations in 1,291 communities. In general, cities had both VHF and UHF stations assigned to them.

The freeze lift led to an enormous rush to obtain stations. Within six months, six hundred applications had been received and 175 new stations had been authorized. By 1954, 377 stations had begun broadcasting and TV could be considered truly national.[13]

rush for stations

Blacklisting

To this fledgling industry came some of the country's best-known talent. They came from radio, from Broadway, and from film—all of which were experiencing downturns as television was burgeoning.

Unfortunately, many of these people became caught in the **blacklisting** mania of the 1950s led by Senator Joseph R. **McCarthy,** a Republican from Wisconsin. A 215-page publication called *Red Channels: The Report of Communist Influence in Radio and Television* listed 151 people, many of whom were among the top names in show business, most of them writers, directors, and performers. Listed after their names were citations against them. Some of these charges were proved to be totally false, such as associating people with "leftist" organizations to which they had never belonged. Others were

Red Channels

true, but were "leftist" only by definitions of the perpetrators of *Red Channels*. These included such wrongdoings as belonging to an End Jim Crow in Baseball Committee and signing a cablegram of congratulations to the Moscow Art Theater on its fiftieth birthday.

Although many network and advertising executives did not believe these people were Communists or in any way un-American, they were unwilling to hire them, in part because of the controversy involved and in part because sponsors received phone calls that threatened to boycott their products if programs employed these people.

For the better part of a decade some well-established writers found that all the scripts they wrote were "not quite right," and certain actors were told they were "not exactly the type for the part." Many of these people did not even know they were on one of the "lists" because these were circulated clandestinely among executives.

In time the blacklist situation eased. Ironically, broadcasting was influential in exposing the excesses of the Communist witch-hunt, which had spread beyond the entertainment industry.

McCarthy's downfall

Edward R. Murrow and Fred Friendly presented a "See It Now" program entitled "The Case Against Milo Radulovich, A0569839." Radulovich had been asked to resign his Air Force commission because his sister and father had been anonymously accused of radical leanings. The program exposed some of the guilt-by-association tactics and was instrumental in helping Radulovich retain his commission. Later Murrow and Friendly prepared several programs on Senator Joe McCarthy, who had alarmed the country by saying that he had a list of hundreds of Communists in the State Department. The Murrow-Friendly telecasts helped reveal this as a falsity.

Later there was network coverage of the 1954 hearings of McCarthy's dispute with the Army. As the nation watched, McCarthy and his aides harassed and bullied witnesses. Public resentment built up against McCarthy and the Senate voted 67 to 22 to censure him.[14]

The Live Era

Programming of the 1950s was predominantly live. "I Love Lucy" continued to be filmed and several other programs jumped on the film bandwagon as foreign countries began developing broadcasting systems. Americans could envision the reuse of their products in other countries and, hence, the possibility to recoup film costs. **Reruns** in the United States itself were as yet unthought of, although some programs were **kinescoped** so they could be shown at various times. These kinescopes were low-quality, grainy-film representations of the video picture. However, most of the popular series of the day originated in New York and were telecast as they were being shot.

kinescopes

This live aspect created problems for writers, actors, and technicians. Costume changes needed to be virtually nonexistent. The number of story locales was governed by the number of sets that could fit into the studio. Timing

production problems

was sometimes an immense problem. In radio, scripts could be fairly accurately timed by the number of pages, but television programs contained much action, the time of which often fluctuated wildly in rehearsals. One writing solution was to plan a search scene near the end of the play. If the program was running long, the actor could find what he was looking for right away. If the program appeared to be moving too quickly, the actor could search the room for as long as necessary.[15]

The programming of the early 1950s is often looked back upon as the **golden age** of television. This is mainly because of the live dramas that were produced during this period. One of the most outstanding plays was Rod **Serling's** *Requiem for a Heavyweight,* the psychological study of a broken-down fighter. Another was Paddy **Chayefsky's** *Marty,* the heartwarming study of a short, stocky, small-town butcher who develops a sensitive romantic relationship with a homely schoolteacher.[16]

dramas

All of these productions made abundant use of close-ups—the real emotion and action of the plays took place on actors' faces. Cameras were usually placed in the center of the studio with sets arranged in a circle on the periphery so the cameras could have easy access to each new scene. Framing shots was a little more difficult in early TV because cameras had **fixed lenses** rather than zoom lenses. The fixed lenses could only frame shots at one specific point. For example, a fixed 25 mm lens mounted on a camera in the middle of a studio would show a wider shot than a 50 mm lens mounted on the same camera. Usually four or five of these fixed lenses were mounted on a turning device called a turret. When the camera picture was not on the air, the camera operator could switch from one lens to another thus obtaining a closer or longer shot for the next on-air picture.

Besides conventional drama, some innovative formats were tried, many of them the brainchild of Sylvester L. "Pat" **Weaver,** the president of NBC from 1953 to 1955. One of his ideas was the **spectacular,** a show that was not part of the regular schedule but designed to expand the horizons of creativity. The most outstanding of the spectaculars was probably "Peter Pan," starring Mary Martin, which was viewed by some 165 million Americans.

Weaver's innovations

Weaver was also involved in developing the "Today" and "Tonight" shows. "Today" started in Chicago as "Garroway at Large," an informal variety series. Its host, Dave Garroway, became host on the daily morning "Today" show and the format changed somewhat to include both news and variety. Another star of the show was J. Fred Muggs, a baby chimpanzee. The late-night show "Tonight" originally starred Steve Allen, and then picked up steam as Jack Paar became its controversial host.

Pat Weaver felt that programs should be network controlled rather than advertising agency controlled. Most of the early TV and golden era radio program content had been controlled by the advertising agencies. Weaver developed a magazine concept whereby advertisers bought insertions in programs and the program content was supervised and produced by the networks. This was used for "Today," "Tonight," and many of the spectaculars, somewhat to the chagrin of the advertisers.[17]

FIGURE 3.12

A "Playhouse 90"
show entitled "The
Country Husband."
Here Barbara Hale
(*left*) can't understand
why Frank Lovejoy
insists on personally
escorting Felicia Farr
home. *(Courtesy of
Columbia Pictures
Television)*

other programs

Other highlights of this early programming were "Victory at Sea," a documentary series that recreated the naval battles of World War II; Elvis Presley performing on the Ed Sullivan show—from the waist up; the introduction of the crime-drama "Dragnet," which was the first show to bump Lucy from the top of the ratings; Jackie Gleason's comedy sketches, the most famous of which was "The Honeymooners"; and "Omnibus," a potpourri of cultural elements.[18]

Color TV Approval

The problems connected with color TV were not resolved until 1953. In 1950 the FCC finally accepted the CBS system of color instead of the RCA system

CBS system accepted

even though the CBS system was incompatible with existing black-and-white receivers. This was a surprise to many engineers who felt an incompatible system would not be chosen. However, the FCC was pressured by Congress and the companies involved to make a decision and felt the CBS system provided higher quality color pictures.

A great deal besides compatibility and consumer cost and inconvenience was at stake. The company that won the color battle stood to gain enormous financial benefits through the building and licensing of TV sets. RCA con-

RCA objections

tinued to fight the CBS system by refusing to program in color and by gaining allies among other set manufacturing companies and TV stations.

There were valid reasons why stations did not want to purchase CBS color cameras and other color gear. If a station were to transmit with the

FIGURE 3.13

A 1952 camera with turret lenses. *(Courtesy of RCA)*

FIGURE 3.14

Dave Garroway with chimp J. Fred Muggs (*left*) and Phoebe B. Beebe. Muggs was discovered in classic show business style by one of "Today's" producers, who spotted him in an elevator. *(Courtesy of NBC)*

CBS color system, it could not simultaneously transmit with the present black-and-white system. This meant that a station could not program to people with old sets and people with new color sets at the same time.

CBS also ran into difficulty manufacturing its color sets. Certain metals that were needed for the sets were also needed for the Korean War effort, and CBS had difficulty obtaining them. As a result, the CBS color system project essentially came to a halt, and a general state of confusion concerning color reigned in the TV industry.[19]

end of CBS system

To help solve the problem, the National Television System Committee (NTSC), the same committee of engineers that had decided on the 525-line system, volunteered to study the problem. Because this committee included more members who favored the RCA system than the CBS system, very few people were surprised when it recommended the compatible system. However,

FIGURE 3.15

"Dragnet," starring Jack Webb (*right*) as Sergeant Friday and Harry Morgan as Officer Cannon. In this episode, called "The Gun," a Japanese widow has been murdered and Friday and Cannon are assigned to find the assailant. *(Courtesy of NBC)*

significant improvements had been made in the RCA system, some of which the NTSC initiated, and its color picture was now much better than it had been when RCA originally demonstrated it to the FCC. The FCC took the NTSC recommendation and in 1953 adopted RCA's electronic compatible system, which is still used in this country today. At the time, even CBS supported the adoption.

NTSC recommendation

However, for a long time RCA-NBC was the only company actively promoting color. NBC constructed new color facilities and began programming in color, but both CBS and ABC dragged their heels, and most local stations did not have the capital needed to convert to color equipment. Even more important, consumers were reluctant to purchase color TV sets because they cost twice as much as black-and-white sets and the limited color programming did not merit this investment. The vicious cycle was eventually reversed. As more color sets were sold, the prices were lowered and programming in color increased, causing even more sets to be sold. But not until the late 1960s were all networks and most stations producing color programs.

slow growth

Films for TV

The days of live programming for other than news and special events began to disappear in the mid-1950s for several reasons. One was the introduction of **videotape** in 1956. The expense of the equipment prevented it from taking hold quickly, but once its foot was in the door, it revolutionized TV production techniques. Programs could now be performed at convenient times for later airings. As the equipment became more sophisticated, stops could be made to allow for costume changes, scene changes, and the like. As the equipment became even more sophisticated, mistakes could be corrected through editing procedures. Scenes could even be taped out of sequence and assembled in order at a later time, in much the same manner as film.

videotapes

FIGURE 3.16

The CBS color TV set. The rotating color wheel disc was built right behind the outside of the picture tube. *(Courtesy of Ed Reitan)*

However, the live era had begun to yield to film even before tape took hold. Film companies had been antagonistic toward TV, not even allowing TV sets to appear in movies. The 1953 merger of United Paramount Theaters and ABC opened the film door, and ABC soon convinced several major film companies to begin producing film series for TV. Some of the early filmed TV series were "Cheyenne," "Colt 45," "Death Valley Days," "Wagon Train," and "December Bride." As the list demonstrates, westerns predominated. By 1959 thirty-two western series were on prime-time TV. The one with the greatest longevity was **"Gunsmoke,"** which revolved around Dodge City's Matt Dillon, Chester Goode, Doc, and Miss Kitty.[20]

westerns

Other types of programs took hold also, again mainly on film. The old crime-mystery formula surfaced on TV in such programs as "77 Sunset Strip," "M Squad," and "Perry Mason." Dancer Fred Astaire presented a widely acclaimed 1958 special; Dick Clark appealed to the teenagers with "American Bandstand"; and Leonard Bernstein conducted his first "Young People's Concerts" in 1958. For the younger audience there was such fare as "Captain Kangaroo," and "The Mickey Mouse Club."

other programs

Several public service programs also aired at this time. Fidel Castro appeared on "Meet the Press" and Nikita Khrushchev, amid much controversy, appeared on "Face the Nation." Two young reporters, Chet **Huntley** and David **Brinkley,** teamed up for NBC's coverage of the 1956 political convention and found themselves a berth as nightly anchorpersons for the fifteen-minute network news.[21]

The TV boom continued in the late 1950s—more TV sets, more viewers, more stations, more advertising dollars. But to this rising euphoria came a dark hour.

The Quiz Scandals

Quiz programs on which contestants won minimal amounts of money or com-
pany-donated merchandise had existed on both radio and television. However,
in 1955 a new idea emerged in the form of **"The $64,000 Question."** If they
beat out challengers over a number of weeks, contestants could win huge cash
prizes. Sales of Revlon products, the company that sponsored "The $64,000
Question," zoomed to such heights that some were sold out nationwide. The
sales success and high audience ratings spawned many imitators. Contestants
locked in soundproof booths pondered, perspired, and caught the fancy of the
nation.

advertiser success

From time to time there were rumors that some quiz programs had been
fixed. Then in 1958 a contestant from the daytime quiz show "Dotto" claimed
to have evidence that the show was rigged, and a contestant from "Twenty-
One" described that program's irregularities. The networks and advertising

FIGURE 3.19

M.C. of "The $64,000 Question" was Hal March (*right*). Gino Prato, an Italian-born New York shoemaker, mops his brow while listening to a four-part question on opera. *(From United Press International)*

agencies denied the charges, as did Charles **Van Doren,** the most famous of the "Twenty-One" winners. A grand jury and a House of Representatives sub-committee conducted hearings and found discrepancies in the testimony.

Charles Van Doren

In the fall of 1959 Van Doren appeared before the House subcommittee and read a long statement describing how he had been convinced in the name of entertainment to accept help with answers to questions in order to defeat a current champion who was unpopular with the public. Van Doren was also coached on methods of building suspense and when he did win, he became a national hero and a leader of intellectual life. After several months on the program, he asked to be released and finally was allowed to lose. He initially lied, he said, so that he would not betray the people who had invested faith and hope in him. Other witnesses from various programs followed Van Doren to testify to the means by which dull contestants were disposed of so that lively personalities could continue to win and hence hold audience attention.

In retrospect the **quiz scandals'** negative effect on TV was short lived. The medium was simply too pervasive a force to be permanently afflicted by such an incident. Congress did amend the Communications Act to make it unlawful to give help to a contestant, but for the most part the networks rectified the errors by canceling the quiz shows and reinstating a higher percentage of public service programs.

They also took charge of their programming to a much greater extent. The presidents of all three networks decreed that from then on most program content would be decided, controlled, and scheduled by networks, which would then negotiate time sales with advertisers. This was a further extension of Weaver's "magazine concept." Beginning in 1960, most program suppliers contracted with the networks rather than advertising agencies. This made life more profitable for the TV networks, too, because they established profit participation plans with the suppliers.[22]

network programming control

The UHF Problem

technical problems

When the FCC ended its freeze and established stations in the UHF band, it intended that these stations would be equal with VHF stations. In reality, they became first-class and second-class stations. UHF was technically not as effective as expected. Its weaker signal was supposed to be compensated for by higher towers, but this did not prove to work in practice.

converters

People did not have sets that could receive UHF, so in order to tune in UHF stations, they needed to buy converters. Many were unwilling to do this, and UHF found itself in a vicious circle. In order for people to buy UHF converters, it had to offer interesting programming material; in order to finance interesting program material, UHF stations had to prove to advertisers that they had an adequate audience.

programming

Because VHF stations were established first, they were the first to obtain network affiliations and hence capture the best programming. Consequently, UHF stations had to depend on syndicated material and local talent.

The FCC tried to help the fledgling UHF stations in a number of ways. In 1954 it changed the number of TV stations that one company could own from five to seven, provided that no more than five of those seven were VHF. This meant that organizations such as the three networks were free to buy two UHF stations each. And this is what NBC and CBS did—each purchased

network ownership

UHF stations and placed their network programs on them. The theory was that with network programming on UHF, people would be willing to buy converters. In reality people did not buy the converters and UHF penetration increased so slightly in the markets where NBC and CBS had their stations that both networks abandoned the UHF stations after several years.

deintermixture

In 1957 the FCC proposed a procedure known as **deintermixture.** The intent was to make some markets all UHF and some all VHF so that in the all-UHF markets people would be forced to buy converters if they wanted to receive any television. This plan did not succeed either, however, mainly because established VHF stations fought any efforts to convert them to UHF.

all-channel receiver bill

A third attempt to give UHF a boost was the passage of the **all-channel receiver** bill that Congress authorized in 1962. By amending the Communication Act, the FCC was given the authority to require both a UHF and a VHF tuner on all TV sets. All sets manufactured since 1964 include both UHF and VHF, but most UHF stations still play second fiddle to VHF stations.

From the 1950s to the 1960s

success

The 1950s was a decade of great growth and experimentation for television. Its technical base was devised by the FCC during the freeze period, the color television problem was resolved, and attempts had been made to resolve the UHF problem. Networks and stations established programming concepts and relationships with advertisers. The video tape recorder added more flexibility to programming, and film and TV began a cautious marriage that further enhanced programming possibilities.

FIGURE 3.20

A Huntley-Brinkley
newscast. Chet
Huntley broadcasts
from New York, while
David Brinkley, seen
on the television
screen, broadcasts from
Washington, D.C.
(Courtesy of NBC)

The TV industry had the strength to survive both the external attack of blacklisting and the internal scandal of quiz program rigging. By the end of the 1950s even the greatest of skeptics were conceding that TV was not just a passing fad. TV had not established itself as a primary news and information medium as yet, but that accomplishment awaited the turn of the decade.

status

Reflections of Upheaval

Television journalism gathered force and prestige during the 1960s—the decade of civil rights revolts, the election of John F. Kennedy, space shots, satellite communications, assassinations, Vietnam, and student unrest.

The networks encouraged documentaries and increased their nightly news from fifteen to thirty minutes in 1963, thereby assigning increased importance to their news departments. Anchoring on camera for NBC were Huntley and Brinkley and for CBS, Walter **Cronkite,** the person who became known as the most trusted man in America. Network news departments scored points over their print counterparts when President **Kennedy** agreed to have news conferences televised live.

news

The quiz scandal had helped precipitate a rise in documentaries. To atone for their sins, networks increased their investigative fare. Documentaries were now easier to execute because technical advances included 16mm film to replace the bulky 35mm. Film and sound could now be synchronized without an umbilical cord between two pieces of equipment; and wireless microphones were becoming reliable, enabling speakers to wander freely without having to stay within range of a mike cord.

documentaries

As ongoing documentaries, ABC established "Close-Up," and the other networks offered "CBS Reports" and "NBC White Paper." Some of the notable documentaries of the era included "Biography of a Bookie Joint," for which concealed cameras oversaw the operation of a Boston "key shop"; "The Real West," an authoritative report narrated by Gary Cooper that countered the westerns; a 1962 airing of "The Tunnel," for which NBC secured footage

FIGURE 3.21

FIGURE 3.21

Filming preparations for "The Tunnel." NBC news correspondent Piers Anderton (*left*) reviews the building of this tunnel under the Berlin Wall with student Dominico Sesta (*center*) and cameraman Peter Dehmel. The 450-foot passage began in this basement of a West Berlin home. (*Courtesy of NBC*)

of an actual tunnel being constructed by young Berliners to bring refugees from East to West Berlin; "The Louvre," for which cameras were allowed inside the famous building for the first time; and "D-Day Plus 20," a reminiscence between Eisenhower and Cronkite filmed on Normandy beaches.

Many documentaries reported on the racial problem. "Sit-In" dealt with resistance to restaurant segregation; "Crisis: Behind a Presidential Commitment" chronicled the events surrounding Governor George Wallace's attempted barring of the schoolhouse door to prevent blacks from attending the University of Alabama; "The Children Are Watching," dealt with the feelings of a six-year-old black child attending the first integrated school in New Orleans.

civil rights

Civil rights actions were covered live—including the riots in the Watts section of Los Angeles and the death of Martin Luther King.

Television reacted to the civil rights movement in another way—it began hiring blacks. Radio had also been lily-white, but not so visibly. Both media began hiring blacks in the 1960s. Jackie Gleason's chorus included a black dancer. Scriptwriters began including stories about blacks, which often were not aired in the South. Networks and stations hired black on-camera newspeople. A black family moved to the nighttime soap opera "Peyton Place," and Diahann Carroll in "Julia" became the first black heroine.[23]

Great Debates

Television has been credited, through the **"Great Debates"** between John F. Kennedy and Richard M. Nixon, with having a primary influence on the 1960 election results. Kennedy and Nixon met for the first debate at the studios of Chicago's CBS station. Kennedy, tanned from campaigning in California, refused the offer of makeup. Nixon, although he was recovering from a brief illness, did likewise. Some of Nixon's aides, concerned about how he looked on TV, applied Lazy-Shave, a product to cover five o'clock shadow. Some people believe that the fact that Kennedy appeared to "win" the first debate had little to do with what he said. People who heard the program on radio felt Nixon held his own, but those watching TV, especially the reaction

Communication Systems

FIGURE 3.22

One of the so-called "Great Debates." John F. Kennedy has his turn speaking while Richard Nixon and moderator Howard K. Smith look on. In the foreground are the newspeople who asked the questions. *(From United Press International)*

FIGURE 3.23

The Kennedy funeral. This off-monitor shot shows John-John saluting the flag covering his father's caisson. *(Courtesy of* Broadcasting *magazine)*

shots of Kennedy and Nixon listening to each other, could see a confident, attentive Kennedy and a haggard, weary-looking Nixon whose perspiration streaked the Lazy-Shave. Three more debates were held, and Nixon's makeup and demeanor were well handled, but the small margin of the Kennedy victory at the polls has often been attributed to the undecided vote swung to Kennedy during the first debate.[24]

Three years later television devoted itself to the coverage of the assassination of John F. Kennedy. From Friday, November 22, to Monday, November 25, 1963, there were times when 90 percent of the American people were watching television. As a New York critic wrote, "This was not viewing. This was total involvement." From shortly after the shots were fired in Dallas until President Kennedy was laid to rest in Arlington Cemetery, television kept the vigil, including the first "live murder" ever seen on TV as Jack Ruby shot alleged assassin Lee Harvey Oswald. Many praised television for its controlled, almost flawless coverage. Some felt TV would have made it impossible for Lee Harvey Oswald to receive a fair trial and that it was the presence of the media that enabled the Oswald shooting to take place.[25]

Kennedy assassination

Commercial Television 91

FIGURE 3.24

Walter Cronkite
visiting Vietnam.
*(Reprinted with permission
from the September 14, 1987
issue of* Broadcasting*)*

space launches

Through the medium of television, Americans were able to witness the nation's race into space. In 1961 Alan Shepard was launched into outer space within the view of network television cameras. Subsequently, televised pictures were sent from outer space and the moon and television viewers witnessed Neil Armstrong's first step onto the moon.[26]

Vietnam

For the first time in history, war came to the American dinner table. As the troop buildup began in Vietnam in the mid-1960s, the networks established correspondents in Saigon. Reports of the war appeared almost nightly on the evening news programs. This resulted in an inner-circle battle at CBS. Fred Friendly, then president of CBS news, resigned when his network carried reruns of "I Love Lucy" rather than the Senate hearings on the escalation of war activities in Vietnam.[27] In 1968, amid rising controversy over the war, Walter Cronkite decided to travel to Vietnam to see for himself and returned feeling that the United States would have to accept a stalemate in that country.

student unrest

Much of the controversy surrounding the war originated on the country's campuses, where students were becoming increasingly dissident regarding various issues. This, too, was covered by the media, as was the 1968 Democratic convention in Chicago, where youths protested the steamrolling nomination of Hubert Humphrey outside the convention hall. The media became embroiled in the controversy and were accused of inciting the riot conditions. The demonstrators seemed to be trying to attract media coverage. On the other

(a)

(b)

FIGURE 3.25

(*a*) George Maharis in "Route 66." In this series, two adventurers, George and costar Martin Milner, tour the country (in an automobile produced by the show's sponsor). (*b*) Robert Redford in his first TV appearance. This scene is from an episode of "Rescue 8" entitled "Breakdown," aired in February 1960. Redford was hired by director Larry Stewart to guest star with leads Jim Davis and Lang Jeffries and from this Redford launched a well-known career. *(a courtesy of Columbia Pictures Television; b courtesy of Larry Stewart)*

hand, many people inside the convention hall learned of the protest by seeing it on a TV monitor and might not otherwise have known of this show of discontent.[28]

The role of the media in the creation of news was debated frequently in the 1960s and beyond as their image as a source of information increased.

A Vast Wasteland?

In 1961 Newton **Minow,** Kennedy's appointee as chairman of the Federal Communications Commission, spoke before the annual convention of the National Association of Broadcasters. During the course of his speech, he stated the following to the broadcasting executives:

> I invite you to sit down in front of your television set when your station goes on the air and stay there without a book, magazine, newspaper, profit and loss sheet, or rating book to distract you—and keep your eyes glued to that set until the station signs off. I can assure you that you will observe a vast wasteland.[29]

Minow's speech

FIGURE 3.26

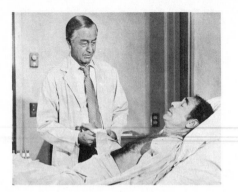

Robert Young as Dr. Marcus Welby. In this episode, guest star Gary Merrill plays a lawyer who refuses to step down from an important case after he has been told he has a terminal disease. *(From MCA Publishing)*

The term **"vast wasteland"** caught on as a metaphor for television programming. Needless to say, the executives were not happy with Minow's phrase.

During the early 1960s the dominant fare was violence. "The Untouchables," "Route 66," "The Roaring 20s,"—all featured murders, jailbreaks, robberies, thefts, kidnappings, torture, blackmail, sluggings, forced confessions, and dynamiting. Saturday morning children's programming was also replete with violent cartoons.

violence and antiviolence

A surge against violence, aided by Minow's challenge, brought forth a change to doctors' shows, such as "Ben Casey" and "Dr. Kildare" and more comedies, such as "The Dick Van Dyke Show," "Gilligan's Island," "Bewitched," and the "Beverly Hillbillies." The last, a series about an Ozark family that struck oil and moved to Beverly Hills without changing their mountain character, headed the ratings but was heralded as the supreme example of the vast wasteland.

Spy shows, such as "Mission: Impossible" and "I Spy," were plentiful for several years and brought with them a returning encroachment of violence. A reaction to this violence brought a new wave of medical shows, comedies and variety shows—"Marcus Welby M.D.," "Here's Lucy," "The Andy Williams Show," and "Laugh-In." This latter show was a fast-paced, free-form series that featured many one-liners. It was the first highly edited program, but the editing was done by physically cutting the videotape, a very painstaking process that is no longer used.

movies

During the 1960s movies were better than ever—on TV. In 1961 "Saturday Night at the Movies" began a prime-time movie trend that, by 1968, saw movies on all seven nights of the week. This rapidly depleted Hollywood's supply of films, so in 1969 ABC introduced a **made-for-TV movie** series, "The Movie of the Week." Its films had not been previously shown at theaters, but were conceived for television viewing.

talk shows

Talk shows with such hosts as Steve Allen, Dick Cavett, David Frost, and Merv Griffin also became dominant in the 1960s. Johnny Carson took over "Tonight" from Jack Paar and proved equally controversial.

Specials of note included "Julie and Carol at Carnegie Hall," featuring the singing of Julie Andrews and Carol Burnett; "My Name is Barbra" with Barbra Streisand; "Frank Sinatra: A Man and His Music;" Vladimir Horowitz playing the piano at Carnegie Hall; and Hallmark's presentation of "Macbeth."

specials

In addition, the science fiction series "Star Trek" aired for three years and developed a cultlike following. NBC was going to cancel the show after its second season but received such a barrage of letters from outraged "trekkies" that it renewed it for one more year.[30]

Instant replay became an instant hit with sports fans. As remote equipment made it easier to leave the studio, the quantity of sports broadcasting increased, and sports themselves changed to accommodate the new medium. Teams changed their playing times in order to avail themselves of larger audiences and added a few more playoffs to increase the excitement.[31]

sports

The decade of the 1960s was a period of great prosperity for television. Despite objections from TV's critics, the commercialism of the medium grew. When Herbert Hoover, who as secretary of commerce had decried advertising in the 1920s, died in 1964, NBC broadcast a tribute that was followed by a beer commercial, a political commercial, and a cigarette commercial.[32]

Government Actions and Reactions

During the 1970s government regulators and broadcasters played a cat-and-mouse game that often engendered hard feelings. The cigarette commercials disappeared in 1972 as the result of the passage of a congressional bill that barred the advertising of **cigarettes** on the broadcast media. This action was motivated by the surgeon general's report that linked cigarettes and cancer.[33]

cigarette ads barred

Two experiments of the 1970s involving the government were **prime-time access** and the family hour. Prime-time access was promoted primarily by Westinghouse Broadcasting. It pointed out that the networks monopolized prime time by programming between 7:00 and 11:00 P.M. and suggested that some of this time should be programmed by the stations themselves in order to meet community needs, as well as to allow more room for syndicated programs. In 1971 this idea was adopted by the FCC, which ruled that networks would be allowed to program only three hours a night in stations in the top fifty markets, leaving the other hour to the local stations. During this hour there were to be no old network programs or old movies; all programs had to be newly created. The rule was modified, however, so that, in actuality, prime-time access became 7:30 to 8:00 P.M. Monday through Saturday. Although prime-time access was established to allow stations to broadcast significant local programming and independently produced programs of high quality, this did not happen to any significant degree. The stations filled the time with the cheapest thing they could find—game shows and remakes of old formats, although some developed local magazine shows.[34]

prime-time access

Family hour began in 1975 as an attempt to curtail sex and violence before 9:00 P.M., the time when children are usually in the audience. The family-hour concept originated with the code of the National Association of Broadcasters, an internal broadcasting organization. However, there were those who believed it came about through subtle pressure by the FCC and, hence, represented an abhorrent attempt by the commission to regulate program content. The family-hour idea was widely opposed by writers, producers, and directors, who took the concept to court. It was decided that the FCC had overstepped its powers and that the family-hour restrictions should be removed from the NAB code. However, most of the networks and stations continued to program that slot with material of a family nature.[35]

Another government action that engendered controversy was the **financial interest-domestic syndication** rule. When the networks reacted to the quiz scandals by exerting a stronger hand over their programming, they often acquired a financial interest in the programs supplied to them by independent production companies. The networks would contribute money to the production of the program and when the program was syndicated to local TV stations, would reap some of this syndication profit. In 1970, the FCC adopted a rule barring networks from acquiring subsidiary interests in independently produced programs and from engaging in domestic syndication of these programs. This was done primarily because the FCC felt the networks were too powerful in the programming business and that financial interest and domestic syndication increased this power. The networks disagreed but during the 1970s accepted the ruling somewhat complacently.[36]

The Equal Employment Opportunity Commission kept a watchful eye on broadcasters during the 1970s. The stations were still hiring and programming for blacks and Mexican-Americans. In addition, spurred on by the women's movement of the 1970s, they added female newscasters and stories about women in responsible positions.

A major dispute developed between broadcasting and the Nixon administration. It arose primarily as a result of the media's disapproval of the administration's Vietnam policies. What particularly annoyed the administration was the **"instant analysis"** that network commentators launched after presidential speeches. After one particular November 1969 speech in which Nixon vowed to hold firm on Vietnam, some commentators expressed their displeasure and supported disengagement from the war. Shortly thereafter, Vice-President Spiro Agnew addressed a fund-raising dinner and blasted the media for its hostility toward the administration and its biased point of view.[37]

The next several years were ones of undeclared war between the media and the administration. Very little of a positive nature was reported to the citizenry concerning Nixon or his appointees. In return, administration sources accused the media news personnel of being eastern-establishment elitist liberals who were out of touch with the "silent majority" of Americans who, the administration claimed, supported Nixon and his policies.

The administration-media war came to an end with the Watergate hearings. The media expressed little sympathy for Nixon and other administration personnel who were accused of masterminding or covering up the break-in at the Democratic party offices in the Watergate complex. Eventually most of these accused people resigned, and the media could claim an undeclared victory.

Programming of the 1970s

The invention of lightweight, portable TV cameras in the 1970s allowed TV news coverage to become even more pervasive and mobile than it had been during the 1960s. The news itself changed as the United States ended its involvement in Vietnam and entered a period during which domestic issues were more newsworthy than international events. Documentaries on pollution, energy, the bicentennial, the economy, and crime gradually replaced those on integration, Vietnam, and student protests.

news

Sports coverage increased greatly during the 1970s, including "Monday Night Football," nighttime world series games, and Olympic coverage. In 1972 sportscasters were transformed into newscasters as Arab terrorists captured and killed Israeli Olympic athletes.

sports

Among comedy series, **"All in the Family"** struck a new chord by probing the previously taboo topics of politics, ethics, and sex. The program, with its bigoted resident of New York's borough of Queens, Archie Bunker, became a ratings leader.

comedy

By the mid-1970s there were fewer movies on TV than at any time during the preceding decade, but some of the movies—including *That Certain Summer*, which dealt with homosexuality—showed that censorship was becoming less restrictive. Spin-offs became popular. "The Mary Tyler Moore Show" begat "Rhoda" and "Phyllis;" from "The Six Million Dollar Man" came "The Bionic Woman;" and "Happy Days" resulted in "Laverne and Shirley." The nighttime soap opera "Mary Hartman, Mary Hartman," refused by the networks, became a hit on independent stations and also some network affiliates. Novels dramatized for TV and shown on a series of nights, often referred to as miniseries, became a popular phenomenon. Among these were programs derived from Irwin Shaw's *Rich Man, Poor Man* and Alex Haley's *Roots*. The late night show, "Saturday Night Live," became a hit particularly with the college crowd.[38]

other programs

The Government and Deregulation

Television, like radio, felt the impact of the deregulatory mood of the 1980s. TV station licenses were lengthened from three years to five, and TV stations in small communities no longer needed to keep logs or ascertain community needs. In addition, restrictions governing the amount of commercial time stations could program were lifted.[39] The number of TV stations one entity could

station rules

FIGURE 3.27

The cast of "All in the
Family" in their
Queens home. Top row
left to right: Rob
Reiner as Mike Stivic,
Sally Struthers as
Gloria Stivic, and
Mike Evans as Lionel
Jefferson. Bottom row
left to right: Jean
Stapleton as Edith
Bunker and Carroll
O'Connor as Archie
Bunker. *(Photo/Viacom,
Hlwd.)*

own was increased from seven to twelve, provided the twelve stations were not
in markets that collectively contained more than 25 percent of the nation's TV
homes. UHF stations are assessed for just half the homes in their reach, and
group broadcasters that are more than half owned by minorities can own up
to fourteen stations and reach 30 per cent of TV homes.[40]

TV stereo

In other government action, the FCC authorized **TV stereo** broadcasting
in 1984. This concept caught on quickly, as technologies go. Although the cost
to stations to convert from mono to stereo ran between $50,000 and $100,000,
over sixty stations did so within the first year and many more followed. Gradually networks and stations produced more and more programs in stereo and
more people purchased or installed stereo-receiving systems with their TVs.
Stereo TV appeared to be a technological change that would happen without
the trauma associated with other changes, such as color TV and UHF.[41]

Two of the 1970s regulations, prime-time access and financial interest,
were thrashed through again but with virtually no change occurring in the
end.

Mergers and Acquisitions

The 1980s were not particularly good times for commercial television financially. Stations and networks were faced with competition from many newer

hard times

media that splintered their audience and their advertising dollar. Profits were
down almost universally, and some stations even went bankrupt.[42] The independents had grown explosively in the 1980s, both in terms of numbers and

(a)

(b)

(c)

FIGURE 3.28

(*a*) Mary Tyler Moore. After appearing on Dick Van Dyke's show for several years, Mary launched her own comedy show, which went off CBS voluntarily after eight years. (*b*) Lead Lindsay Wagner (*left*), director Larry Stewart (*center*), and Andrew Prine prepare for a scene of "The Bionic Woman" episode "Rodeo," which aired in October 1977. (*c*) A scene from the drama *Roots,* a David L. Wolper Production depicting one of Alex Haley's ancestors who was sold into slavery in the New World. *(a courtesy of MTM Enterprises; b courtesy of Larry Stewart; c courtesy of The Wolper Organization, Inc.)*

audience. In fact, they were one of the main competitive forces causing the erosion of network and affiliated station dominance. But by the late 1980s, some of them found that they, too, had overproliferated and overextended.[43]

This downturn in broadcasting in general, combined with a national phenomena of company takeovers, led to major changes in the management and ownership of all three networks and was instrumental in creating a fourth network.

The first network affected was ABC, which was purchased in 1985 for $3.5 billion by **Capital Cities** Communications, a company one-third the size of ABC. Capital Cities, known for its lean economic policies, did not hesitate long in pruning back the number of employees at ABC. Although the takeover was termed "friendly," it was the first time in history that a major network had been purchased and it was referred to as "the minnow swallowing the whale."[44]

ABC

FIGURE 3.29

One of the Fox
network's first series,
"Married. . .with
Children." Here Peggy
(Katey Segal) and Al
(Ed O'Neill) postpone
going to a favorite
restaurant so Al can
assemble one of Bud's
(David Faustine) toy
cars. *(Courtesy of Fox
Broadcasting Company)*

NBC

In less than a year, a second major network, NBC, was purchased as part of an overall takeover of its parent company RCA for $6.28 billion.[45] This acquisition brought with it an irony of history because the purchaser was **General Electric,** a company that had been involved in the formation of NBC and then eased out of ownership by RCA. This, again, was called a "friendly takeover." The GE man who was put in charge of NBC was Robert Wright. He became the president and chief executive officer when Grant Tinker, who had run the network since 1981, resigned.[46] Tinker had replaced Fred Silverman, who had been dubbed "the man with the golden gut" because he had been largely responsible for putting CBS and then ABC at the top of the ratings. His magic did not work at NBC, however.[47] When Tinker took over NBC, he was able to lead the network from third place to first place.[48]

CBS

Things were not so "friendly" in the CBS executive suites. In 1985 Ted Turner of Turner Broadcasting attempted an unfriendly coup of the network that CBS was able to thwart only by buying back $1 million of its own stock.[49] This greatly damaged CBS' financial footing and it began laying off employees. At one point the chairman of the board, Thomas Wyman, tried to persuade Coca-Cola to buy CBS without notifying the other members of the board. This cost Wyman his job and brought back William Paley who had retired from CBS in 1983. Paley formed an alliance with Lawrence **Tisch** who had also been nibbling at the edges of CBS in what some thought was a takeover bid. Tisch was appointed chief operating officer and continued layoffs, which became particularly controversial within the news division. The division felt that its broadcast journalism tradition was being severely compromised.[50]

Even though the three major networks were suffering from losses of both money and morale, a brand new network, **Fox Broadcasting Company (FBC),** appeared on the scene. This was promulgated primarily by Rupert **Murdoch,** an Australian media baron who had recently obtained U.S. citizenship. In 1985, Murdoch purchased 50 percent of 20th Century Fox, which was owned primarily by oil magnate Marvin Davis. Davis and Murdoch then purchased six TV stations owned by Metromedia.[51] Shortly thereafter, Davis sold out his share of the businesses to Murdoch who proceeded to build a network using his six stations and other independent stations throughout the country.[52] The first show offered by the network was "The Late Show Starring Joan Rivers." This caused quite a stir because Rivers had been substitute hosting for Johnny Carson and left him for a show that directly competed with him. Several months later Rivers wound up being "relieved" of the show by Fox. FBC built up the number of programs offered to the affiliated stations slowly but in a manner that was very similar to the three traditional networks.[53]

FBC

Many production companies and stations were also affected by mergers and acquisitions. Ted Turner, rebuffed by CBS, bought MGM then turned around and sold all of it except the film library.[54] Several companies that owned groups of stations had to go private to avoid unfriendly takeovers—Lorimar and Telepictures merged, and Coca-Cola bought Embassy Communications, just to mention a few.[55]

production companies

This merger mania, which affected virtually all aspects of American business, was particularly visible within the media area. The media, by their very nature, were able to give their trials and tribulations a great deal of publicity.[56]

Programming Trends of the 1980s

Programming added a bit more glitter primarily because digital special effects and multifaceted graphic systems were well developed.

As the elderly population accelerated, TV began paying more attention to older demographics and programs such as "The Golden Girls" (three older divorced or widowed women who share a house in Florida) and "Murder, She Wrote" (a mystery writer in her fifties who solves crimes) became quite popular.[57]

older programming

An all-time high prime-time ratings record was set in 1983 with the airing of the final episode of the popular series "M*A*S*H."[58] Also high on the ratings was a 1980 episode of the evening soap opera "Dallas," which answered the question, "Who shot J. R.?" The fate of J. R., the series' cunning, ruthless lead, had been left hanging over the summer and became a burning question both nationally and internationally.

ratings leaders

Comedies and dramas remained the primary prime-time TV fare. NBC's "Cosby Show," a well-written and well-acted family sit-com garnered both high ratings and accolades. Two other popular comedies were "Cheers" and "Family Ties."

comedy

FIGURE 3.30

The stars of "The
Golden Girls." Betty
White (*left*), Bea
Arthur (*center*), and
Rue McClanahan
(*right*) discuss a
problem around the
kitchen table of their
Florida home. *(Photo ©
The Walt Disney Company)*

FIGURE 3.31

The Huxtable family
of "The Cosby Show."
Clockwise from center:
Cliff (Bill Cosby),
Rudy (Keshia Knight
Pulliam), Clair,
(Patricia Rashad),
Theo, (Malcolm-Jamal
Warner), Vanessa
(Tempest Bledsoe),
Sondra (Sabrina
LaBeauf), and Denise
(Lisa Bonet). *(Photo/
Viacom, Hlwd.)*

drama

NBC took a chance on "Hill Street Blues," a multicharacter, complex series about San Francisco police and left it on the air even though it had weak ratings. It was awarded a number of Emmys and gradually developed a large loyal audience. "Cagney and Lacey" plowed new ground by starring two women police officers; "L. A. Law" projected a not-too-flattering impression of lawyers; and "Miami Vice" was glitsy enough that some felt it represented the triumph of style over substance.

news

In the news area, the 1986 *Challenger* space shuttle explosion was covered thoroughly for several weeks and the sale of arms to Iran occupied a great bulk of airtime for months during 1986 and 1987. The ability to deliver news

stories by satellite from anywhere in the world began to change the dependence of local stations on network news.[59]

Satellites also played a crucial role in the 1985 "Live Aid" concert to help the starving in Africa. This was described as a global juke box that originated primarily in Philadelphia and London but also in cities in eight other countries. The world's top musicians performed via satellite for a worldwide TV audience of 2 billion and radio audience of 1.6 billion.[60]

music

A new type of programming came to TV in the 1980s in the form of home shopping. Goods were shown over the air and people called in to order them. This became enough of a success that a proliferation of home shopping services developed—individual programs, series, and entire station programming.[61]

home shopping

Commercial television has undergone many structural and programming changes during its forty-year life, but it remains a dominant pastime for the American public.

Commercial TV Chronology

1884	Nipkow invents mechanical scanning
1927	Farnsworth transmits electronic pictures
1935	Sarnoff pledges millions to RCA's development of TV
1939	TV displayed at world's fair
1941	FCC authorizes black-and-white 525–line TV
1942	TV development halted because of World War II
1948	Many facets of TV grow 4000 percent in one year
1948–	
1952	TV "freeze"
1950	FCC accepts CBS's color system
1951	"I Love Lucy" is filmed
1953	FCC adopts RCA's color TV system
1954	Televising of McCarthy-Army hearings
1954	Number of TV stations one company can own revised
1956	Videotape introduced
1957	FCC proposes deintermixture to help UHF
1959	Charles Van Doren confesses to the quiz scandals
1960	"Great Debates" between Kennedy and Nixon
1961	Newton Minow makes "vast wasteland" speech
1962	Congress passes all-channel receiver bill
1963	Networks nightly newscasts increased to thirty minutes
1963	Telecasting of John F. Kennedy funeral
1968	Cronkite travels to Vietnam
1968	Youths protest at Democratic Convention
1969	ABC introduces made-for-TV movies
1970	FCC adopts financial interest-domestic syndication rule
1971	Prime-time access rule adopted
1972	Cigarette commercials taken off radio and TV
1975	Family-hour concept begun
1981	Grant Tinker replaces Fred Silverman at NBC
1983	Paley retires from CBS
1983	Final episode of "M*A*S*H" aired
1984	TV deregulated
1984	TV stereo authorized
1985	Capital Cities Communications purchases ABC
1985	"Live Aid" broadcast
1986	*Challenger* space shuttle explodes
1986	GE takes over RCA
1986	Paley returns to CBS and joins forces with Lawrence Tisch
1986	Fox Broadcasting Company begins programming

Conclusion

Commercial television, born into a fast-paced society, has been forced into the role of an early bloomer. Within forty years it has progressed from hot lights and green makeup to instantaneous worldwide communication.

In some ways television technology seems to prove that necessity is the mother of invention, but in other ways it seems to suggest that invention is the mother of necessity. Because the original mechanical scanning techniques might never have been adequate for a popular medium, the electronic techniques were indeed necessary. On the other hand, matters such as lines of resolution and color were hotly debated by various industry groups, but only after the technology was approved did the need for its existence become evident. The invention of the video tape recorder altered TV production techniques in ways not envisioned. Similarly, portable equipment was developed before its full range of possibilities, particularly in terms of newsgathering, was explored.

Television's greatest boosters could not have predicted how quickly and thoroughly TV would be accepted by the American public. The medium's growth in terms of sets, programs, and advertising during the late 1940s and early 1950s is a phenomenon unto itself. Performers, such as Milton Berle and Lucille Ball, latched onto early TV and became instant celebrities. More reluctant "TV stars," such as Frank Costello and Senator Joseph McCarthy, found that TV could also create notoriety.

The elements of government that had dealt with radio suddenly found a new medium featuring both sight and sound upon their doorstep. Taking a cue from radio, the FCC imposed a freeze to work out technical allocations. The choice of UHF to resolve the channel shortage was not the most fortuitous decision, and it is one that still engenders government patch work. Other controversial government actions have included eliminating cigarette ads, instituting prime-time access, influencing the family-hour concept, imposing the financial interest rule, and deregulating.

Private business has also been predominant in TV from the early founding of networks to the recent rash of takeovers.

Programming on TV traditionally has cycles in terms of content and direction. Doctors' shows, detective stories, anthology dramas, feature films, westerns, and social-commentary comedies all have had their moment or moments of glory.

Successful programs, such as "Your Show of Shows," "Playhouse 90," "See It Now," "Tonight," "Gunsmoke," "Mary Tyler Moore," "Roots," "Hill Street Blues," and "The Golden Girls" are dependent upon group efforts. Many individuals have also made significant contributions, among them: David Sarnoff, who laid the groundwork and financial base for the original development of TV; and Pat Weaver, who developed the spectacular and the magazine concept of advertising.

Television has survived dark moments, such as blacklisting and the quiz scandals, and celebrated bright moments, such as the telecasting of space shots. It has been accused of creating public opinion concerning the Vietnam War and praised for unifying a nation during the Kennedy assassination. Its future will undoubtedly change, but commercial television—which has taken the country by storm in a historically short period of time—seems bound to continue.

Thought Questions

1. Should the FCC have waited longer to approve the beginning of commercial television so that the technical quality would be better than it is?
2. What, if anything, could have been done to help TV personnel who were blacklisted?
3. Could a phenomenon such as the quiz scandals reoccur in TV today? Why or why not?
4. What are the merits and demerits of deregulation?

Public Broadcasting

Overview

The public radio and television system of the United States has been struggling beside its more glamorous commercial cousin—floundering and fluctuating, arguing and achieving, staggering and starring. Its history is marked by victory and defeat and its present is replete with insecurity, but the system has nonetheless emerged as a viable aspect of American broadcasting.

   The following topics chronicle these ups and downs:

The early development and partial demise of educational radio

The establishment of radio networking, culminating with NPR and APR

The status of nonaffiliated radio stations

Financial problems of public radio

Reservation of educational television channels

The role of the Ford Foundation in early educational TV

The Educational Broadcasting Facilities Act

The Carnegie Commission and ensuing legislation that led to the formation of CPB and PBS

TV programming from the Station Program Cooperative and other sources

Controversies during the Nixon administration

The lack of action on Carnegie II

Financial problems of public television

Programming milestones of public TV

4

I think public television should be the visual counterpart of the literary essay, should arouse our dreams, satisfy our hunger for beauty, take us on journeys, enable us to participate in events, present great drama and music, explore the sea and sky and the woods and hills. It should be our Lyceum, our Chatauqua, our Minsky's, and our Camelot. It should restate and clarify the social dilemma and the political pickle.

E. B. White
writer and philosopher

Early Educational Radio

early experiments

The roots of **public broadcasting** go back to radio and TV operations that were originally referred to as **noncommercial broadcasting** or **educational broadcasting.** In fact, broadcasting was born noncommercial, and many of the early stations were started by educational institutions. For example, as early as 1917, Professor Earle M. Terry of the University of Wisconsin transmitted voice and music to listeners in farm areas around Madison. The professor and his students experimented with various forms of music and particularly favored Hawaiian music because its twang carried well. During World War I the station broadcast weather and crop information and was the only radio station not under Navy control.[1]

radio rush of the 1920s

The early 1920s witnessed a radio rush by colleges and universities around the country. Seventy-four such institutions were broadcasting by the end of 1922. Colleges used radio stations primarily to aid extension activities, raise funds, and offer college credit courses that people could listen to in their homes. The main problem facing these stations was the same one that plagued other early stations—they all had to broadcast on the same frequency of 360 meters. As more and more stations added their signals to the airwaves, home-study students often could not hear their lessons because of interference. As a result, faculty members became disillusioned with the effectiveness of radio and many colleges ceased their broadcasts. In 1925 thirty-seven educational stations left the air and only twenty-five new ones began broadcasting.

commercial station takeover

As commercial stations became more firmly entrenched, they overpowered the educational stations both in wattage and dollars. During the 1920s radio stations generally shared time with one another, so it was not uncommon for an educational and commercial station to alternate hours. If the commercial station decided it wanted a larger share of the time, it would petition the Federal Radio Commission. Both the commercial and the educational station were required to go to Washington for the appeal. This was an expensive and time-consuming process that the educational station could not easily afford. Usually the educational stations found themselves coming away with the short end of the stick on disputes involving time, power, and position to the extent that they were unable to use their broadcasting facilities effectively for any type of continuing programming.[2]

reservation threats

After 1925 the secretary of commerce strongly urged people who wished to enter broadcasting to buy an existing station rather than add one to the already overcrowded airwaves. As a result, many educational facilities were propositioned by commercial ventures interested in buying the stations. Frequently the institutions succumbed to the pressure because the financial drain of the stations outweighed the dwindling public service value.

The result was a downward spiral for educational radio. In 1928 thirty-three stations gave up, followed by thirteen more the following year.[3]

In 1929 the National Committee on Education by Radio was organized. It suggested that 15 percent of all radio stations be reserved for noncommercial educational use. The group tried to establish this reservation policy in the

1934 Communication Act. It was unsuccessful, largely because commercial interests so touted their own cultural contribution as to negate the need for strictly educational stations.

reservations defeated

Despite defeat, educational organizations such as the National Association of Educational Broadcasters (NAEB), the National Education Association (NEA), and the U.S. Office of Education (USOE) kept this issue alive while the educational stations continued to dwindle to approximately thirty.[4]

FM Educational Radio

With the advent of FM broadcasting, educators, through perseverance, began to taste victory. In 1945 the FCC reserved the twenty FM channels between 88.1 and 91.9 exclusively for noncommercial radio. These channels at the lower end of the FM band are reserved for educational use everywhere across the country, so it can be assumed that any station broadcasting below 91.9 on the FM band is noncommercial.

FM reservations

By the end of 1945 there were six FM educational stations on the air.[5] This number grew to forty-eight by 1950 and passed twelve hundred by 1987.[6] Part of this FM increase was due to a 1948 ruling authorizing low-powered, ten-watt educational FM stations that generally reached only a two- to five-mile radius and were easily and inexpensively installed and operated. For a short time, approximately half the FM educational stations were of the ten-watt variety, generally operated by educational institutions as training grounds for students.

growth of FM

In fact, the majority of applications for FM noncommercial radio stations were from educational institutions—colleges, universities, high schools, public school systems, and boards of education. Others were from state or municipal authorities or religious groups and a few were owned by nonprofit community groups such as the Pacifica Foundation.[7]

Networking

As the number of educational radio stations increased, the need for some form of network grew. Actually, the need had been evident as far back as the 1930s. Several attempts were made then to exchange program material so that stations would not be burdened with producing all their programs themselves. However, the attempts were not very successful and little was accomplished except for a few isolated instances of station program exchange.

program exchanges

The National Association of Educational Broadcasters (**NAEB**), organized in 1934 as a spokesgroup for the educational radio field, formed the first workable duplication and distribution operation. This so-called bicycle network, begun in 1949, was not a network in the same sense as the commercial networks. Programs were not sent through wires to all stations simultaneously; they were taped and then sent from one station to another by mail on a scheduled round-robin basis.[8]

NAEB bicycle network

The Advent of NPR

In 1967 a Public Broadcasting Act, which formed the Corporation for Public Broadcasting (CPB), was passed by Congress primarily to serve the needs of public television, but radio was also included. At that time, the NAEB ceded its radio networking duties to National Public Radio (**NPR**), an arm of CPB that began operations in 1970.

NPR was formed to upgrade the quality of public radio by providing excellent programming. In order to help with the upgrade, NPR set up certain qualifications that stations had to meet in order to become affiliates. For example, they had to have at least five full-time paid employees, and they had to program at least eighteen hours a day every day of the year. Over three hundred stations have become affiliated.[9]

Under NPR's original organization, qualifying stations wanting to receive programming from NPR paid a fee of several thousand dollars per year and received about seventy hours of programming per week.

This included news, senate hearings, concerts, dramas, self-help material, documentaries, and programming from foreign countries, particularly the British Broadcasting Corporation and the Canadian Broadcasting Corporations. Probably the most popular of the early programs was an in-depth evening news program called "All Things Considered" that dealt with issues, events, and people. Later two similar programs, "Morning Edition" and "Weekend Edition" were offered. The latter was hosted by Susan Stamberg, who had been very instrumental in the development of "All Things Considered."[10]

The fees from stations did not cover the costs of programming. About half of the budget came directly from the Corporation for Public Broadcasting and other monies came from companies that underwrote programming.[11]

When stations were not airing NPR material, they programmed locally, primarily with classical music. Some stations also aired jazz, rock, swing, or other forms of music. Many local stations also produced their own dramas, children's programs, sports, news, and health programs.

Originally NPR material was sent to the stations over phone wires from the Washington headquarters. Although some of the programming distributed nationally was produced by local stations, most of it was produced in Washington, partly because the phone lines led from Washington to the local stations. Later NPR distributed by satellite, but most of the programming was still Washington based.[12]

More recently, NPR has **"unbundled"** its programming so that stations can air certain blocks of programming (e.g., morning news only) without having to pay for the rest of the programs.[13]

FIGURE 4.1

Susan Stamberg, host of "Weekend Edition." *(Courtesy of National Public Radio)*

FIGURE 4.2

The NPR satellite center. *(Courtesy of National Public Radio)*

The Formation of APR

Some of the structural elements of NPR were not popular with many of the stations. For one thing, NPR received most of its money from the federal government through CPB and government cutbacks led to cutbacks in NPR programming, particularly in the cultural and concert areas. In addition, most of the programming was produced in Washington and local stations felt their creative programs deserved wider dissemination.

reasons

FIGURE 4.3

A feature called "Buster The Show Dog" was a regular part of "A Prairie Home Companion." Host Garrison Kiellor (*left*) performs with Kate MacKenzie, Dan Rowles, Stevie Beck, and Tom Keith.
(Courtesy of American Public Radio/Rob Levine, photographer)

structure

As a result, a group of public stations joined together in 1982 to form American Public Radio (**APR**), an independent, nonprofit, private network headquartered in St. Paul, Minnesota. Its qualifications for affiliates are similar to those of NPR. The purpose of the network is to acquire radio programming from various local stations and disseminate it to other stations. Part of what made this possible is satellites that can pick up programs from any station and beam them to any other station.[14]

finances

American Public Radio has not received federal funding. It is supported by fees paid by its members and based, to some degree, on the size of their market. These fees support the national office and the satellite distribution system. Programs are donated by many of the local stations. Usually these have been funded locally by underwriting or contributions from listeners, or they have been funded by a special program fund administered by APR that is designed to stimulate development of new ideas.

programs

APR programs material similar to NPR, but it programs a much higher percent of classical music (55 percent).[15] One of its most popular programs was "A Prairie Home Companion" produced in Minnesota and hosted by Garrison Keillor. It included music, interviews, and homespun philosophy. The program aired from APR's inception until 1987 when Keillor decided to leave the program.[16]

Over three hundred stations have affiliated with APR, but most of those are also affiliated with NPR. This is an arrangement that is possible within public broadcasting but not commercial broadcasting.[17]

Nonaffiliated Stations

That leaves close to a thousand stations that are not affiliated with either NPR or APR. Most of these are small stations operated by colleges to train students in broadcasting. Many are not interested in network programs because they want to produce their own programming for training purposes. In general, they do not qualify for NPR affiliation because they do not have five full-time paid employees and they do not broadcast every day of the year.

At first most of these stations were ten watts. However, during the mid 1970s, NPR instituted a campaign to increase its number of affiliates by trying to get these stations to upgrade in power, hours, and personnel. One of the results was that in 1978, the FCC adopted new policies for the **ten-watters** stipulating that they upgrade to at least one hundred watts (Class A) or switch to a frequency somewhere in the commercial band or to a newly created frequency of 87.9, which was not on most radios. Also, all public radio stations were required to be on the air on a full-time basis or be subject to sharing their frequency with other interested entities.

ten-watters

Most of the stations opted to increase power; very few increased hours of operation, mainly because they did not have requests to share the frequency; and most remained uninterested in national network affiliation. NPR lost its interest in most of these stations and the whole issue of ten-watt station upgrade simmered down or was treated with benign neglect.[18]

present state

Financial Stress

None of the public radio entities are rich. All are constantly scraping for money. The 1980s were particularly difficult because the government cut back support for public broadcasting. Educational institutions also had their own financial problems that led to decreased support for the stations they owned.

For a short time NPR teetered on the brink of bankruptcy and was saved only by a 1983 loan from the Corporation for Public Broadcasting.[19] In 1985 NPR voted to rearrange its financial structure so that federal funds would go directly to the stations instead of NPR. As a result of this reorganization, stations will pay more dues to NPR.[20]

reorganization

Public radio stations have had to rely increasingly on support from listeners and from companies who were willing to underwrite or supply grant money. Both stations and networks have undertaken some commercial fundraising ventures, such as leasing satellite time to other users, establishing a paging system, entering the cable audio business, and transmitting digital data. Aside from leasing satellite time, most of these commercial ventures have been unsuccessful.[21]

fund-raising

However, public radio continues to survive and provide excellent programming. Those involved with it find it worthwhile in terms of recognition, prestige, and service. One report has indicated that 13 percent of the adult population is aware of public radio and 4 percent listens regularly. This listening audience is skewed toward people with high incomes and extensive education.[22]

Overall, the development of public radio has been slow but steady. Conflicts arise because of its diverse purposes and because it does not receive the prestige given public television, but it does make a substantial contribution to the media structure.

Early Educational Television

Early television was dominated by commercial interests with a little activity from educational institutions. Several universities offered college credit courses over local commercial channels, and in 1950 Iowa State University built a TV station that it ran as a partially commercial venture.[23]

Educators, while recognizing the value of television, also realized that they would face a losing battle akin to the early radio experience if they attempted to compete with commercial companies for stations. As a result, the same groups that fought for the allocation of noncommercial FM radio stations attempted to convince the FCC that there should be **channel reservations** for noncommercial television stations. These groups formed an organization called the Joint Committee for Educational Television (JCET), hired a lawyer, and raised money to lobby for their cause.

station reservations

JCET was aided in its cause by FCC commissioner Freida Hennock, the first woman appointed to the FCC, who did not hesitate to say that she favored educational stations. When the freeze was lifted in 1952, the FCC had authorized the reservation of 242 noncommercial channels—80 VHF and 162 UHF. Unlike FM, these channels were scattered around the dial, as channels were reserved in different cities. For example, channel 13 was reserved in Pittsburgh, channel 11 was reserved in Chicago, and channel 28 was reserved in Los Angeles.[24]

Obtaining the channels and activating the stations were two different matters, mainly because of the huge sums of money that television stations demand. Fortunately for educational television, the **Ford Foundation** became interested in its cause and provided much of the money for the early facilities and programs. In 1953 the nation's first educational TV license, KUHT, was granted to the University of Houston in Texas. By 1955 there were nine stations on the air and by 1960, forty-four stations.[25]

Ford Foundation

The Ford Foundation also became involved in programming by helping to establish the National Educational Television and Radio Center in Ann Arbor, Michigan, which acted as a distribution bicycle network center. This helped stations acquire enough programming to fill the hours, but because tapes were mailed to stations, nothing of a timely nature could be exchanged. To compound this problem, the early programs were reproduced on kinescopes, which gave them a very grainy quality. Eventually this center in Ann Arbor dropped some of its functions (including radio), changed its name to **National Educational Television (NET),** started producing programs, moved to New York, and allied itself with the New York public TV station for production.

Chicago TV College and MPATI

The Ford Foundation also funded general programming concepts, such as **Chicago TV College,** a fully accredited set of televised courses that enabled students to earn two-year college degrees through at-home viewing or a combination of at-home viewing and on-campus class attendance.[26] Another Ford-funded experiment of the late 1950s was the Midwest Program on Airborne Television Instruction (**MPATI**), by which programs for schoolchildren were broadcast onto two UHF channels by an airplane that circled two states.[27]

FIGURE 4.4

People being taken on a tour of KUHT, Houston, shortly after it went on the air. *(Courtesy of KUHT, Houston)*

FIGURE 4.5

Professor Harvey M. Karlen conducting a lesson on national government for Chicago's TV College during the midsixties. *(Courtesy of Great Plains National Instructional Television Library)*

FIGURE 4.6

Bill Nixon conducting a science lesson used by MPATI. *(Courtesy of Great Plains National Instructional Television Library)*

Despite its largess, the Ford Foundation could not underwrite all of educational television, so educators turned to the government for the additional funds to build stations. In 1962, after a year of debate, Congress passed the **Educational Broadcasting Facilities Act,** which authorized $32 million for five years. These monies were to be made available on a matching-fund basis to states to assist in the construction of educational television facilities. At this point, government funds could be used only for facilities and not programs— a satisfactory arrangement to the educators, who were leery of strings that the government might attach to programming money. This act led to an increased interest by many groups in establishing educational TV stations. Because of the many requests, the FCC in 1966 revised its assignment table upward and set aside 604 channels in 559 communities. By the end of 1965, the number of educational stations on the air had doubled from the 1960 number of 44.[28]

The ownership of these educational television stations was somewhat akin to radio station ownership in that educational institutions owned many stations. However, many more TV stations than radio stations were owned by states and community groups. State ownership became common, particularly in the South. Alabama, the first state to own and operate TV stations, built nine stations that cover the state in a mini-network. About fifteen other states followed Alabama's lead of owning one or more stations. About one-fourth of educational television stations were licensed to nonprofit community corporations. These corporations were usually formed by community leaders and supported by the communities they served.[29]

Educational TV was not without its growing pains. In many cities, the channel allocation for the educational television station was in the UHF band, making it difficult for the station to gain either audience or community financial support. What is more, educational programming was, in a word, dull. Most of it was produced on a shoestring budget and consisted of local free talent discussing issues or information. As a result, educational television became known for its talking heads and did not attract large audiences even in VHF cities. The Ford Foundation, feeling that it alone was supplying too much of the support to educational television, began withdrawing some of its financing. Innumerable educational broadcasting organizations and councils appeared and disappeared, some because of financial problems and some because of lack of clear focus or because of political infighting. The result was a system so loosely organized that it impeded the impact of the medium.

The Carnegie Foundation therefore set up the **Carnegie Commission** on Educational Television. This group of highly respected citizens spent the better part of two years studying the technical, organizational, financial, and programming aspects of educational television. In 1967 it published its report, "Public Television: A Program for Action."

The Carnegie Commission changed the term "educational television" to "public television" to overcome the pedantic image the stations had acquired. It also recommended that "a well-financed and well-directed system, substantially larger and far more pervasive and effective than that which now exists in the United States, be brought into full being if the full needs of the American public are to be served."[30]

The Public Broadcasting Act of 1967

Most of the Carnegie Commission's many recommendations were incorporated into the **Public Broadcasting Act of 1967.** However, there were two major changes between the Carnegie recommendation and congressional passage of the act. One was that radio was added to the concept. The Carnegie Commission had occupied itself only with TV, but the radio interests that led to NPR were included by Congress. The Carnegie group had recommended that public broadcasting be given permanent funding, perhaps through a tax on TV sets, but Congress opted for one year's funding of $9 million with additional funding to be voted at a later time.

Congressional changes

The Public Broadcasting Act of 1967 had three main sections. Title I of this act authorized an additional $38 million for the construction of facilities. Title II provided for the establishment and funding of the **Corporation for Public Broadcasting (CPB)** to supply national leadership for public broadcasting and to make sure that it would have maximum protection from outside interference and control. Title III authorized the secretary of health, education, and welfare to make a comprehensive study of educational and instructional broadcasting.

provisions

The heart of the bill, Title II, stipulated that the fifteen members of the board of the Corporation for Public Broadcasting be appointed by the president with consent of the Senate. This corporation was in no way to be considered an agency or establishment of the U.S. government. The main duties of the CPB were to help new stations get on the air, to obtain grants from federal and private sources, to provide grants to stations for programming, and to establish an interconnection system for public broadcasting stations.[31]

CPB

The Corporation for Public Broadcasting was specifically forbidden from owning or operating the interconnection system. Therefore, the corporation created the **Public Broadcasting Service (PBS),** an agency to schedule, promote, and distribute programming over a wired network interconnection, which later became a satellite interconnection. PBS also had a governing board consisting of station executives. This service was not to produce programs itself, but rather was to obtain them from such sources as public TV stations, production companies, and foreign countries. Hence a three-tier operation was established: (1) the stations produced the programs; (2) PBS scheduled and distributed the programs; and (3) CPB provided funds and guidance for the activities.[32]

PBS

FIGURE 4.7

Bert (*right*) and Ernie,
two regulars on
"Sesame Street."
*(Courtesy of © Children's
Television Workshop)*

1970s Programming Processes

"Sesame Street"

The advent of the Corporation for Public Broadcasting and its accompanying funding allowed public television to embark upon innovative programming of high quality. The first series to arouse interest in virtually every public TV station was the successful 1970 children's series **"Sesame Street,"** produced by a newly created and newly funded organization, the Children's Television Workshop (CTW). This series helped strengthen PBS as a network because it was in demand throughout the country. In the same year, the public television drama "The Andersonville Trial" won the Emmy for Best Program of the Year.

other programs

Other early PBS series that met with sustained popularity were "The French Chef," Julia Child's cooking show produced in Boston; "Mr. Rogers' Neighborhood," a children's program produced in Pittsburgh; "Black Journal," a public affairs series dealing with news and issues of importance to blacks produced in New York; "Civilisation," a British import on the development of western culture; and "Firing Line," a debate series hosted by William F. Buckley produced in Washington.[33]

Station Program
Cooperative

In 1974 PBS established a **Station Program Cooperative,** a unique method by which many of the PBS programs are selected. First PBS conducts an audience research survey to determine national program needs; then it solicits program proposals from stations and production companies that theoretically should meet the needs determined by the survey. These proposals are catalogued and sent to the various stations, which begin voting on those programs they would like to carry. The first few rounds of voting allow stations to express their interest in certain programs, but do not bind the stations to purchase the program. Occasionally, several major stations will agree in the first round to

FIGURE 4.8

Fred Rogers (*right*) with Mr. McFeely, one of the many characters who visited regularly on "Mister Roger's Neighborhood." *(Courtesy of Family Communications, Inc.)*

FIGURE 4.9

The French Chef, Julia Child about to prepare an unusual dish. *(Courtesy of WGBH Educational Foundation/Paul Child, photographer)*

support a particularly popular program or series and its future is, therefore, secure before additional rounds of voting occur. But more frequently, after a number of proposals have been eliminated for lack of interest, actual purchase rounds are conducted and stations are required to help pay the production costs of those programs they wish to air. As a result of this process, each program produced for PBS under the Station Program Cooperative has a slightly different group of stations paying for and airing it. Often the money that stations contribute to production costs come from the CPB community service grants by following this route: CPB to local stations to PBS to producing stations.[34]

Not all PBS programs come through the station program cooperative. Some are underwritten by foundations, corporations, or the government and some are purchased from foreign governments. In fact, there have been those who claim that PBS stands for "primarily British shows" because of the large amount of fare produced by the BBC. Others feel PBS stands for "Petroleum Broadcasting Service" because of the large number of shows underwritten by oil companies.

other program methods

PBS's concept of cultural programming was under competitive attack during the late 1970s and early 1980s because several cable TV services were established for cultural programming, including BBC programs. However, most of these cultural services died prematurely for lack of adequate financing. Thus, public broadcasting programming suffered little long-term damage.[35]

For a period of time PBS led the way in programming for the **hearing impaired.** Encoded captions were placed on the TV signal so that hard-of-hearing people, by purchasing a decoder, could see both the picture and written captions. Some public TV programs had a person, seen in the corner of the screen, signing for the hearing impaired. At one point, during the 1970s, public TV was essentially obligated to provide captioning because a court had ruled that public stations, because they receive federal funds, have more of an obligation to serve the hearing impaired than commercial stations. However, in 1983 the Supreme Court overturned that decision.[36]

Government Conflict

Although the organization for public broadcasting set up by the Public Broadcasting Act of 1967 looked good on paper, it had problems in actual operation. A major controversy surfaced during the Nixon administration when the CPB, which had been formed to insulate public broadcasting from government, became something of an arm of the government. The conflict centered around programs that were critical of the government and around the concept of localism.

The programs that caused ire were primarily nationally aired documentaries and public affairs programs, such as "Washington Week in Review," which were against Nixon and his policies. As more and more of these programs aired, members of the administration began stating that public television should emphasize local programs, not national ones. Many within public television felt that this localism policy was espoused because the administration felt local programming would not be as influential as national programming and, therefore, any criticisms of the administration that did creep in would not become significant issues. The administration's rebuttal stated that since the three commercial networks were national in scope, there was no need for a fourth network, but there was need for local programming.

The administration carried its localism philosophy into the budgeting area. After Congress passed a bill that would have given public broadcasting $64 million for 1972–73 and $90 million for 1973–74, Richard Nixon vetoed the bill on the grounds that the national CPB organization exerted too much control over the local stations.

With this action the chairperson and president of CPB resigned and Nixon appointed new people to the posts who were more in line with his way of thinking. With CPB now more in tune with the administration, a schism developed between CPB and PBS over who should control the programming decisions for public broadcasting. PBS wanted to continue news and documentary programs having a national emphasis. CPB pressed for local control and an emphasis on cultural programming rather than on documentaries.

The controversy continued but gradually abated as Watergate consumed the administration and as new political forces took hold. In addition, PBS reorganized its board slightly so that some of the members were laypeople instead of public broadcasting station people. These laypeople tended to communicate more evenly with the lay members of the CPB board. Monitoring committees were also set up to resolve disputes whenever PBS or CPB felt that particular programs or series were not balanced or objective.[37] In addition, public TV licensees formed a trade organization, the National Association of Public Television Stations, to do some of their lobbying and planning.[38]

PBS board reorganization

All of this helped settle down the CPB-PBS government conflict, but it did not thoroughly contain it. Occasional resignations and unpleasant words still fill the air from time to time.[39]

Carnegie II

Because of the constant unrest in the public broadcasting arena, the Carnegie Commission decided to review the whole subject. This review started in 1977 and culminated in 1979 with a report entitled "A Public Trust: The Landmark Report of the Carnegie Commission on the Future of Public Broadcasting."

In this report the commission recommended abolishing CPB, which, despite the best of intentions, had not been insulated from political pressure. The commission also recommended that CPB be replaced with a Public Telecommunications Trust. The trust was to be a private, nongovernmental, nonprofit corporation that would provide leadership and long-range planning for public broadcasting and would administer government funds to local stations and to a Program Services Endowment that, in turn, would fund innovative projects.

provisions

The commission also recommended that funding be more than doubled for public broadcasting and that a spectrum fee be assessed to help raise this additional funding. This spectrum fee would be levied on users of the electronic spectrum—commercial broadcasters, CBers, AT&T, doctors' call services, and the like.[40]

However, this time the Carnegie recommendations did not go into wholesale effect. They came at a time when financial cutbacks were the order of the day and when a recommendation proposing double spending was not well received.[41]

lack of implementation

The structure of public broadcasting remained essentially as it was in the 1960s and 1970s. CPB's influence was cut slightly in that decisions on specific grants were usually made by independent panels made up of program producers.[42]

Financial Problems

Like public radio, public TV underwent difficult financial times during the early 1980s. Financing public television is much more complicated than financing public radio, however, because of the greater expense.

sources of funds

In the late 1970s public TV was supported as follows: 70 percent federal, state, and local governments; 8 percent foundations; 3 percent corporate underwriting; 11 percent community subscribers and auctions; and 8 percent other.[43] By 1985, government support was less than half at 47 percent. Other percentages were; foundations, 4 percent; underwriting, 16 percent; subscribers, 23 percent; and other, 10 percent.[44]

government

The federal government reduced its support to public broadcasting and hard-pressed state and local systems followed suit, not only in terms of percentages but in hard dollars. Much of the government cutback affected CPB and its activities. Local stations suffered also from direct cuts and from the cuts affecting CPB since many CPB activities involve some form of grant to local stations. In addition, federal funding was only authorized for two years in advance, making it difficult for public TV to plan for the future.[45]

foundations

Foundations also reduced their contributions to public broadcasting, mostly because their money became tighter. One foundation to come to public broadcasting's aid, however, was the **Annenberg Foundation** established by Walter Annenberg, the founder of TV Guide. The purpose of Annenberg's grant ($150 million over a ten-year period beginning in 1982) was to develop collections of materials for college-level courses that use electronic media.[46]

corporate underwriting

Some of the slack from government funding and foundations was picked up by corporate underwriting from companies such as Xerox, Mobil, and 3M. Congress sweetened the pie for corporations in 1981 by approving the airing of company logos by public stations. Prior to this the stations could merely acknowledge that "This program was made possible by a grant from XYZ Corporation." With the change in policy, stations could show corporate logos and mention the product line of the contributing company. This was referred to as **enhanced underwriting.**[47]

citizens

People from the community also tried to give public broadcasting more of a helping hand. Monies from individuals paid a yearly fee to a local public station, and in return received a magazine that listed the station's programs, increased percentagewise. So did money from auctions, which had become an institution, particularly at community-owned stations. Donations of unusual objects from members of the community were auctioned off over the air with people calling in to place their bids.[48] The auctions became so profuse and consumed so much air time, however, that negative reaction set in and many public TV stations began curtailing their auctions.[49]

ownership changes

The change in finances was also reflected in changes in ownership. Government agencies, such as public schools and colleges and states, had been the predominant owners of early TV stations, but as their fortunes waned, nonprofit community groups became the more predominant owners. By 1987, 43 percent of licenses were held by community organizations, as compared to 32 percent by colleges, 16 percent by state authorities, and 9 percent by local educational or municipal authorities.[50]

Because the total amount of money from traditional sources was down, public TV, like public radio, tried to undertake commercial fund-raising ventures. It considered establishing a pay-cable network, going into subscription TV, providing data services, and leasing its satellite time.[51] Except for the leasing of satellite time, most of these ideas did not materialize. The leasing of satellite time, however, enabled PBS to recover most of its own costs for operating a satellite system.

fund-raising

The most controversial of PBS's fund-raising proposals, however, was the airing of actual commercials. In the spring of 1982, Congress authorized ten public TV stations to experiment with advertising in order to obtain funds.[52] The experiment, which lasted about eighteen months, had ambivalent results and, at the end of the period, the stations appeared to conclude that direct advertising was not something public broadcasting should embrace. Commercial TV networks and stations were pleased with this decision because they were unhappy with the thought of public TV stations seeking ad money from the same companies they were approaching. Also, the spectre of a public broadcasting system overly dependent upon the business community seemed unhealthy to some who feared that big business could dictate programming content.

commercials

However, PBS revised its enhanced underwriting rules to allow more commercial information to be given about companies that donate for programs. It determined that logos, slogans, locations, description of product lines or services, and trade names could be included in attractively designed announcements.[53] By the late 1980s these enhanced underwriting announcements were running as long as thirty seconds and were looking very much like commercials.[54]

By the late 1980s, funding in general had improved for public television, but it still felt itself scraping to make ends meet.[55]

Programming of the 1980s

Despite financial problems, public television could be very proud of its programming record by the 1980s. It had "Frontline," which was the only regularly scheduled prime-time documentary, and "The MacNeil/Lehrer News Hour," which had begun in 1983 and was often acclaimed for its in-depth look at news issues.[56] These programs, and other public affairs shows, were not without their critics who accused them of left-leaning biases.

Two of the most controversial programs of the decade were "Death of a Princess" and "The Africans." The former was a docudrama, co-produced by the BBC and WGBH in Boston, which graphically depicted the execution of a Saudi Arabian princess and her common lover, both of whom had been accused of adultery. The Saudi's objected to the protrayal, complaining it was

controversy

FIGURE 4.10

Robert MacNeil (*left*) and Jim Lehrer of "The MacNeil/Lehrer Report," an in-depth examination of the day's top news stories produced jointly by WNET in New York and WETA in Washington. *(Courtesy of MacNeil/Lehrer Productions)*

FIGURE 4.11

A scene from "Death of a Princess." *(Courtesy of WGBH Educational Foundation, Boston)*

an attack on the Islamic religion.[57] "The Africans" was a point-of-view miniseries about the influence (mostly negative) of western countries, such as England and the United States, on the African continent. It had been underwritten in part by the National Endowment of Humanities, which labeled it an "anti-Western diatribe" and demanded that the NEH credit be removed from the series.[58]

FIGURE 4.12

A scene from one of the "Wonderworks" programs called "Mighty Pawns." Paul Winfield plays the principal of an inner city school where student Alfonso Ribiero leads a chess team to victory.
(Courtesy of Wonderworks)

Public broadcasting stations won the right to editorialize during the 1980s. Previously they had been prevented from this on the premise that federally funded stations could take on the appearance of government propaganda organs. However, in 1984 this was declared unconstitutional and public stations were given the right to editorialize.[59]

editorializing

The National Geographic Specials were the most-watched of the PBS programs with a program called "The Sharks" garnering a cumulative audience of 17.4 percent of the population.[60] Programs about the universe such as "Planet Earth" and "Cosmos" also received critical acclaim.

In the drama department, many of the BBC's finest productions came to public television, particularly on the long-running "Masterpiece Theatre." Newly written American drama was also encouraged on such series as "American Playhouse," "Hollywood Television Theatre," and "Visions." "Great Performances" was the longest running performing arts series in TV history.[61]

drama

An area where public television shown most brightly was children's programming. In the dramatic series "Wonderworks," it had the only regularly scheduled prime-time show designed for children. "Sesame Street," which by now was as old as most college students, was still one of the most popular public TV offerings. It had been followed by other noteworthy soft-teaching programs such as "Electric Company," "3–2–1 Contact," and "Square One TV."[62]

Frequently these children's programs, which were designed primarily for children at home, were used within the schools. Public TV also produced some programs designed more for in-school use than at-home use, such as "Newscasts from the Past" and "The Challenge of the Unknown."[63]

children's programming

Public Broadcasting   125

In the area of formal education, PBS has created Adult Learning Service (ALS). This is a cooperative effort between public TV stations and over 1200 colleges and universities to offer college credit courses that people can watch in their homes. ALS researches courses that are available and then helps stations and colleges schedule and promote the courses that fit their needs. Over 100,000 students a year enroll in these courses.[64]

Public TV also decided to tackle illiteracy and set up a project called Project Literacy US (PLUS) that will include programs aimed at teaching reading to adults.[65]

Over 100 million people (60 percent of the population) watch public television at least once a week. Although this translates to ratings that are only in the 3 to 4 percent bracket, many of the programs have more impact than those found on commercial TV.[66]

Public Broadcasting Chronology

1917 Professor Terry of the University of Wisconsin transmits signals
1925 Many educational radio stations taken over by commercial interests
1929 National Committee on Education by Radio asks for station reservations
1945 FCC reserves twenty FM channels for education
1949 NAEB begins bicycle network for radio
1950 Iowa State University operates commercial TV channel
1950s Ford Foundation funds various TV stations, NET, Chicago TV College, and MPATI
1952 FCC reserves 242 noncommercial TV channels
1953 First educational TV license granted to University of Houston
1962 Educational Broadcasting Facilities Act passed
1965 Carnegie Commission formed
1966 FCC sets aside 604 educational TV channels
1967 Public Broadcasting Act authorizes CPB
1970 "Sesame Street" begins
1970 NPR incorporated
1970s Public TV and President Nixon clash
1974 Station Program Cooperative established
1977 Second Carnegie Commission formed
1978 FCC adopts new rules for ten-watters
1981 Congress approves public broadcasting's airing of company logos
1982 First Annenberg grants awarded
1982 APR established
1982 Ten public TV stations experiment with commercials
1984 Public broadcasting given right to editorialize
1984 PBS revises underwriting rules
1986 PLUS begins to tackle illiteracy

Conclusion

Public broadcasting is caught between the best and worst of democracy, free enterprise, and culture. America's democratic traditions make governmental financial support of broadcasting in any form a risky undertaking. The spectre of possible government influence poses a threat, and yet to date, government seems to be the only source able to finance a culture-oriented programming service. The free enterprise system allows the marketplace to determine what products and services should survive, but public broadcasting, if it is to maintain even a partial noncommercial nature, will probably continue to be unable to compete in this commercial marketplace. Although culture and education are supposedly highly revered in America, they cannot draw the audiences and finances necessary to ensure their survival.

All of this has influenced the history and operation of public broadcasting. The early radio stations, which were operated by colleges, were gradually overpowered by commercial interests. Through concerted effort, educators persuaded government to reserve stations in the FM band for educational broadcasting. Most of these stations exist on modest budgets provided by educational institutions and on endowments, grants, and public donations. They do provide a hallmark of cultural programming, including classical music and some highly rated programs from National Public Radio and American Public Radio.

Public television began with stations specifically reserved for it, but educational institutions were unable to bear as large a percentage of the costs as they had for radio. Fortunately, the Ford Foundation underwrote much of the early development of educational TV, but it could not shoulder the entire weight. Therefore, the government, amid much controversy, began in 1962 to provide grant money for facilities only.

Still, educational broadcasting was not able to compete in the free enterprise system with its commercial counterparts. The Carnegie Commission spent several years studying the situation and recommended a public television system heavily supported by government funds for both facilities and programming. The Carnegie plan was quickly passed by Congress. By recommending the Corporation for Public Broadcasting to act as a buffer and to administer funds to the Public Broadcasting Service and those producing programs, the Carnegie Commission hoped to avoid the pitfall of government influence over programming. But this was not to be—both internal and external schisms over programming philosophy rocked the CPB-PBS-NPR family during the Nixon administration. Following that the government withdrew much of its funding, leaving public broadcasting to find other methods of financing, turning more sharply to the leaders and concepts of free enterprise.

The purpose of financial recruitment, of course, is to find support for the programming services that have improved greatly since the Carnegie plan went into effect. The Station Program Cooperative programs, foreign programs, underwritten programs, and local programs have moved public broadcasting from "talking heads" to dynamic culture.

The triangle formed by democracy, free enterprise, and culture has created many controversies as public broadcasting tries to maintain an equilibrium among government influence, corporation influence, and audience demographics. Just where public broadcasting is headed is unclear, but this is nothing new for an industry that has spent most of its life in a state of uncertainty.

Thought Questions

1. How can public broadcasting be insulated from government influence over program content?
2. In your opinion, what is the proper balance for local versus national programming within the public broadcasting system?
3. Should the recommendations of Carnegie II be implemented? Why or why not?
4. What are some ways public broadcasting could ease its financial strain?

Cable Television

Overview

Although the emergence of cable television as a well-known industry is a relatively recent phenomenon, it is almost as old as television broadcasting. Broadcast TV became an overnight success while cable struggled for several decades before it received any widespread recognition.

This chapter details the struggle, recognition, and problems of cable television discussing the following topics:

The beginnings of cable television

Cable's early days as a retransmission and distant import service

Broadcaster objections to cable

Pole attachment controversies

The ambivalent role of the FCC in cable's development, particularly in relation to must-carries, syndicated exclusivity, distant signal importation, local origination, and channel designation

Early cable local origination and public access

The controversy over cable's copyright payments

Home Box Office's breakthrough in satellite distribution

The phenomenal growth of cable in the late 1970s and early 1980s

The franchising boom

The growth of MSOs, advertising, and union participation

The proliferation of pay-cable and basic cable services

The varying forms of local programming

Experiments in interactive services

The retrenchment of cable as the boom subsided and reality took hold

The cutback in programming services, nationally and locally

Signal theft

Cable deregulation and government interaction

5

Cable television should be the supermarket of the mind.

Milt Gelman
TV writer

The Beginnings of Cable

transmission pattern

Cable television is almost as old as commercial broadcast television. It is, in fact, an outgrowth of the broadcast industry in that it developed because people in poor reception areas wanted to receive TV. They were unable to receive regular TV because broadcast television is dependent upon antenna systems in **"line of sight"** with the transmitter or with a clean reflection of the signal from a hill or building. If the receiving antenna cannot be lined up with the station transmitter, then TV reception is poor or nonexistent. Cable TV signals, on the other hand, are carried in wire rather than through the open airwaves so cable reception is not dependent on placement of an individual antenna in relation to the transmitter. Rather, a master antenna is erected for one community and the signal received is strengthened and distributed by wire to any home in the area that subscribes to the service by paying a fee. For this reason, cable TV has also been referred to as Community Antenna Television (**CATV**).

first systems

There are many different stories about how cable TV actually began. One is that it was started in a little appliance shop in Pennsylvania by a man who was selling television sets. He noticed that he was selling sets only to people who lived on one side of town. Upon investigation, he found that the people on the other side of town could not obtain adequate reception, so he placed an antenna at the top of a hill, intercepted TV signals, and ran them through a cable down the hill to the side of town with poor reception. When people on that side of town would buy a TV set from him, he would hook their home to the cable.

Another story on the origin of cable service tells of a "ham" radio operator in Oregon who was experimenting with TV just because of his interest in the field. He placed an antenna on an eight-story building and ran cables from there to people's homes. The initial cable subscribers helped pay for the cost of the equipment and after that was paid off they charged newcomers $100 for a hookup.[1]

Whether either of these stories represents the true beginning of cable TV is hard to say. In many remote and mountainous places, friends and neighbors gathered together with the intent of providing television reception for themselves.

freeze effect

One factor that helped cable TV in its beginning was the **freeze** on television station expansion from 1948 to 1952. The only way that people could receive TV if they were not within the broadcast path of one of the 108 stations on the air was to put up an antenna and, in essence, catch the signals as they were flying through the air.

multichannel capability

Most of the early cable TV systems were capable of handling three signals. This is another way in which cable TV differs from broadcast TV. Any particular cable service is concerned with more than one channel, while a particular broadcast service is concerned with only one channel. Cable TV provides a viewer with many channels of programming, whereas a particular network or particular station provides only one channel.

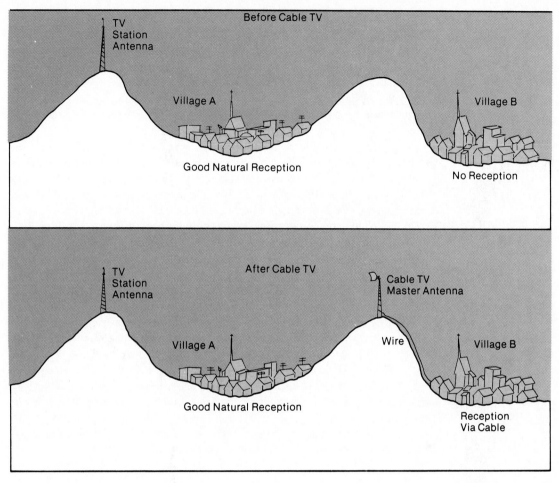

Before Cable TV

TV Station Antenna

Village A

Good Natural Reception

Village B

No Reception

After Cable TV

TV Station Antenna

Cable TV Master Antenna

Village A

Wire

Village B

Good Natural Reception

Reception Via Cable

FIGURE 5.1

Cable TV reception process.

early importation

FCC and broadcaster opinions

As early as 1949 a multichannel antenna system was developed in Lansford, Pennsylvania, as a moneymaking venture. With no local signal available, the system carried the three network signals by importing them from nearby communities, a practice that became known as **distant signal importation.** Fourteen such signal importation companies were in operation by the end of 1950 and the number swelled to seventy by 1952.[2]

At this point the FCC was convinced that as the number of stations increased, the need for CATV would diminish and gradually vanish entirely. The one factor that eluded its attention was that many communities were too small to support the expense of station operation. If the basic philosophy of "mass communication for an informed public" was to be a reality, CATV would have to grow. Signals would have to be imported from cities where stations could find support for their operations.

However, even with 65,000 subscribers and an annual revenue of $10 million in 1953,[3] cable TV was still only a minor operation. Most broadcasters felt no concern about this business that was growing on the fringes of their signal contour. Some, however, were becoming alarmed by the attitude of permanence growing in some cable systems. **Coaxial cable** was replacing the open-line wire of early days and space was leased on telephone poles for line distribution instead of the house-to-house loops augmented here and there by a tree.

telephone pole problems

Placing the cable on **telephone poles** was a source of major resentment for the cable industry. The cable companies had to rent space on the telephone poles for their cable because the telephone companies naturally had a monopoly on all the telephone poles in town. The telephone companies could charge rates that the cable companies felt were unfair. The cable companies also felt that the phone companies often gave them an unnecessarily hard time about where, how, and when the cables were to be attached on the poles. Much discussion and lobbying revolved around pole attachments, but the issue was never satisfactorily resolved.

more importation

As cable TV grew, it became more sophisticated. In addition to importing signals into areas where there was no television, it began importing distant signals into areas where there was just a little television. For example, if a small town had one TV station, the cable system would import the signals from two TV stations in a large city several hundred miles away.

In general, during the decade of the 1950s, cable TV grew from a few very small operations run by friends and neighbors to a system of television reception for small and medium-sized areas that had poor reception or a very limited number of TV stations. Most of these systems were operated by small, locally run organizations that charged subscribers an initial installation fee and monthly fees compatible with modest profits.[4]

Cable's Muddled Growth

objections to importations

The importing of distant signals brought about the first objections to cable TV. Existing TV stations in an area would find that the size of their audience had shrunk because people were watching the imported signal. In fact, sometimes the signal imported would be the same as the one on the local station. For example, a local station might be showing a rerun of "I Love Lucy" and find that the imported station was showing the same rerun, splitting the audience for the show in half. As a result, local stations could no longer sell their ads for as high a price as they had before the importation.

legislation failure

Some of the stations in areas affected by cable TV appealed to Congress and the FCC to help them in their plight. In 1959 the Senate Commerce Subcommittee suggested legislation to license CATV operators. The actual extent of CATV operations, even at this late date, was impossible to identify because the operators were not required to report to any governmental agency—in spite of the fact that in many areas of the country the audience served by

CATV ran as high as 20 percent of the available viewers. Attempts to draft legislation were filled with arguments and debates that lasted until 1960 and ended with the defeat of a bill.[5]

With the failure of federal intervention came a rash of state and local attempts to assert jurisdiction over CATV. In most areas the local city council became the agency that issued cable **franchises** and placed stipulations on how the cable system was to conduct business. Competing applicants for a cable franchise would present to the council their plans for operation of the system, including such items as the method of hookup (for example, telephone poles or underground cable), the speed with which hookups would be made, the fees to be charged to the customer for installation and for the regular monthly service, and the percentage of profit that the company was willing to give the city for the privilege of holding the franchise. Based on this information, the council would award the franchise to the one company it felt was most qualified.[6]

local franchising

During the 1960s this was a very calm process; usually only one or two companies applied for a franchise in a given area. Areas with good reception did not even bother with franchising because no companies would have applied.

Although the number of cable systems doubled between 1961 and 1965, cable TV was still small business. In 1964 the average system served only 850 viewers and earned less than $100,000 annually. **Multiple system ownership**—ownership of a number of cable TV systems in different locations by one company—was less than 25 percent because of the lack of economic incentives.[7]

early 1960s figures

The FCC maintained a policy of nonintervention in cable TV matters during the early 1960s. It was hoping that the problems between the operators and broadcasters would be settled by court decision. Unfortunately, the situation only became more confused as court case piled upon court case.

The early 1960s then saw the beginnings of a rift between broadcasters and cable TV people over distant signal importation, a lack of action on the part of the federal government, and the beginning of actions on the part of local governments.

The FCC Acts

In April of 1965 the FCC did act and issued a notice that covered two main areas: (1) All CATV-linked common carriers from this time forward would be required to carry the signal of any TV station within approximately sixty miles of its system; and (2) No duplication of program material from more distant signals would be permitted fifteen days before or fifteen days after such local broadcast.[8]

FCC rules

The rule of local carriage, which became known as **must-carry,** caused little or no problem. Most CATVs were glad to carry the signals of local stations. However, the thirty-day provision caused bitter protest from cable operators because it limited their rights in relation to what they could show on their distant imported stations. This rule, which became known as **syndicated exclusivity,** meant that if a local station was going to show an "I Love Lucy" rerun on January 15, a cable TV operator would have to black out that rerun on a distant imported station if it were on any time during the month of January.

The cable TV industry marshalled its forces and in May of 1966 succeeded in having the thirty-day provision reduced to only one day. This, of course, angered the broadcasters.

As background to this conflict, one must visualize the frustration of broadcasters who had spent large sums of money. Many were barely able to survive on the revenue generated by their station while cable operators, who had invested far less money, were realizing profits by carrying the broadcasters' signals and importing distant signals. During this period there were both filed and pending applications for cable coverage of areas that would account for at least 85 million people. It is easy to understand the fears of station operators. Their basic desire was for the FCC to provide them with full administrative protection.

The attitude of the FCC during this time was strongly in favor of local TV stations. The second report issued by the FCC in 1966 restricted cable service in the top one hundred markets to existing services. This order came when 119 systems were under construction, 500 had been awarded franchises, and 1,200 had applications pending. All of these systems would be required to prove that their existence would not harm any existing or proposed station in their coverage areas. By making no increase in staff to handle this load, the FCC was in essence freezing the growth of cable operations in the top one hundred markets. This made station owners very happy.[9]

The effect of this ruling on cable operators during the remaining years of the decade was the reverse of what the FCC had intended. Cable systems that were unable to expand were sold to large corporations that could withstand the unprofitability of the freeze period. Multiple system ownership was quite prevalent by 1970.

In 1972 the FCC issued another policy on cable TV that changed the must-carry rule from all stations within sixty miles of the cable TV to all stations within thirty-five miles plus other local stations that, as shown by polls, were viewed frequently by people in the area. The policy also reconfirmed syndicated exclusivity and set up guidelines delineating how much distant signal importation a cable system could undertake, depending primarily upon the size of the market in which it was located and the number of stations already in the market.

During the remainder of the 1970s, the FCC withdrew further and further from cable regulation, leaving it more and more in the hands of local municipalities.

Communication Systems

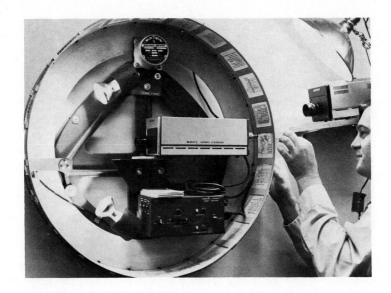

FIGURE 5.2

A simple system whereby local announcements are placed on the TV screen one after the other as the drum rotates. *(Courtesy of TeleCable of Overland Park, Kansas)*

Early Programming

When cable TV first began, it was a common carrier similar to the telephone company; that is, it picked up signals and brought them into homes for a hookup fee and regular monthly fees. This meant that the programming was only that which was picked up from local stations or from distant signal importation.

The very early systems had only three channels for the three networks. As television and its resulting technology grew, cable systems provided as many as twelve channels of programming. The cable system could use each channel from 2 through 13 because its signals were on wires that were not subject to the same interference that makes it impossible to use all twelve channels of broadcast TV in a particular area. Different cable systems placed different programming on these twelve channels, but this usually consisted of converting all the local VHF stations plus all the local UHF stations to VHF space on the dial. If there was not a large number of local stations, then the cable system would bring in stations from nearby communities.[10]

twelve channels

Under this early system there was no **local origination** of programs. Gradually, however, some of the cable facilities began to undertake their own programming. The most common "programming" in the beginning involved unsophisticated weather information. Cable TV operators would place a thermometer, barometer, and other calculating devices on a disc and have a TV camera take a picture as the disc slowly rotated. This would then be broadcast on one of the vacant channels so that people in the area could check local weather conditions. Some systems had news of sorts. This might just involve a camera focused on a bulletin coming in over a wire service machine or on three-by-five-inch cards with local news items typed on them. At any rate, it was a simple, inexpensive, one-camera type of local origination.

A cable TV facility covering a remote sports event. *(Courtesy of TeleCable of Overland Park, Kansas)*

early local origination

Gradually, studio-type local origination was used, usually in the form of local news programs, high school sports events, city council meetings, local concerts, and talk shows on issues important to the community. The first regularly scheduled cable local origination took place in 1967 in Reading, Pennsylvania, and shortly after that local programming appeared in San Diego, California. The FCC became involved and gave the San Diego cable system the authorization to engage in local programming. After this, other cable systems began such programming.[11]

1970 rule

In October of 1969 the FCC issued a rule that required all cable TV systems with 3,500 or more subscribers to begin local origination no later than April of 1970. The purpose was to promote local programming in areas where it did not previously exist.

By April of 1970 many of the cable TV operators who were not engaging in local origination claimed hardship, telling the FCC that they did not have the funds to build studios, buy equipment, and hire crews. As a result, the FCC order was not enforced and later was modified to say that systems with over 3,500 subscribers only had to make equipment and channel time available to those who wished to produce programs.

early public access

This brought about a different type of local programming that became known as **public access.** Individuals or groups would use the equipment provided by the cable operator to produce programs that did not have the operator's input or sanction and then cablecast those programs over one of the system's channels. This differed from local origination, which was programming planned by the cable system.

In most areas public access was not a success, and the equipment provided by the cable company went unused for lack of interest on the part of the local community. In other areas, particularly where cable systems had shown an active interest in local programming, some exciting and innovative projects were undertaken in the public access area. These programs included public affairs programs and video art. Unfortunately, in several places the people who made use of the public access time were would-be stars who were using cable for vanity purposes or people from fringe groups who wished to propagate various causes or even lewd modes of behavior. Such individuals and groups made it difficult for the cable operators to present balanced opinions and also gave the concept of public access an unsavory reputation.[12]

Another form of local programming that was instituted by some cable systems was the showing of movies. A cable company would use one of its channels, or perhaps only the evening hours of one of its channels, to show movies that only subscribers who paid an extra fee could receive. These movies were leased from film companies and were shown without any commercial interruptions. Regulations prevented cable systems from showing the well-known films that broadcasters wanted to show during prime time, but because the small cable systems only had a small number of subscribers who would pay for the movies, they could not afford to pay for blockbuster movies anyway. The quality and vintage of most cable movies were similar to what could be seen on late, late broadcast TV. The channels for movies were not a huge success, but they did bring extra income to some cable systems.

early movie channels

By the mid-1970s only 20 percent of the cable companies were engaging in any type of local programming, and the most popular type of such programming was local sports.[13]

However, throughout the 1960s and 1970s promises were made, broken, and remade concerning the potential services and programs that would be available through cable. Cable lived on the edge of a promise that "within the next five years" cable TV will perk your coffee, help your kids with their homework, balance your checkbook, bring you the top sports events, secure your home, do your shopping, and teach the handicapped. But except for a few isolated experiments, cable remained, until the mid-1970s, primarily a medium to bring broadcast signals to areas with poor reception.

promises made

The Copyright Controversy

Another point of contention between broadcasters and cable operators was payment of **copyright fees.** When cable systems rebroadcast network, local station, or distant station signals, they did not pay any fees to those who owned the copyrights to the materials. In some instances, the networks or stations had paid for and created the material, so they owned the copyright. In other cases, the networks or stations had purchased or rented program material from film companies or independent producers and had paid copyright fees.

The stations, networks, film companies, and independent producers felt that cable companies should pay copyright fees for the retransmission of material because these retransmission rights were not included in the broadcast TV package. Cable operators felt that copyright had already been taken care of by the TV stations and networks, and cable systems were merely extending coverage.

producers versus cable companies

The basic fight over copyright began in 1960 when United Artists Productions, holders of license rights to a large library of feature films purchased by a network at a cost of $20 million, sued two cable operators for copyright

United Artists's case

infringement of their performance rights. The initial ruling that was handed down in 1966 favored United Artists. However, on appeal to the U.S. Supreme Court, the decision was overruled in 1968 and the controversy once again boiled over.

Further complicating the issue was the fact that the copyright law under which the United States then operated had been written in 1909. Congress had been trying for many years to update the copyright situation with new legislation, but the many controversies and special interests surrounding the issue, including the cable feud, made this one of the most lobbied bills of history.

1976 law

Finally, a new copyright law was passed in 1976 and went into effect in 1978. Under this law, cable TV systems paid a compulsory license fee to a five-member, newly created government body called the **Copyright Royalty Tribunal.** This body would then distribute the money to copyright owners. The amount of this compulsory license fee was 0.7 percent of a cable operator's revenue from basic monthly subscriptions. The copyright law gave the Tribunal authority to adjust the rate for inflation. Both cable operators and copyright holders seemed happy with this plan.[14]

claims on money

However, two major problems arose. One involved the distribution of the money collected from the compulsory license. Many interests claimed a share of the money because their program material was retransmitted over cable, and it was the Tribunal's job to decide who got how much. In 1979 it decided that the $20.6 million collected should be distributed as follows: 70 percent to movie producers and program syndicators; 15 percent to sports; 5.25 percent to public television; 4.5 percent to broadcasters; and 0.25 percent to National Public Radio. Hardly any of the entities were satisfied with their pieces of the pie and most went to court in an attempt to gain increases.[15]

changes in industry

The other problem arose because the cable industry changed greatly after 1976 and the $20.6 million collected through compulsory licensing seemed like a paltry sum to copyright holders. Cable systems with expanded channel capacity were now retransmitting signals from stations all over the country. However, according to copyright holders, they were paying as though they were only retransmitting the local must-carries and a few distant imports.[16]

By the early 1980s there was seething discontent with the infant copyright law. Congressional lobbyists from many entities were trying to initiate bills to better their lot in life.

The Beginnings of Satellite Cable

early HBO

In 1975 the stage was set for a dramatic change in cable programming when a company called **Home Box Office (HBO)** began distributing movies and special events via **satellite.**

Home Box Office was actually formed in 1972 by Time, Inc., as a movie/special pay service for Time's cable system in New York. The company decided to expand this service to other cable systems and so set up a traditional broadcast-style microwave link to a cable system in Wilkes-Barre, Pennsylvania. In November of 1972 it sent its first programming from New York to Wilkes-Barre.

During the next several years HBO expanded its microwave system to include about fourteen cable companies in two states with over eight thousand subscribers. This was not an overly successful venture and it was not profitable for Time.

Then in 1975, shortly after domestic satellites were launched, Time decided to bring two of its cable systems the Ali-Frazier heavyweight championship fight from Manila by satellite transmission. The experiment was very successful and HBO decided to distribute all of its programming by satellite.[17]

This distribution method was easier and cheaper than the system that required huge microwave towers, and it meant that as soon as HBO sent its signals to a satellite, they could be received throughout the country by any cable system that was willing and able to buy an earth station satellite receiving dish.

HBO on satellite

HBO then began marketing its service to cable systems nationwide, which was no easy chore at first. The original receiving dishes were ten meters in diameter and cost almost $150,000, a stiff price for cable systems, many of which were just managing to break even. But the technology of satellites moved quickly enough that by 1977 dishes as small as 4.5 meters sold for under $10,000.[18]

Another problem HBO encountered centered on the rules that had been established, mainly for over-the-air subscription TV, that prohibited pay services from bidding on movies and sports events that conventional broadcasters wanted to show. These rules had been established to protect broadcast TV stations from a phenomenon known as **siphoning.** The fear was that pay-TV might simply take over, or siphon, the TV station's programming by paying a slightly higher price for the right to cablecast the material. HBO and several cable TV system owners took these rules to court. In March of 1977, the court set aside the pay-cable rules restricting programming, leaving HBO free to develop as it wanted.[19]

marketing problems

When HBO first began marketing its service, it offered the cable system owners 10 percent of the amount they collected by charging subscribers extra for HBO programming. Approximately 40 percent of the fee was to go to HBO and 50 percent to the program producers. When cable owners indicated that they were not happy about their percentage, HBO raised it so that the systems retained about 60 percent of the money and HBO and the program producers split the other 40 percent.[20]

With receiving dishes manageable in terms of both cost and size, with programming of an appealing nature, and with financial remuneration at a high level, cable systems began subscribing to HBO in droves. Likewise, HBO became very popular with individual cable subscribers and by October of 1977, Time was able to announce that HBO had turned its first profit.

HBO success

FIGURE 5.4

A "live" in-concert taping by HBO of singer Diana Ross before a regular nightclub audience at Caesar's Palace in Las Vegas. This was one of HBO's early programs. *(Courtesy of Home Box Office)*

Turner's superstation

Shortly after HBO beamed onto the satellite, Ted **Turner,** who owned a low-rated UHF station in Atlanta, Georgia, decided to put his station's signal on the same satellite as HBO. This meant that cable operators who had bought a receiving dish for HBO could also place Turner's station on one of their channels. This created what became referred to as a superstation because it could be seen nationwide. A company transmitting the station charged cable operators a dime a month per subscriber for the signal,[21] but they in turn did not charge the subscriber as they did for the HBO pay service. The economic rationale for the superstation was that the extra program service would entice more subscribers. The charge to the cable companies did not cover the station's costs, but it was able to charge a higher rate for its advertisements once it had a bigger audience spread out over the entire country.

With two successful program services on the satellite, the floodgates opened and cable TV took on an entirely new complexion.

Cable's Gold Rush

late 1970s and early 1980s figures

In the late 1970s and early 1980s cable TV experienced a phenomenal growth that was sparked mainly by the development of satellite-delivered services. In 1975, 265,000 homes subscribed to pay cable, but that number grew to eight million by 1981.[22] Similarly, revenues from pay cable increased. In 1979, pay revenues grew 85 percent over 1978 figures. The industry predicted that this was a peak that would subside to about 50 percent growth.[23] However, in 1980 pay revenues grew 95.5 percent over 1979 revenues.[24] The number of cable subscribers also grew. During the 1970s cable added subscribers at the rate of about 1.1 million per year, but in 1980 3.1 million subscribers were added.[25] Between 1975 and 1980 the percentage of homes subscribing to cable jumped from 17 to 23 percent,[26] and during this same period of time, the cable TV profits grew 641 percent.[27] With figures such as these, it is no wonder that cable experienced a veritable gold rush.

One of the main areas in which this gold rush manifest itself was franchising. Cities that could not have given away franchises in earlier years because their reception of TV signals was so clear suddenly became prime targets for cable and its added programming services. In general, cabling moved from rural areas into major cities such as Pittsburgh, Boston, Los Angeles, Dallas, and Cincinnati, where numerous companies applied for the franchises.

In most areas the local city council made the decision regarding which company would win the franchise, although in some areas state regulatory bodies also had a voice. Most city governments were not accustomed to dealing with such matters. Gradually, through the use of consultants, they established lists of minimal requirements that they wanted from the cable companies. These requirements pertained to number of channels, length of time to complete the wiring job, subscriber rates, local input into programming, and similar issues.[28]

Cable companies, in their fervor to obtain franchises, usually went well beyond what the cities required. They, too, hired consultants who contacted city leaders to learn the political structure and needs of the city and to decide how the company should write its franchise proposal to ensure the best possible chance of winning. Since only one company could receive a franchise for a particular area, cries of scandal sometimes accompanied these procedures as the cable companies tried to gain influence. A practice dubbed "rent-a-citizen," whereby a cable company would give blocks of its free stock to influential citizens in return for their support, came under sharp attack.[29]

The development of local programming became an important part of the franchising process. Cities usually requested that cable companies set aside a certain number of their channels for access by governmental, educational, community, and/or religious groups. In order to program these channels, cable companies promised to provide equipment and sometimes personnel.

As franchising competition became more intense, cable companies began promising cheaper rates, shorter time to lay the cable, more channels, more equipment, more local involvement, and generally more and better everything. All of this was written in a bid (usually several volumes thick) and submitted to the city council, which then evaluated all bids, usually with the help of consultants. When the winner was announced, the losers might well call foul play over the procedures used by the winner and the decision could wind up in court or back in the city council's lap. Often the winning cable company was unable to meet all the requirements stipulated in the bid, especially in regard to the speed with which the system was to be built. This led to court cases and fines for breach of contract.[30] But cabling did move forward at a rapid rate.

As the cable companies promised more and more, they realized that they might not recover their investment for about a decade. This hastened a process that was already in full swing in the cable industry—the takeover of small

franchising

franchising problems

MSOs

"ma and pa" cable operations by large multiple system owners and then the consolidation of these MSOs with other large companies.[31] Large companies emerged in cable, partly because they wanted to be part of the gold rush and partly because only large companies had the resources to withstand the expenses of the franchising process and the other start-up costs of laying cable, marketing, and programming.

advertising

Other groups also began to stake their claims in cable's gold rush. Advertisers who saw cable reaching close to 30 percent penetration of the nation's households in the early 1980s became interested and began placing ads on cable's programming channels, both local and national. Sports programming was the first to attract advertising, but soon corporations were committing to advertising in areas such as news, health, and entertainment.[32]

unions

Members of the various unions and guilds that operate in the broadcasting industry were not involved in cable programming when it first began because the cable companies did not recognize the unions or abide by their regulations. Original cable programming was written, directed, produced, and crewed by nonunion members. However, after several long, bitter strikes in the early 1980s, the unions won the right to be recognized and to receive residuals from cable TV. Thus, the same people who worked in broadcast programming began to work in cable programming.[33]

equipment manufacturers

Perhaps the biggest winners in the cable gold rush were the equipment manufacturers who supplied the materials needed to build the cable systems. The suppliers of the converters that enable a regular TV set to receive the multitude of cable channels, the earth station dishes, and the cable itself found their order desks piled high. Space on a satellite became a precious commodity as more and more companies wanted to launch national programming services. Satellite transponder time that had rented for about $200 an hour or $1 million a year took on a new dimension in 1981 when RCA auctioned off seven **transponders** (parts of a satellite that carry particular program services) on its Satcom IV satellite for figures ranging from $11.2 to $14.4 million for six years. The FCC later disallowed the auction method of leasing transponders and RCA had to settle for charging all successful bidders $13 million for seven years and nine months of use.[34]

Although large companies, advertisers, unions, and equipment manufacturers all flocked to cable during the late 1970s and early 1980s, the most noticeable type of cable growth was in the area of programming and special services. One after another new services were started in pay-TV in what became known as basic cable, in local programming, and in interactive services.

Pay-Cable Services

movies and specials

Shortly after Time became successful with its HBO service, Viacom launched a competing pay-TV service on satellite called **Showtime.** Viacom, like Time, owned various cable systems throughout the country and had been feeding them movies and special events through a network that involved bicycling and microwave.[35]

(a)

(b)

(c)

FIGURE 5.5

(*a*) Jason Robards starring in Eugene O'Neill's *Hughie*, one of Showtime's Broadway productions. (*b*) A scene from *The Greek Passion* performed by the Indiana University Opera Company and taped by Bravo. This was the first opera to be taped for cable television. (*c*) A scene from "Everything Goes," an adult quiz show on Escapade. *(a courtesy of Showtime; b courtesy of Bravo; c courtesy of Escapade)*

Following the launching of Showtime, Warner Communications in conjunction with American Express (Warner-Amex) began **The Movie Channel,** which consisted of movies twenty-four hours a day.

Time, buoyed by its success with Home Box Office, began another pay service called Cinemax that consisted mostly of movies that were to be programmed at times complementary to HBO.

Galavision went on the satellite as a Spanish-language movie service, Times-Mirror established Spotlight as a movie and special service to its cable companies, and several companies entered the field of cultural pay programming in the form of Bravo (owned by Rainbow) and The Entertainment

culture

Channel (owned by Rockefeller Center and RCA). Disney began a family-oriented service called The Disney Channel, which offered old Disney films and newly developed materials.[36]

"adult" programming

Several companies also entered the area of satellite-delivered pay "adult" programming, including R-rated movies, skits, and specials. The primary one was Playboy, which joined forces with an already established program service called Escapade.

Premiere

In what will probably remain only as an interesting footnote to the history of pay-cable, the Getty Oil Company and four film production companies joined together in April of 1980 to form Premiere, a company that was intended to be a pay-TV movie service that would compete with HBO, Showtime, The Movie Channel, and Cinemax. The difference was that the Getty venture, in conjunction with Columbia Pictures Industries, MCA, Paramount Pictures Corporation, and 20th Century-Fox Film Corporation, was going to show the films from those companies and then not allow them on any other pay-TV service for nine months. In essence, Premiere would have had half of the industry's top motion pictures nine months before any of the other pay services could exhibit them. Needless to say, the other pay services fought this in the courts. After approximately a year of battling, Premiere closed its doors, the victim of two negative court decisions.[37]

Basic Cable

The late 1970s and early 1980s also saw a proliferation of new satellite-delivered programming services that became known as basic cable. Some of these were supported by advertising, some were supported by the institutions that programmed them, and some were supported by small amounts of money that the cable companies paid to the programmers. These were lumped together and referred to as basic cable because the subscriber did not usually pay any substantial additional fee for them.

services

Some of the representative basic cable services that were born within the timespan of only a few years were:

ESPN—Established by Getty Oil and based in Connecticut, this service provided twenty-four hours a day of advertising-based sports programming, featuring every conceivable type of sport.

CBN—The Christian Broadcasting Network, headquartered in Virginia, provided free religious programming with the cost being underwritten by donations from those who wished to contribute to their religion. Trinity Broadcasting Network (TBN) operated in a similar manner.

Nickelodeon—The programming of this Warner-Amex service was at first totally commercial-free children's programming, some of it produced at cable systems owned by Warner-Amex. Cable systems paid about fifteen cents a subscriber a month in order to receive the service.

Black Entertainment Television—This programming was only on several hours a week and was designed to provide for the special interests of the black community. Cable systems were charged for it.

Satellite Program Network—An unusual concept, this Oklahoma-based network served as a distribution vehicle for independently produced programs, some of which had their roots in public access. It charged the producers who, in turn, secured their own advertising to cover their costs. The programming was quite varied including interviews, sports, gardening, and women's programs.

CNN—Owned by the same Ted Turner who put his Atlanta UHF station on satellite, Cable News Network programmed news twenty-four hours a day. It charged the cable systems that cablecast it and also solicited advertising.

Nashville Network—Westinghouse started this country-western format channel that included not only music but also game shows and dramas.

USA Network—Supported by advertising, this network provided a variety of programming with a heavy emphasis on sports. It also included children's shows and health programs.

WTBS, WGN, and WOR—These were all "superstations" that carried commercials and also charged cable operators about ten cents a subscriber a month. WTBS was the Atlanta-based original superstation; WGN was from Chicago; and WOR was from New York.

SIN—The Spanish International Network was supported by advertising and provided Spanish-language programs. It paid the cable system for each Spanish surname on its subscriber list. As a result its slogan was "SIN pays."

C-SPAN—Based in Washington, D.C., the Cable Satellite Public Affairs Network, a nonprofit organization, delivered live coverage of the House of Representatives. It charged cable systems a modest fee per subscriber per month.

ARTS—Owned by ABC, this advertising-based service provided cultural programming, much of it from Europe.

CBS Cable—This was an advertising-supported cultural service owned and operated by CBS. It featured a great deal of originally produced American material.

MTV—Another Warner-Amex product, Music TV featured music videos and news and information about the musical scene. This became very successful quickly and a number of competing music video services were launched.

Satellite News Channel—Designed to compete with CNN, this was a Westinghouse-ABC joint venture of twenty-four hour news.

(a)

(b)

(c)

FIGURE 5.6

(*a*) An ESPN taping of a college basketball game. (*b*) Host Reggie Jackson (*right*) interviews high school rodeo champ Mike Esposito for a Nickelodeon program. (*c*) MTV's taping of the rock group .38 Special as they perform at Denver's Rainbow Music Hall. (*d*) Paul Ryan interviews Phyllis Diller for the SPN program "The Paul Ryan Show." This program started as a public access show on Theta Cable in Los Angeles and moved to the SPN network. (*e*) The "Bobby Jones Gospel Show" on Black Entertainment Network. *(a courtesy of © ESPN; b, c courtesy of Warner-Amex Satellite Entertainment Company; d courtesy of David Sarro, producer/Rita Bozyk, photographer; e courtesy of Black Entertainment Network)*

(d)

(e)

Cable Health Network—This service was first owned by Viacom and was devoted to information about physical and mental health.

Disco Network—On a different track, this was an audio service fed over satellite. Although rarely talked about, cable is also capable of delivering radio stations and audio material to the home. Several other audio services were Home Music Store, WFTM Chicago, and National Classical Network.

And on and on. There were also services that provided words and numbers on the screen. For example, Associated Press, United Press International, and Reuters News Service all had cable news services that printed out the latest news. Several companies had stock market and other financial reports and some had weather information available twenty-four hours a day.

text services

For several years both new pay and basic services were announced at a rapid rate—sometimes several in one day. Some of these never got off the ground, others existed only for short periods, and others showed signs of longevity.[38]

Local Programming

Because the franchising process placed so much emphasis on the local community in which the cable TV system originated, local programming took on an entirely new dimension in the late 1970s.

The older systems that still only had twelve-channel capacity usually allocated only one channel to local programming. But the newly franchised systems that were able to take advantage of improved technology to provide twenty, then thirty-seven, then fifty-four, then over a hundred channels usually promised an entire complement of local channels.

increase in channels

At least one of these channels was generally reserved for local origination—programming that the cable system itself initiated. The others were some combination of access channels: public access for the citizenry at large; community access for such community groups as the Girl Scouts and the United Way; government access for local officials; educational access for schools and colleges; religious access for religious groups; and leased access for businesses, newspapers, or other individuals interested in buying time on a cable channel to present their messages, sometimes with commercials included. Often these channels were shared (for example, by the local government and the schools), and sometimes there were several channels for the same type of organization (for example, one channel for the local college, one for the public school system, and one for private schools).

types of channels

How these channels were to be organized and operated differed widely from community to community. Sometimes cable companies provided equipment, studio space, and professional personnel to help the various groups create their programming and then were responsible for seeing that the programs were cablecast over the system. At other times, cable companies donated equipment to a community nonprofit access organization, the city council, or

channel operation

a local school district and the organization then produced its own programming with its own hired or volunteer staff. Some cable companies merely made a channel available for programming and the local organizations or individuals interested in doing the programming used their own resources for equipment and crew.

Of course, not all the local programming planned by cable systems and local groups actually materialized, but the quality of access programming improved greatly from the early days when access consisted mainly of vanity TV.[39]

Interactive Cable

Interactive cable was highly touted in franchise applications. Once again, promises were made that cable would perk the coffee, help the kids with homework, do the shopping, protect the home, and teach the handicapped. In some areas of the country a fair amount of interactive cable was actually undertaken. Cable signals that are wired from a central location to a home also have the capability of going from the home to a central location, either on the same or a different wire. This enables cable to take on a two-way capability and allows for interaction between the subscriber and whatever entities the cable system wishes to provide.

The pioneer and most publicized interactive system was **Qube,** which was operated by Warner-Amex cable. Qube was initiated in Warner's Columbus, Ohio, system in 1977 and later was incorporated into several of its other systems.[40] The basic element of the original interactive Qube was a small box with response buttons that enabled Qube subscribers to send an electronic signal to a bank of computers that could then analyze all the responses. An announcer's voice or a written message on the screen asked audience members to make a decision about some question, such as who was most likely to be a presidential candidate, whether a particular congressional bill should be passed, what play a quarterback should call, whether an amateur act should be allowed to continue, or whether a city should proceed with a development plan. Audience members made their selection by pressing the appropriate button in multiple-choice fashion. A computer analyzed the responses and printed the percentage of each response on the participant's screen.[41]

In 1981 Qube added to its interactive capability the ability for subscribers with computers to access a computer data bank to obtain a wide range of information, such as consumer tips, weather, news, video games, and articles from magazines. No doubt, some of this was helpful for students' homework.[42]

Another quasi-interactive service offered by some cable systems was **pay-per-view,** whereby subscribers who paid an extra one-time-only sum could see one-time-only events, such as championship boxing matches or special movies. Most pay-per-view plans asked subscribers to phone the cable company to order the event that would then be sent through the wire only to those homes requesting and paying for it. Special addressable equipment was needed in order

process

Qube

pay-per-view

to service part of the cable system, while preventing the event from entering the homes of people who did not want the pay-per-view event.[43]

Home security was another interactive area that cable entered. Again, because of the two-way capability, various burglar, fire, and medical alarm devices could be connected to the cable system. A computer in a central monitoring station could send a signal to each participating home about every ten seconds to see if everything was in order. If any of the doors, windows, smoke detectors, or other devices hooked to the system were not as they should be, a signal was sent back to the cable company monitoring station, which then notified the police.[44]

For the most part these interactive services did not become profitable, but proposals for them caught the attention of city councils and were often responsible, at least in part, for decisions regarding franchise awards.

Consolidation and Retrenchment

The mid-1980s saw a defoliation of the cable industry. Perhaps the promises had been too lavish and the anticipation too great, but the bloom fell off the rose. The rate of new subscriber growth leveled while the rate of disconnects from old subscribers increased.[45] Programming services consolidated and went out of business and those that were left sported much of the same type of play-it-safe programming that had traditionally been found on ABC, CBS, and NBC. MSOs drowned in red ink as they tried to live up to the promises they had made in regard to wiring big cities. Advertisers did not respond to the cable lure as quickly or profusely as expected. Even a magazine produced by Time about cable TV programming failed.[46]

Companies took actions to stem financial woes. Warner-Amex, for example, sold several of its cable systems so that it could concentrate resources on a few cities. Group W sold all of its systems to five cable TV companies who then divided the systems among themselves. Storer and Times-Mirror traded several systems so that the systems each owned would be more geographically contiguous. In general, a few companies, such as Tele-Communications and American Television and Communications, acquired more systems while other companies folded or retrenched.[47]

Programming Changes

In the area of pay programming, Showtime and The Movie Channel merged in 1983. At first they were co-owned by Viacom and Warner-Amex, then Viacom bought out Warner to become the sole owner.[48] HBO, the perennial leader, remained profitable but suffered a slowdown that led to changes in the executive suites.[49]

Spotlight went out of business and Galavision converted from a pay service to basic. Bravo changed owners, programming emphasis, and marketing strategy several times trying to find its niche. The Entertainment Channel,

FIGURE 5.7

Setting up for a movie
filmed for HBO called
*Man Who Broke 1,000
Chains*. *(Courtesy of HBO)*

the cultural programming service owned by Rockefeller Center and RCA, turned its programming over to the basic service ARTS, which then became the Arts and Entertainment Channel.[50]

The Playboy Channel hardened and softened its programming several times caught between angry citizens' groups who did not want nudity and a small but loyal group of viewers who did. This churn in philosophy angered its partner, Escapade, and the two parted company with Playboy paying a $3 million divorce settlement to Escapade.[51]

The pay services also began making exclusive deals with various movie studios so that certain films could appear only on one company's channels. These services also began making their own exclusive movies and programs.[52]

basic services

Problems occurred in the basic services also. The most highly touted failure was that of the cultural service, **CBS Cable,** which stopped programming after losing $50 million. The service had programmed ambitious, high-quality television, but did not receive sufficient financial support from either subscribers or advertisers. Because it had touted its service so aggressively, its demise was almost heralded by some of the cable companies who resented the encroachment of the broadcast networks into their cable business.[53]

Another well-publicized coup occurred when Ted Turner's Cable News Network slew the giants, ABC and Westinghouse, by buying out their Satellite News Channel. This meant less competition for CNN, which was then able to proceed on less tenuous financial footing because it did not have to compete for advertisers with SNC.[54]

MTV survived but under new ownership (Viacom instead of Warner-Amex) and with less of the hoopla that accompanied its early days. Most of the imitative music video services failed, however, including one launched by Ted Turner that lasted less than a month.[55]

FIGURE 5.8

Dr. Ruth Westheimer in a characteristic pose she uses on her popular sex-oriented talk show on Lifetime. *(Courtesy of Lifetime)*

On other fronts, USA Network was bought by Time, MCA, and Paramount, who did not always agree on the network's direction. Getty Oil was bought out by Texaco, who did not have any interest in maintaining the as yet unprofitable ESPN—it wound up being purchased by ABC. Daytime and Cable Health Network merged to form Lifetime. The children's service, Nickelodeon, began accepting commercials and changed ownership from Warner to Viacom. CBN changed from a religious format to a family-oriented format that included very old reruns. Satellite Program Network changed its name to Tempo Television and changed its format from public access programs to a potpourri of programs.[56]

The rash of new program services stopped, but The Discovery Channel was launched in 1985 to provide nonfiction programming about such areas as nature and history.[57] C-Span also added a second service because it gained access to covering the Senate.[58]

Both basic and pay services suffered from **piracy** in the form of signal theft by people who had not paid for the programming. Some of this was within the cable systems where people figured out methods of obtaining the services by tapping off a neighbor's feed or altering technical devices so that the signals would enter their homes. In addition, an increasing number of people were buying their own satellite dishes and receiving the signals directly from the satellites. The cable industry figured it was losing $500 to $700 million annually because of service theft.[59] As a result, many of the services began **scrambling** their signals so they could not be received by the home dish owners. This irritated the dish owners and one of them, who called himself Captain Midnight, became irate enough to figure a way to break into the national scrambled transmission of HBO long enough to decry the service's scrambling policies. The cable services did not necessarily want to lose the viewers watching

piracy

through home dishes; they merely wanted them to pay for the service. Thus, a number of methods for marketing the cable networks to home dish owners began to surface.[60]

local programming

Activity on the local access and local origination channels also slowed during the mid-1980s. Cable systems that had promised truckloads of production equipment to local organizations tried as best they could to drag their feet on these obligations because they were so costly. Local programming departments that had included five or six employees dwindled down to a precious few. In some instances, the director of marketing took over programming "on the side." On some systems the only local programming became messages typed on the screen about local news and events. They had, in essence, reverted to a sophisticated version of the early three-by-four-inch cards. Local programming did not by any means disappear, but it did not fill the multitudes of channels promised in many of the franchise agreements.[61]

interactive

In the area of interactive services, Warner-Amex killed its highly touted Qube system by 1984, mainly because it was not profitable. No other similar interactive services rushed in to fill the void.[62] Cable essentially removed itself from the home security business, leaving it to companies who specialized in just that.

home shopping

A new breed of programming, **home shopping,** proliferated on cable and a multitude of national and local services began. Many of these disappeared as soon as they began—or sooner. The market could not bear all of them. They were not truly interactive in the sense of actually ordering items through the TV. They showed items on the screen and then relied on the tried and true technology of the old-fashioned telephone in order to arrange purchases.[63]

pay-per-view

Pay-per-view became a big topic during the mid-1980s and, in a manner similar to the shopping networks, pay-per-view networks were proposed, announced, launched, reorganized, renamed, and discontinued. Some of these were founded by companies already in the pay-cable business such as Playboy, Viacom, and Time. Others were formed by ex-cable executives or companies new to the field. Most offered several movies and/or sports events per week. Cable systems offered these services to their subscribers on a slow but steady basis and, although profits were rarely in the picture, revenues grew to an estimated $70 million in 1986. Enthusiasm remained high for the concept of pay-per-view.[64]

Deregulation and Regulatory Issues

Cable TV, like its cousin broadcasting, has been affected by the deregulatory mood of government. In 1980 the FCC abolished both the syndicated exclusivity and distant signal importation rulings, opening the way for cable systems to import as many stations as they wished and to play the programming of those stations without having to worry about whether the same programming would be on a local TV station.[65] Broadcasters, motion picture producers, and sports interests, concerned about the effects of this deregulation on their businesses, appealed the FCC ruling. The Supreme Court ruled in favor of the FCC action and, therefore, in favor of the cable TV interests.[66]

syndicated exclusivity and
distant signal importation

FIGURE 5.9

Jewelry being offered for sale on the home shopping service, Teleshop. *(Courtesy of Financial News Network)*

However, in 1988 the FCC reinstated syndicated exclusivity, so the issue continued to see-saw.

Another major piece of deregulation legislation, the Cable Telecommunications Act of 1984, was passed by Congress and delineated the role of cable systems and local governments. The main positive change for cable operators was that cities that had at least three broadcast stations could no longer regulate the rates that cable systems charged their customers for the basic service. Instead, these prices were to be determined by the marketplace. The most positive aspect for the cities was that they were allowed to raise their franchise fee from 3 percent to 5 percent.[67]

local governments

The old issue of must-carries surfaced again for several reasons. Some cable practitioners began promoting the abolishment of must-carries so that cable systems could fill their channels with satellite programming rather than with local station programming. They reasoned that local stations could be received over the airwaves, so cablecasting them constituted an unnecessary duplication of services to the consumer. Local stations of course did not agree because they wanted to be carried on the cable systems.

must-carries

In a series of legal maneuvers, the FCC and the courts outlawed and reinstated must-carries a number of times, each time changing the provisions slightly as to which stations needed to be carried. Heading into the 1990s, this issue was unstable.[68]

Another issue to rear its head again was copyright. The Copyright Tribunal, acting under its right to adjust fees, ordered cable systems to pay 3.75 percent of their gross receipts for each imported distant signal that was above the number of distant signals allowed under the 1972 rulings. This ruling had set up a complicated formula for distant signal importation that depended on the size of the cable system, the size of the market, and the number of stations in the market. The systems most affected were fairly large ones in densely

copyright

populated areas, but these systems did serve about ten million people.[69] The cable companies, of course, objected and appealed. At first the courts ruled in favor of the Copyright Tribunal. This made the program producers very happy because it meant more money for them if the cable companies paid the fee. In 1983, the first year the 3.75 percent fee was in effect, cable operators paid $25 million more to the Copyright Tribunal than they did in 1982 for a total of $69.1 million. Cable companies continued to fight the fees in both the courts and Congress and did succeed in getting a court to rule that cable systems had been overpaying copyright fees by at least 40 percent. In general, the whole issue of copyright still dangled.[70]

overbuild

Another legal issue to surface was called **overbuild** and involved whether or not a city could give one company exclusive rights for cabling the city or a part of the city. After several trips to the courts, this issue was decided in favor of the companies that wished to overbuild, (i.e. set up second systems). In other words, cities had no right to refuse to allow a company to wire a city just because it did not win the franchise. This idea is, of course, opposed by the established cable companies.[71]

pole attachments

Even the old issue of pole attachments recycled. This time the Supreme Court ruled that the rates the phone companies charged the cable systems could be regulated by the FCC.[72]

Outlook for the Future

recovery

Heading into the 1990s, cable TV looked like it was on the road to recovery. Companies were not making the 98 percent profits once envisioned; in fact, many companies had totally fallen by the wayside. Most of the ones that remained had undertaken their shakeouts and were plodding along nicely in a nonsensational manner.

The expensive part of cable operation (laying the cable) was almost complete, so companies had less expense to balance against income. The income also rose because of deregulation. Now that the cities could not regulate rates, cable companies could charge customers more. They did charge an average of 6.7 percent more during the first six months after deregulation.[73] Cable systems were selling for higher prices than in the past,[74] and cable was even called "the best business in America."[75]

Given cable TV's checkerboard history, anything could happen. The future is uncertain but optimistic.

Cable TV Chronology

1940s First cable TV begins
1949   Distant signal importation begins
1960   United Artists sues cable TV for copyright infringement
1965   FCC issues rules on must-carry and syndicated exclusivity
1966   FCC publishes report to restrict cable service
1970   Cable systems are ordered to have local origination
1972   Revision of must-carry rule from sixty to thirty-five miles

1975   HBO begins distributing movies by satellite
1977   Qube initiated
1977   HBO turns a profit
1978   New copyright bill goes into effect
1978   Franchising boom begins
1978–
1981   Numerous cable programming services started
1980   Pay-TV revenues grow 95.5 percent
1980   FCC abolishes syndicated exclusivity and distant signal importation rules
1981   RCA auctions off satellite transponders
1982   Copyright Tribunal increases distant signal copyright fees to 3.75 percent
1983   Showtime and The Movie Channel merge
1983   Daytime and Cable Health Network merge
1983   CNN buys SNC
1984   ABC buys ESPN
1984   Qube is ended
1984   Congress passes cable bill concerning cities' rights
1984   Warner-Amex sells some cable systems
1985   Several pay-per-view services launched
1985   Group W sells all of its cable systems
1985   Overbuild sanctioned
1985   Discovery Channel launched
1985   Must-carry rules temporarily abolished
1986   Captain Midnight breaks into HBO scrambled transmission
1987   Home shopping networks proliferate
1988   FCC reinstates syndicated exclusivity

Conclusion

Cable TV has undergone many changes, especially in recent times. What started as one- to three-channel systems in the 1950s promised in excess of one hundred channels in the 1970s and settled for about fifty channels in the 1980s. A technology that involved only wires has now wholeheartedly embraced satellites for distance distribution.

An industry that maintained stable, modest, consistent numbers for decades suddenly boasted figures that jumped hundreds of percents in one year and then settled back to modest, slow-growing numbers just as quickly.

Ownership of cable systems changed greatly, too, from the early community-oriented, ma-and-pa owned systems to multiple system owners. These MSOs, in turn, merged and acquired more cable-related properties and then began a consolidation and withdrawal process because they were overextended.

The whole regulatory scene has undergone evolution. Early cable TV was essentially unregulated with rules arising only as broadcasters convinced federal governmental bodies of the need to protect them from encroachment by the cable systems. The government's reaction, primarily from Congress and the FCC, was one of confusion and abstention; however, during the 1960s and early 1970s rules were enacted dealing with must-carry, distant signal importation, syndicated exclusivity, pole attachments, and copyright. As the government developed a deregulation policy during the 1980s, cable TV became even less regulated by both federal and local agencies than it had been. However, all the old issues from the 1970s continued to rear their heads.

The franchising of cable systems was picked up by local governments almost by default and was a sleepy process until cable boomed in the late 1970s. Then municipalities were besieged by cable companies eager to wire the cities. Franchise proposals became huge documents and the franchising process bred large stakes, complete with a certain amount of accompanying corruption. As the wiring of major cities became complicated and expensive, the fervor for obtaining additional franchises abated, although some companies did become interested in overbuild.

More than anything else, there was an immense change in cable TV programming. Original cable systems retransmitted broadcast programming of both a local and imported nature. A few systems originated their own programming through public access, local origination, or movie channels, but the original attempts at this were largely unsuccessful. Promises, not performance, were the hallmark of early cable. But the tide turned with the late 1970s success of the HBO satellite venture, followed closely by the beginning of superstations. The floodgate opened to a raft of programming services produced exclusively for cable TV as opposed to broadcast TV. Cable programming became a glamour business with material distributed on both a national and local basis. A multitude of pay services and basic services emanated from the satellites and the cable systems themselves revived the concept of local programming, often with the promise of numerous local channels to meet various needs within the community. The interactive services that had been long on talk and short on action began to undergo experimentation, most notably in the Qube system. The programming concepts of the late 1970s and early 1980s changed the entire face of the cable industry. A shakeout occurred in the mid-1980s and many of the national programming services merged, were bought out, or went out of business. Local programming was greatly cut back and interactive services were virtually nonexistent except as they related to pay-per-view. Cable programming is still abundant but experimentation and growth have stabilized.

The cable industry is still in a state of flux, making it an exciting phenomenon. Its future course has yet to be determined.

Thought Questions

1. What should be the policies regarding cable copyright payments, distant imports, must-carries, and syndicated exclusivity?
2. What types of programming, both national and local, do you think will have long-term success on cable TV?
3. What do you think will be the state of cable TV ten years from now?

Other Electronic Media

Overview

During the 1920s, 1930s, and 1940s, radio was the only really significant electronic medium. It was joined by broadcast television in the 1950s and 1960s. Then, in the mid-1970s and 1980s, a myriad of media appeared on the scene. This led to the possibility that people might have an enormous choice in how they received their telecommunications entertainment and information. However, the market did not seem to be able to bear the huge proliferation and a shakeout period occurred. No one knows as yet which of these media will survive or in what form they will survive.

This chapter chronicles these media's ups and downs by considering the following topics:

The overall status of the newer media

Subscription TV's experiments, setbacks, growth, and demise

The success of videocassette recorders

The shaky existence of the laser videodisc and the birth and death of the stylus disc

MMDS's halting route to becoming "wireless cable"

SMATV's competition with cable

The grandiose nature of DBS and its more mundane applications

The enthusiasm for and realities of LPTV

The experimentation and varying standards of teletext

The possible uses and ramifications of the telephone

6

New media have meant new values. Since the dawn of history, each new medium has tended to undermine an old monopoly, shift the definitions of goodness and greatness, and alter the climate of men's lives.

Eric Barnouw in
A Tower in Babel

Status

Although commercial broadcasting, public broadcasting, and cable TV are the dominant forms of electronic media, many other ideas have sprung up over the years as ways of bringing entertainment and information to the masses through devices that employ electrons.

Some of these came and went quickly, such as the EVR, developed by CBS during the 1960s, which employed electronic images placed on photographic film. The film did not record images as does a motion picture camera. Rather, it contained electronic impulses. The advantage of putting the electronic impulses on film rather than tape was that prints could be made of the film quickly and inexpensively, in contrast to duplications of videotape, which were slow and expensive. The film was placed in the EVR machine and the output was displayed on a TV monitor. CBS predicted that EVR would replace film in classrooms, homes, and in the broadcasting industry. But such never happened. Both technical and marketing problems plagued the EVR and it died aborning.

status of newer media

Most of the newer electronic media were born or had their heyday in the late 1970s and early 1980s, and some were considered a threat to the established media. However, most are now in an unsettled state. Some are essentially moribund and others are undergoing reconditioning. Only videocassettes appear to be an unqualified success.[1]

Subscription TV

Subscription TV (**STV**) is the oldest of the alternative broadcast media, having been tested in the early 1950s. Some of the original subscription TV systems delivered the programs to the home TV set through wires, but most sent it over the airwaves in a **scrambled** form so that home viewers needed to use a special decoder in order to see the signal.

wire and airwaves

The early STV experiments were low-key and failed primarily because of equipment malfunctions.[2] In 1957 a high-profile motion picture exhibitor, Henry Griffing, set up an operation in Oklahoma that was supposed to deliver first-run movies by wire. Subscribers paid $9.50 per month no matter how much they watched. Griffing's plan was to have first-run film, but unfortunately he was undercapitalized. This presented difficulties in wiring Bartlesville and in securing films. His project failed and proved a major setback for subscription TV.[3]

Oklahoma system

Because of the publicity connected with Bartlesville, both motion picture exhibitors and broadcasters began lobbying for congressional action or FCC action against these kinds of systems. Motion picture exhibitors feared that people would not go out to the movies anymore if they had them in the home and broadcasters feared that these systems would be able to bid more for available programming and talent and thus threaten their audiences for free TV.

restrictions

What followed was a shaky decade for STV, during which the FCC allowed it to operate but saddled it with so many pro-broadcasting restrictions that it was stifled. For example, no more than one STV operation was allowed

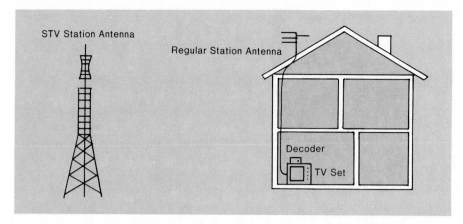

FIGURE 6.1

STV configuration.
*(From Gross, Lynne Schafer,
The New Television
Technologies. 2d ed. © 1986
Wm. C. Brown Publishers,
Dubuque, Iowa. All Rights
Reserved. Reprinted by
permission.)*

in any one community, and STV systems were prevented from **"siphoning"**—outbidding broadcasters for programming. At one point STV was even outlawed in California through a referendum passed by voters who were influenced by a very strong anti-STV campaign launched by motion picture exhibitors.[4] The few companies that did try STV were largely unsuccessful.

Ironically, a positive turning point for subscription TV was pay-cable, its competitor in many areas. When Home Box Office succeeded in having some of the FCC pay-TV rules rescinded (including the antisiphoning rule and the rule that limited pay services to one per community), subscription TV received a shot in the arm.

For several years during the late 1970s and early 1980s the bloom was on the rose and STV systems were installed in many cities, primarily on UHF stations. Predictions were made that STV was on "the threshold of an era of expansion comparable to that of television itself in the 1950s."[5] Since the STV operations were garnering profits at 20 to 25 percent of revenues (in contrast to conventional television, which was less that 18 percent), these predictions seemed justified. In 1982 the entire STV industry encompassed 1.4 million subscribers and Oak Media, the largest company, had 600,000 customers in five cities.[6] STV even did well on the regulatory front and most of the remaining restrictions that had choked it were lifted.[7]

success

Then in the mid 1980s the bottom fell out because STV could not hold its own against the rapidly expanding cable systems, which enticed subscribers with not just one, but several pay-TV services on top of an array of other channels.

failure

Subscribers defected in droves and seven systems shut down in 1983.[8] Even Oak Media, the biggest, closed down or sold out all its systems by 1985.[9] By 1987 only one subscription TV service, SelecTV in Los Angeles, was still on the air and it was operating primarily because it had a satellite feed to other media.[10]

Videocassette Recorders

Betamax

Home **videocassette** recorders are the biggest success story among the newer media. The first videocassette machine specifically designed for the home consumer market was the Sony **Betamax** introduced in 1975. This machine used half-inch tapes that could record for one hour. It sold for $1300, but the price was actually $2300 because the recorder could only be purchased with a new Sony color TV set.[11]

VHS

The price went down and the features went up as Sony encountered competition from Matsushita, which in 1976 introduced a cassette recorder that it called video home system (**VHS**) and marketed through its Japan Victor Company (JVC). This JVC recorder had a marked advantage in that it could record for up to two hours. People who bought the recorders were using them primarily for recording feature films off the air, so the two-hour format enabled them to record an entire feature on one cassette.[12]

With the advent of competition, the technological war was on. Sony altered its Beta format so that it could record longer too. Both Beta and VHS adopted such features as pause buttons that enabled consumers to eliminate commercials while they were recording, timers that allowed recordings to be made while the owner was not at home, devices that allowed one program to be recorded while another was being viewed, still frame and variable speed that allowed the owner to hold a particular picture on the screen or view the picture either in slow motion or fast speed, and, of course, taping methods that allowed for longer and longer recordings.[13]

Many other companies also began manufacturing cassette machines, with more of them adopting the VHS method than the Beta method. The two methods were not compatible, so a tape recorded on a VHS machine could not be played back on a Beta machine even though both used half-inch tapes. The main difference between the two was the method by which the tape threaded itself in the machine.[14]

growth

Once these videocassette machines became somewhat sophisticated, their sales soared beyond expectations. More than one million recorders were sold in 1979,[15] and more than three million in 1981.[16] People used them primarily for recording off the air, but consumers also began purchasing prerecorded videocassettes that contained movies, self-help programs, and educational material. A whole new business of video rental sprang up. Some people also bought cameras to use in conjunction with their videocassette recorders to make "home movies."

lawsuit

Into this success story came a lawsuit. In 1976, a year after Sony introduced its Betamax, Universal Studios and Walt Disney Productions brought suit against Sony claiming that any device that could copy their program material was being used in violation of **copyright** and should not be manufactured. The case was dragged through the courts and in 1979 a federal judge sided with Sony, saying that copying TV program material for use in the home comes under the "fair use" doctrine and is not an infringement of copyright.

FIGURE 6.2

A Sony Betamax videocassette recorder. *(Courtesy of Sony Corporation of America)*

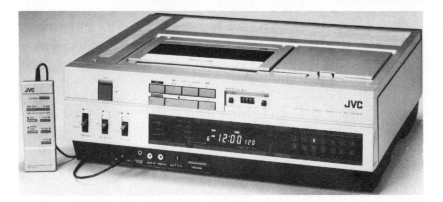

FIGURE 6.3

A VHS videocassette recorder. *(Courtesy of JVC Industries Company)*

However, two years later an appeals court sided with Universal and Disney, saying that home recording was indeed a violation of copyright and that Sony, which obviously knew its recorders were being used for this purpose, was at fault.[17] Fortunately for Sony and other VCR manufacturers, in 1984 the Supreme Court reversed this decision and ruled that home taping does not violate copyright laws.[18] This decision was fortunate for consumers, too, for by 1984 over thirteen million recorders had been sold.[19]

Ironically, after Sony won the copyright battle, it lost the crucial marketing war with VHS, which became the preferred choice of consumers. In fact, Sony itself began selling the VHS, which many felt was the death knell for the Beta format.[20]

By the late 1980s, VCRs had become a fairly dominant device, populating half the American homes. Sony introduced an **8mm** (quarter-inch) VCR that it hoped would cut into the market.[21] Several companies introduced camcorders, single units that contained both camera and tape recorder.[22] Movie studios were also giving VCRs significant attention as a distribution method,[23] rating services were reporting their use,[24] researchers were studying their patterns and effects,[25] and, unfortunately, the film industry was losing about a billion dollars a year through piracy.[26]

VHS choice

success

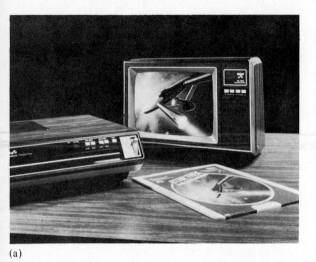

(a)

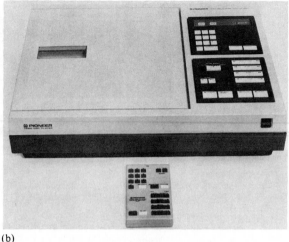

(b)

FIGURE 6.4

(*a*) RCA's
Selectavision stylus
videodisc player.
(*b*) Pioneer's LaserDisc
videodisc player.
*(a courtesy of RCA
Corporation; b courtesy of
U.S. Pioneer Electronics)*

difficult start

stylus system

laser system

RCA introduction

Videodisc Players

Another entry into the consumer market, the **videodisc,** has had a stormier existence than the videocassette. Development of the disc began in earnest in the late 1960s and was soon followed by many pronouncements of the potential efficiency, low cost, and flexibility of the disc. Announcement after announcement followed that the videodisc would be on the market "next year," but many next years came and went with the promises unfulfilled.

As with videocassette recorders, there actually were two different disc systems. The less advanced was the **capacitance disc** developed by RCA. It had a diamond stylus that moved over grooves similar to those of a phonograph record.

The second system, the **laser disc,** had a very muddled parentage that included marriages, divorces, and partial custody by a large number of companies, including MCA, Philips, Magnavox, IBM, and Pioneer. The laser disc used a laser beam that read information embedded in a plastic disc, a technology that guaranteed the disc would never wear out and that any frame could be quickly accessed and viewed in freeze-frame almost indefinitely. Unlike VCRs, both were playback-only systems; they could not record programs off the air. [27]

The laser disc, then owned by MCA-Philips and called DiscoVision, was the first on the market, selling for $700. The first models of the machine itself sold out shortly after its 1978 introduction, but the discs were so technically flawed that their sale and manufacture had to be discontinued.[28]

In 1981 RCA introduced its capacitance disc player, called Selectavision, for $500 each. RCA predicted it would sell 200,000 machines in the first year. At first sales were brisk, 26,000 players and 200,000 discs in the first five weeks,[29] but the pace quickly dropped and RCA fell far short of its 200,000 goal for the first year. In fact, after three years only 500,000 machines had been sold and RCA abandoned Selectavision. The company blamed videocassette recorders, which could record as well as play back.[30]

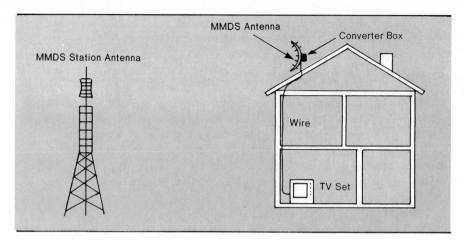

FIGURE 6.5

MMDS configuration.
*(From Gross, Lynne Shafer,
The New Television
Technologies. 2d ed. © 1986
Wm. C. Brown Publishers,
Dubuque, Iowa. All Rights
Reserved. Reprinted by
permission.)*

When RCA threw in the towel in 1984, the laser disc, which Pioneer had bought in 1982 and renamed LaserDisc, was still suffering from technical glitches and videodiscs seemed dead.[31]

However, Pioneer managed to solve the manufacturing problems and began marketing the disc players to education and industry. In these markets, the laser disc's vast storage capability and its ability to freeze pictures and access them quickly seem to be genuine assets.[32] Videodisc technology is also joining with audio **compact disc (CD)** technology to create machines that play either audio or videodiscs.[33]

disc uses

Predictions of the death of the videodisc may have been premature. It has worthwhile features that may yet find a useful purpose.

Multichannel Multipoint Distribution Service

Multichannel Multipoint Distribution Service (**MMDS**) is a concept whose name and focus has changed innumerable times over the years.

Its first name, Multipoint Distribution Service (MDS), was given to it in the sixties when the FCC established frequencies for three services—Instructional Television Fixed Service (ITFS), Operation Fixed Service (OFS), and MDS. ITFS was to be used primarily by educational institutions to transmit televised instructional material to groups of schools. OFS was to be used by corporations for internal communications among different business locations, and MDS was to be used for commercial programming purposes.

MDS

Although all these services are broadcast in a manner similar to commercial TV, the frequencies used are two thousand megahertz higher than conventional broadcast frequencies. As a result, they cannot be received with regular TV sets. In order to receive these signals, a special antenna and **down-converter** system is needed. The MDS systems also have a range of about twenty-five miles, so they cannot be used to send televised material over long distances.

reception

Not much use was made of the MDS frequencies for many years, mainly because no one knew how to make money using the channels. However, with the advent of HBO and other pay-TV services, several companies began to see a way of making money utilizing the MDS frequencies to transmit pay movies to apartments, hotels, and, to a limited extent, homes.

programs

In most instances an MDS company bought a satellite dish, contracted with HBO to use its programming, then retransmitted the HBO programming from its high-frequency antenna to antennas on top of hotels and apartments that downconverted the signal so it could be received on a regular TV. The apartment or hotel paid for its special antenna setup and also paid the MDS operator for the programming. The MDS operator, in turn, paid HBO part of what it received.

apartments and hotels

The cost of the special antenna system was several hundred dollars, so most individual homeowners were not interested in MDS. However, in a few instances, homes, as well as apartments and hotels, became customers.

MDS operations started primarily in areas not serviced by cable that could not receive pay movies. The operation was sleepy at first, with one company, Microband, engaging in most of the MDS activity. But by 1980 business was brisk enough that the FCC began an inquiry into technical standards for MDS, methods for choosing from competing applicants for any particular channel, and possibilities for multichannel MDS.[34]

FCC inquiry

The latter point, multichannel multipoint distribution service, engendered the most interest. This would enable MDS operators to offer a number of different services and, therefore, act more like a cable TV channel with its multitude of channels. In fact, Microband began referring to the multichannel concept as **"wireless cable"** or as multichannel TV (MCTV).[35]

multichannel concept

The main problem with developing a large number of MMDS systems was that there were not enough channels allocated to MDS to accomplish this. However, many of the ITFS channels had not been utilized, so in 1983 the FCC reallocated eight of the ITFS channels to MDS. In addition, ITFS systems were given permission to lease their excess time to commercial ventures.[36]

ITFS reallocation

The FCC decided that two four-channel systems would be available in each market and if there were multiple applicants for the systems, a lottery would be held to determine the recipient. In fact, some thirty thousand applications were received from broadcasters, cable companies, telephone companies, and even a major league baseball team.[37]

The first lottery decisions were made in 1985,[38] but a number of companies did not wait for the lottery and instead established MMDS systems by leasing time from established ITFS channels.[39]

lottery

Neither the lottery-awarded systems nor the systems formed by leasing channels have had an easy time of it. Cable TV, which can offer more channels, has a strong foothold in most areas, making it difficult for this newcomer to attract subscribers. In addition, many of the program suppliers who provide the pay network material for cable TV are reluctant to offer these services to MMDS. HBO, which supplied its services during the single-channel MDS

difficult times

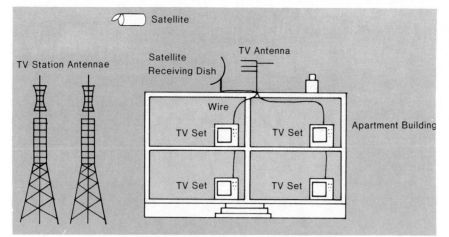

FIGURE 6.6

SMATV configuration.
*(From Gross, Lynne Schafer,
The New Television
Technologies. 2d ed. © 1986
Wm. C. Brown Publishers,
Dubuque, Iowa. All Rights
Reserved. Reprinted by
permission.)*

days when cable was nonexistent in many parts of the country, has refused to contract with MMDS systems. Showtime is seen in Cleveland, but only because of an impending antitrust lawsuit.[40]

MMDS, with about 750,000 subscribers, is in an unstable, infant stage. The future will determine whether or not it can find a niche.

Satellite Master Antenna TV

Satellite Master Antenna TV (**SMATV**) is a system for supplying TV programming to apartments, hotels, and private developments. It became possible in 1979 when the FCC deregulated receive-only dishes by stating that no one needed to apply for a license in order to install and use an earth station that received satellite TV signals.

Prior to this, owners of apartment complexes often had outside contractors install Master Antenna TV (MATV) systems. In its simplest form, this was a regular TV antenna installed on top of the apartment building that was wired to each apartment. In this way, all apartment dwellers could receive the regular broadcast TV signals without each having to place an antenna on the roof of the building.

MATV

After the FCC's 1979 decision, the apartment owners could contract to have satellite signals fed into the Master Antenna system so that cable TV network programming could be played on empty channels not occupied by broadcast TV. This combination of satellite and broadcast signals became known as SMATV.[41]

satellite signals

Although SMATV operations compete with cable companies that may be trying to wire apartment complexes, SMATV does not come under the FCC definition of a cable TV system. The FCC's definition specifically excludes systems that serve fewer than fifty subscribers or serve subscribers in one or more multiple-unit dwellings under common ownership, control, or management. This means SMATV is not regulated and does not need to pay fees to city governments.[42]

lack of regulation

FIGURE 6.7
████████

DBS configuration.
*(From Gross, Lynne Schafer,
The New Television
Technologies. 2d ed. © 1986
Wm. C. Brown Publishers,
Dubuque, Iowa. All Rights
Reserved. Reprinted by
permission.)*

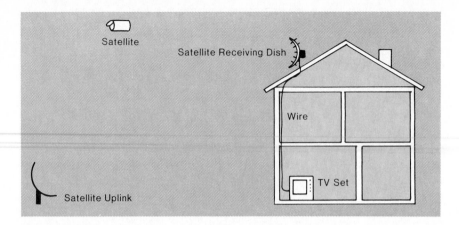

There is no love lost between cable and SMATV, and each takes the other to court frequently to try to solve the problem of who can go after whose customers. Cable systems try to prevent SMATV operations from taking hold in areas they have wired and SMATV operators try to prevent cable representatives from soliciting in their apartment complexes. In addition, most companies supplying pay-cable programming try to avoid contracting with SMATV companies.

cable relationship

Overall, SMATV is a small but rather steady business with between 450,000 and 600,000 customers.[43]

Direct Broadcast Satellite

The term direct broadcast satellite (**DBS**) was coined to describe a grandiose technology that would transmit signals directly from high-powered satellites to two and one-half foot dishes located on roofs of individual homes. To date, this technology has not gotten off the ground, but a more mundane form of direct satellite reception, **backyard dishes,** has been growing—steadily and unsteadily.

receive-only dishes

Both forms of satellite reception were the result of the 1979 FCC ruling that stated no one needed a license in order to have a receive-only satellite dish. At this point, some individuals, particularly those in rural areas where TV reception was poor, bought the same kind of large three-meter dishes that cable TV systems bought and set them up in their backyards. They could then receive the cable programming for free, but they had paid in excess of $3000 for the dish. This caused a minor stir among cable TV program suppliers, but the number of backyard dishes was so small that the cable folk did not aggressively pursue the issue.[44]

STC plan

The more grandiose plan for DBS came from a company called Satellite Television Corporation (**STC),** which in 1979 informed the FCC it wanted to develop a programming service that would go directly from satellites to homes. The idea was opposed by the traditional networks because the establishment of DBS would mean people could receive programming directly, without the intermediaries of broadcast stations.[45] However, the FCC sided with STC and in 1980 invited applications.[46] And who should apply but CBS, RCA, and other familiar names in the broadcast world.

In all, the FCC received thirteen applications and in 1982 approved eight of them, expecting that the services would be operational in 1985. Most of the proposed services planned to program movies, sports, and other entertainment.[47] CBS proposed that the DBS spectrum be used for **high-definition TV.**[48] The FCC mandated that "due diligence" reports be given in 1984 to make sure that the companies were progressing toward DBS service and not sitting on the frequencies for future use.[49]

proposals

With great fanfare, market surveyors proclaimed that 23 percent of consumers would be likely to buy the rooftop dishes and another 20 to 35 percent would be somewhat likely.[50] The whole idea looked so promising that one other company, United States Satellite Broadcasting Company, essentially snuck into the arena.

In 1982 it petitioned the FCC, in a vaguely described proposal, to begin a DBS service, which the FCC approved. It then purchased time on a Canadian midpower satellite that could beam signals to four-foot receiving dishes.[51] United set up five channels of service and offered them in the Midwest for $40 per month and a one-time installation cost of $300.[52] After six months of operation, United folded its tent, having attracted only about a thousand customers.[53]

United

Meanwhile, several of the original DBS applicants flunked their "due diligence" tests and others didn't even bother to fill out the forms.[54] Then in a surprising move, Satellite Television Corporation, the company that had pioneered the DBS concept, announced it was pulling out of the business after having spent five years and $140 million gearing up for it.[55] This left DBS in turmoil, where it remains.

STC withdrawal

On the more mundane front, the backyard dish business continued to grow,[56] and as it did, the companies with programming on satellites became more upset that the dish owners were receiving the programming for free. The result was that many of the services began scrambling their signals but did offer them to backyard dish owners who wanted to pay for them.[57] This caused a temporary dip in satellite dish sales. Now, however, many dish owners have decided the programming is worth the monthly fee and the home satellite business, which claims about 1.7 million dish owners, does not seem to be in danger of disappearing.[58]

backyard dishes

Low-Power TV

Low-power TV (**LPTV**) was first authorized by the FCC in 1980. From the FCC's standpoint, LPTV was an extension of **translators**—stand alone transmitters that had been rebroadcasting, mostly to rural areas, the programs of full-service stations almost since the beginning of commercial television. These translators had never been allowed to originate their own programming, but now the FCC decided that existing translators, as well as new facilities similar to the existing ones, could engage in their own programming.

translators

FIGURE 6.8

LPTV configuration.
(From Gross, Lynne Schafer,
The New Television
Technologies. 2d ed. © 1986
Wm. C. Brown Publishers,
Dubuque, Iowa. All Rights
Reserved. Reprinted by
permission.)

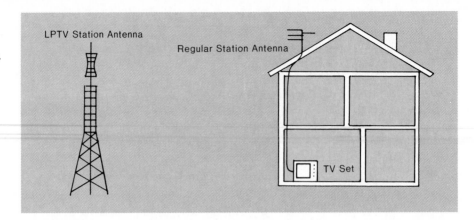

These stations, as their name implies, have very limited power—ten watts for VHF and one thousand watts for UHF. This means they can only cover a radius of twelve to fifteen miles, but they are also inexpensive with start-up costs as low as $50,000 compared to $2 million for conventional stations. They are sandwiched in between regularly operating stations. For example, if a city has a channel 4 and a channel 6, a low-power station could operate on channel 5 provided it did not interfere with either of the other full-power stations. Theoretically, there is room for about four thousand LPTV stations nationwide in addition to the four thousand translators presently in existence.[59]

The FCC's main purposes for establishing LPTV were to add additional channels to small and medium-sized markets and to enable groups not usually involved with TV ownership to have a voice in the local community. The FCC hoped that women and minorities, in particular, would apply. Some did, but a great number of applications were from large companies such as Sears, Federal Express, and even ABC and NBC.[60]

The FCC received over five thousand applications before April of 1981 when it decided to impose a freeze on new applications so it could sort out those already submitted. When it began approving applications late in 1981, first to rural areas and then to cities, it held to its original purpose and made most of the awards to people who were not part of the broadcasting establishment. Although the applications were decided by lottery, special preference was given to applicants who did not own other media and to those with more than 50 percent minority ownership.[61] As of 1987, approximately 1500 construction permits had been awarded and four hundred plus stations were on the air.[62]

Over two hundred of the stations are in Alaska as part of a network to keep people in remote areas in touch with what is happening in the capital, Juneau. All has not been rosy for the other station owners. Some that received construction permits never activated their stations or sold their permits to other groups. In order to reduce this churn in station selling, the FCC extended the time a station could hold a construction permit without going on the air from twelve to eighteen months.[63]

characteristics

purpose

application procedure

status

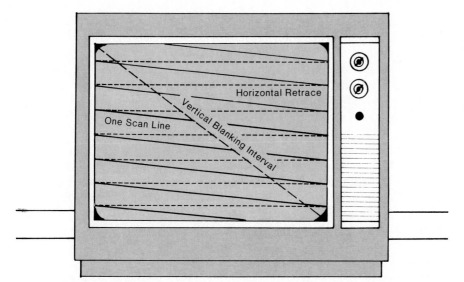

FIGURE 6.9

The TV scanning process. Actually, 525 scan lines are placed on the front of the tube. *(From Gross, Lynne Schafer, The New Television Technologies. © 1983 Wm. C. Brown Publishers, Dubuque, Iowa. All Rights Reserved. Reprinted by permission.)*

Others stations tried to produce local programming but found they could not afford it so instead went to movies, music videos, religious programs, or home shopping services. Some, however, have been able to sell advertising and make modest profits with programs that deal with hunting season, high school band concerts, Yugoslav-American customs, and farm reports.[64]

Teletext

Teletext is a method by which conventional TV stations can display words and graphics on a consumer's TV screen while either the regular picture is on or at a different time. Teletext is possible when the TV station transmits information on an unused portion of the TV signal. This part is called the **vertical blanking interval** and occurs when the video portion of the TV signal is momentarily stopped so that a new frame can be started.

vertical blanking interval

A TV picture is created by a scanning process whereby an electron beam starts in the upper left-hand corner of the screen, proceeds to the right side, turns off momentarily to retrace back to the left-hand side, and then turns on again to scan to the right-hand side. This process continues until the beam reaches the bottom of the screen. At that time, the beam turns off and retraces to the top of the screen to begin the scanning process all over again. This period that the beam is turned off to go from bottom to top is called the vertical blanking interval. Part of this interval is used for information to keep the picture stable, but for many years most of this interval was unused.[65]

The technology for teletext involves digital impulses piggybacking in the vertical blanking intervals, a **decoder,** a **keypad** that the viewers use to select the pages to view, and a **computer.**

The digital impulses create the letters. Either a portion of the screen is on, displaying part of a letter or graphic, or it is off, displaying a background. This on-off phenomenon is much easier to create than the conventional video

process

picture, which depends on many shades of gray and colors. Therefore, the teletext can be transmitted using far fewer electrons than the regular TV picture and can fit in the vertical blanking interval.

decoder

The decoder enables the TV set to read this digital information. Sometimes both digital letters and the video picture are seen on the screen at the same time. The most common use of this is closed captioning for the deaf. The captioning is placed on the vertical blanking interval and deaf people, with special decoders, can access this captioning and place it over the picture. Other people see the programming without the captioning. In a similar manner, foreign language translations can be broadcast through teletext. Most of the uses of teletext do not involve placing the text over the picture, however. They involve the transmitting of information such as weather, news, and financial data, which is seen on the full screen. In order to do this, an A-B switch on the decoder allows either digital words or the TV station's conventional programming to appear.

keypad

The keypad is used to select particular information. Different pages of information are written each time the beam scans from top to bottom, for a total of about two hundred pages. The viewer can use the keypad to capture whichever page he or she desires.

computer

Material to be used on teletext is stored on a computer. The various pages can be called up by teletext employees to be updated, eliminated, or reorganized.[66]

advertising

Generally there is no charge to the consumer for teletext—anyone who buys a decoder can receive it. Most of the companies that have experimented with teletext have attempted to sell advertising, which is incorporated within the information pages in order to cover the costs of programming the service.

British and French origin

Teletext originated in foreign countries, primarily England and France, in the late 1960s. The British and French systems used different methods for creating the words and pictures, so they were not compatible with each other. In general, the British system was less expensive than the French system, but its graphics capabilities were not as good.[67]

The first experimentation in the United States occurred in 1978 at KSL-TV in Salt Lake City. Shortly thereafter several other stations began experimental broadcasts, some with the French system and some with the British.[68]

standards conference

Because these two teletext methods were not compatible, a conference of teletext experts and users was held in 1981 to determine if a standard could be set for the United States so that all teletext transmission would be the same. Employees of the British and French systems spent a great deal of time and money trying to influence American broadcasters that each of their systems was best.

But when the conference was held, AT&T appeared with a new teletext system that was compatible with the French system but not the British system, primarily in the area of graphics.[69] AT&T was hopeful that the FCC would choose its system as the American standard, but the FCC decided to allow the marketplace to select a standard.[70]

FIGURE 6.10

An example of teletext.
(Courtesy of KSL, Salt Lake City)

The French supported the AT&T system because it was compatible with its own, but the British began an aggressive campaign to convince teletext purveyors in the U. S. that they should use the British system.[71]

The following years saw a number of experimental teletext systems inaugurated, some with the British system called World Standard Teletext (**WST**) and some with the AT&T-French system called North American Broadcast Teletext Standard (**NABTS**). In general, the start-up cost for a station using WST was about $30,000 and the decoders, manufactured by Zenith, cost about $300. Start-up cost for the NABTS system was in the neighborhood of $150,000 and decoders were about $1,000—when and if they were available.[72]

WST and NABTS

Many of the companies that experimented with teletext, such as Time, NBC, and Metromedia, found the venture unprofitable and bailed out. Two systems have survived in a limited fashion. One is **Extravision,** which uses NABTS and is broadcast by CBS to all its affiliates with the idea that they can add their own local pages. Only one, KSL, the Salt Lake City station that pioneered teletext, does create its own pages. Some other CBS stations pass through Extravision so that it can be received by people in the area who have decoders.

Extravision

The decoders have been the primary problem. For quite a few years the only decoder available was one from Panasonic that cost $1000 and only worked with Panasonic top-of-the-line TVs. Recently Samson has begun selling a $300 decoder that works with all sets. Proponents of the NABTS teletext system are hopeful this will increase the number of people using their service.[73]

The other surviving system is **Electra,** which uses WST and is an outgrowth of 1983 experiments by the Taft-owned station, WKRC-TV in Cincinnati. Taft distributes Electra from WKRC-TV to all the stations it owns, most of which program local pages. It also sends the material to WTBS and from there it is uplinked to cable systems along with the regular WTBS signal. Some cable systems put the teletext material on a separate channel, some eliminate it from the signal, and some leave it as is so it can be received by people with decoders.

Electra

Decoders have also been a problem with WTS. Zenith manufactured one for $300, but it proved too expensive for most consumers, so Zenith started manufacturing a top-of-the-line TV set that includes the decoder within it for no extra cost. The WTS proponents, in a move almost opposite to the NABTS one, hope this dedicated TV set will increase their subscribers.[74]

Teletext cannot be considered a success at present, but there are still companies fighting to make it economically valid.

The Telephone

Obviously, the **telephone** is not a new technology. Invented by Alexander Graham **Bell** in 1876, it predates both radio and television. But within recent years, it has developed new capabilities that place it in the mass media arena, particularly as a participant in the data information area.

American Bell and AT&T

Bell and two of his friends set up a company that eventually became the American Bell Telephone Company. From the very beginning this company leased phones to customers and charged them for phone calls. Bell first established local switching systems primarily in New England and then developed switching systems that linked various cities in the east. Shortly after it began linking cities, American Bell formed a subsidiary, American Telephone and Telegraph Company (**AT & T**) to handle this long-line business. After a period of time, the parent swallowed the child and AT&T became the main organization, owning American Bell and other local phone companies. AT&T was set up as a monopoly, regulated by the FCC, and for many years maintained the same basic organizational structure.

voice system

During this period of time the telephone business grew into a major national and international service, but it remained primarily a voice service connecting one individual with another. Improvements were made in switching systems and ease of phone installation, but because AT&T was a monopoly making a guaranteed profit, it had little incentive to add new services for the customer. In 1956 a consent decree barred AT&T from entering nonregulated businesses.

outside services

Other companies did begin services that were tied to the phone. For example, some independent companies set up telephone answering services to answer calls for people when they were not at home. Others went into the business of manufacturing and selling phone answering machines.[75]

computer incorporation

Then in the late 1970s, telephone and **computer met.** The computer had its own unique history, dating back to the 1930s when a very bulky **vacuum tube** device was developed at Iowa State University to store and compute data for doctoral candidates. During World War II, computers were used to calculate ballistic tables to produce trajectories for bombing. After the war, these same bulky computers, which often were housed in several rooms, were used on a limited basis by universities and government.

UNIVAC

The first time the general public became aware of computers was in 1952 when UNIVAC, a computer manufactured by Remington-Rand, was used for election coverage. With only five percent of the vote counted, it predicted that Dwight Eisenhower would defeat Adlai Stevenson for president.

172

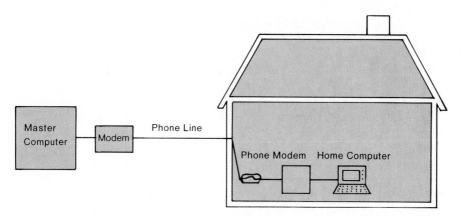

FIGURE 6.11

Telephone-computer configuration. *(From Gross, Lynne Schafer, The New Television Technologies. 2d ed. © 1986 Wm. C. Brown Publishers, Dubuque, Iowa. All Rights Reserved. Reprinted by permission.)*

About this time **IBM** decided to enter the computer business and developed computers designed for business. In 1958 it brought out several computers that used **transistors** instead of the bulky tubes. These computers were bought by businesses at a fairly rapid clip.

During the 1960s the **silicon chip** was introduced to replace the transistor for storage and processing of data. These chips, which were about the size of a fingernail, reduced computer size and cost, increased speed and reliability, and led to a boom in purchasing and renting of computers by the business community.

In the 1970s, the size of computers was further reduced when very large scale integration (**VLSI**) greatly multiplied the number of electronic components that could be placed on a chip. By 1975 electronic hobby magazines were advertising computers that could be assembled from kits and used in the home. Shortly after that, the **Apple** computer was developed and marketed to the home. A number of other companies, including IBM, entered the home computer market which, for several years in the late 1970s and early 1980s, was a booming success.

It was during this period that the **modem** enabled the home and business computers to be connected to the telephone. Several companies, primarily CompuServe and The Source, began compiling data banks of information that users could access through phone lines by paying a fee. Much of the information in these data banks was the same information available through teletext—news, sports, recipes, and games. However, because these services consist of large computers tied to home and business computers through wires, they can be two-way. For example, someone could type the word "Egypt" on a home keyboard; this would go through the modem to a large distant computer that would access news stories about Egypt and send these back to the home computer screen.[76]

Simultaneously with the link-up of computers and telephones, the Justice Department was taking a hard look at AT&T's monopoly situation. Some companies were making noises that they wanted to sell telephones directly to consumers and several independent companies, primarily MCI and Sprint,

IBM

chips

VLSI

data banks

wanted to enter the long-distance phone business. This was all quite controversial because AT&T's long-distance service had been subsidizing its local phone company service in order to keep local phone rates low. This meant other companies could enter the long-distance business and offer much lower rates than AT&T, but they could only do this because they did not need to subsidize the local rates.

All of these issues somewhat coalesced in the Justice Department's 1980 **consent decree** that broke up AT&T. By order of this decree, AT&T had to divest itself of its twenty-two local phone companies. These were spun off into seven new companies serving seven sections of the country. Independent companies were allowed into the long-distance business and customers were to choose which long distance company they wanted. AT&T was also required to notify customers that they could purchase phones instead of renting them from the phone company as they had always done in the past. AT&T kept its long-distance service, was still allowed to operate its research arm Bell Labs, could still manufacture equipment, and was allowed to enter some unregulated fields such as computer sales.[77]

One of the fields AT&T was originally allowed into was **electronic publishing.** However, newspaper publishers and cable TV companies, who were beginning to operate data services called **videotext,** did not like the idea of AT&T entering this field. They felt that because AT&T was so big and powerful and because it already had long lines in place, it could absorb the videotext field before others got to the gate.

The newspapers were particularly loud spoken because they were afraid videotext would replace their newspapers. They voiced powerful enough objections that they were able to get the consent decree modified to prevent AT&T from entering the electronic publishing business until 1989.[78]

Cable TV's history with videotext went back to Qube in the Warner-Amex system in Columbus, Ohio. Qube was not called videotext but it did give subscribers access to data banks and did allow for two-way interaction so that subscribers could answer questions. A few other systems tried two-way videotext, but none of them were economically successful and they dropped it with little fanfare.[79]

The newspaper owners were more aggressive in their pursuance of videotext. Knight-Ridder set up a system called Viewtron in southeast Florida in 1983 and Times-Mirror set up Gateway in southern California in 1984. For both systems, subscribers had to buy a terminal for about $600 and then pay about $12 a month for services that included banking, shopping, news, encyclopedia access, and information about the local area. Unfortunately, both systems collapsed in 1986. Viewtron had over 20,000 subscribers at its peak and Gateway had less than 5,000. In general, subscribers liked the service but did not feel it was worth the cost.[80]

With the primary cable and newspaper systems collapsed, AT&T and the local phone companies began to tiptoe into the data information business. Telephone folks began talking about POTS and PANS—Plain Old Telephone

AT&T break-up

electronic publishing

newspapers

cable TV

viewtron and Gateway

phone company interests

Communication Systems

Service and Pretty Amazing New Service. They also spoke of VANS (Value Added Network Service) and ISDN (Integrated Services Digital Network). The latter is an electronics technology for which a worldwide standard is being devised that theoretically allows two-way processing, storing, and transporting of whatever information anyone desires in simultaneous voice, data, graphics, or video.[81]

Just where all this will go is anyone's guess. CompuServe and The Source have proven successful and theoretically would be incorporated into whatever else comes along. But just how supply, demand, and cost will interact is unclear.

Other Electronic Media Chronology

1952   UNIVAC makes election predictions
1956   Consent decree bars AT&T from nonregulated fields
1957   Henry Griffing sets up pay-TV in Bartlesville, Oklahoma
1958   IBM markets transistor computers
1959   FCC issues rules on subscription TV
1964   California voters pass referendum outlawing subscription TV
1968   FCC issues antisiphoning rule for subscription TV
1975   Sony introduces Betamax half-inch VCR
1975   Hobby magazines advertise computer kits
1976   Matsushita introduces half-inch VHS VCR
1976   Universal and Disney bring copyright suit against Sony
1978   DiscoVision is marketed by MCA-Philips
1978   KSL-TV in Salt Lake City experiments with teletext
1979   SMATV begins because FCC deregulates receive-only earth dishes
1979   Satellite Television Corporation proposes DBS
1979   Videocassette copyright case decided in favor of Sony
1980   FCC authorizes low-power TV
1980   FCC invites applications for DBS
1980   Consent decree breaks up AT&T
1981   RCA begins marketing Selectavision video disc
1981   Videocassette copyright case reversed against Sony
1981   AT&T proposes a teletext standard
1982   Pioneer buys DiscoVision and renames it LaserDisc
1982   AT&T ordered to refrain from engaging in electronic publishing for seven years
1982   FCC lifts rules governing STV
1982   FCC approves eight DBS proposals
1983   Some ITFS channels reallocated to MMDS
1983   United Satellite Communications begins midpower DBS
1983   Knight-Ridder begins Viewtron videotext
1983   WKRC-TV in Cincinnati begins teletext service
1984   Times-Mirror begins Gateway service in Southern California

1984   Satellite Television Corporation bows out of DBS
1984   Supreme Court decides video tape recording does not violate
       copyright law
1984   RCA stops manufacturing its videodisc machine
1984   United folds its DBS operation
1985   Oak Media closes down STV operations
1985   First lottery decisions made for MMDS
1986   Viewtron and Gateway collapse
1986   Sony stops manufacturing Betamax
1987   FCC extends construction time for LPTVs

Conclusion

All of the newer media will have to struggle for their place in the living room. They must compete not only with one another, but with the established broadcast media and with other new media that may develop in the future.

Subscription TV has had a long struggle against established media. This struggle has involved technical, economic, and political problems. STV thrived for a brief period, but its single-channel capability was its nemesis.

The videocassette recorder has been much more successful reaching a 50 percent penetration rate. However, along the way the Beta format fell by the wayside.

The stylus video disc format was discontinued, too, but the disc's other format that involves a laser has not been a big success. Although it is high quality, its inability to record has hampered it.

MMDS, SMATV, DBS, and LPTV have all achieved small niches in the media field. Both MMDS and backyard dishes have their greatest success in areas that do not have cable TV. SMATV encompasses primarily apartment complexes, and LPTV programs for very local areas.

Broadcast stations are the primary experimenters in the teletext field, viewing this media as an adjunct to their regular service. Wired text services have generally floundered but may receive impetus from the telephone-computer alliance.

Technologically, the various new media differ from one another. Subscription TV is an over-the-air system that utilizes broadcast frequencies in a scrambled manner. Both videocassette machines and videodisc players are consumer-controlled devices that hook into the consumer's TV set. MMDS is an over-the-air system, but it transmits at higher-than-broadcast frequencies. SMATV uses a wired system similar to cable. DBS employs satellites and home receivers. LPTV is an over-the-air system, but it covers a much smaller area than regular TV. Teletext uses portions of the broadcast signal, and the telephone-delivered text services use wires.

From a regulation point of view, STV, MMDS, DBS, and LPTV come under the jurisdiction of the FCC and need to fight for regulations that will enhance their situations. Videocassettes, videodiscs, and SMATV do not fall under regulation, but court cases have evolved around them. Because teletext comes from TV stations, it is regulated, but regulation has not affected it to any degree. The wired text services that have been tried to date have not been regulated, but it is a court decision that keeps AT&T out of electronic publishing for seven years.

In the programming area, competition exists in abundance for movies. Not only are broadcast and cable entities utilizing them as primary program fare, but movies are also a mainstay of MMDS, SMATV, STV, DBS, and LPTV. In addition, movies are available for sale and on a rental basis for videocassettes, and they have been a prevalent fare of videodiscs. Other sought-after programming forms include sports and local events. Teletext and videotext compete in the area of data information transmission.

The future path for all these media forms is uncertain, but nevertheless exciting.

Thought Questions

1. Which of the media discussed in this chapter do you think will be most widely accepted by the public in the coming decades? Why?
2. How could videodisc player sales be improved?
3. What role do you feel telephone companies will play in the data field?

Physical Characteristics

Telecommunications is totally dependent upon electronics and mechanics. Eliminate the electron and the total of telecommunications would disappear. For better or for worse, complicated equipment is needed in order to produce and distribute programming. When this equipment functions properly, it can be a great aid to the creative process. When it becomes cantankerous, it can damage or even destroy concepts, careers, and temperaments. Ingenuity and curiosity keep the electronic concepts in a constant state of flux, which leads to a spiral of improvements spiced with problems.

PART

Equipment

7

Television is a triumph of equipment over people.

Steve Allen
TV personality

Overview

Whether it be the coverage of the Olympics or a public access show about mind reading, production involves people using equipment in creative ways to produce programs. Over the years this equipment has changed, mainly in ways that make it smaller and more flexible, although the basic elements of production have remained essentially the same. Even though this chapter is not intended to teach production techniques, it should provide the reader with a general understanding of the production process, which involves the following topics:

Characteristics of microphones in terms of directionality and use

The functions of turntables and compact disc players

Characteristics of open-reel, cassette, and cartridge audiotape recorders

The selection, amplification, and routing functions of audio control boards

The design of radio station studios, control rooms, and master control areas

Uses of portable audio equipment

The overall functions of TV camera viewfinders, lenses, and electronics

The process by which pictures are created in tube cameras and CCD cameras

The possibilities of high-definition TV

Equipment on which cameras are mounted

The incorporation of film equipment into the TV process

Creative uses made of character generators and computer graphics

Effects possible through switchers and digital effects generators

The purpose of monitors

The process by which video material is recorded

The multiple purposes of VTRs and VCRs

Configurations of TV studios, control rooms, and master control areas

The growth and development of portable TV gear

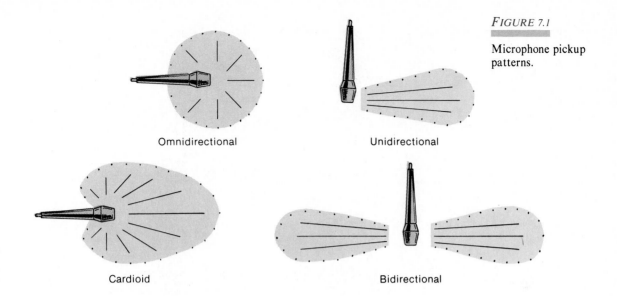

FIGURE 7.1

Microphone pickup patterns.

Omnidirectional

Unidirectional

Cardioid

Bidirectional

Microphones

Similar audio production equipment is used to produce radio and television programs. Some of this equipment has been in operation for years while some seems to change almost daily.

Microphones are among the oldest production equipment. Although their size and construction have changed over the years, they have basically always been devices for converting sound into electricity so that it can be sent through wires as variations in voltage. Different mikes are constructed to convert this sound in different ways. Some methods maximize fidelity at the expense of microphone ruggedness; others sacrifice fidelity for low cost; and still others emphasize the shape and ease of movement, but sacrifice frequency response.

construction

Microphones are also designed to pick up sound in varying ways. The four most common variations are: **unidirectional,** which picks up sound mainly from one side; **bidirectional,** which picks up sound mainly from two sides; **cardioid,** which picks up sound in a heart-shaped pattern; and **omnidirectional,** which picks up sound from all directions.

directionality

Each directional aspect has its own special uses. When only one person's voice is to be heard, such as that of a newscaster or sportscaster, a unidirectional mike is the best choice. It will pick up the person's voice sitting in front of it and minimize distracting noises in the background. Bidirectional mikes had their heydey during the era of radio drama—they enabled actors to face

Types of microphones.
(*a*) An omnidirectional
microphone. (*b*) A
unidirectional
microphone. (*c*) A
cardioid microphone.
(*d*) A bidirectional
microphone. *(a,c courtesy
of ElectroVoice, Inc.,
Buchanan, Michigan; b,d
courtesy of Share Brothers,
Inc.)*

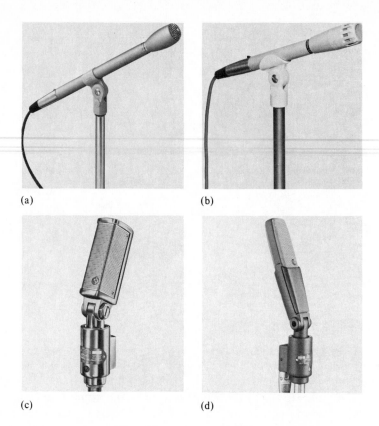

(a)   (b)

(c)   (d)

each other to deliver their lines. Cardioid mikes are very satisfactory for two people seated next to each other and are frequently used on TV talk shows. Omnidirectional mikes pick up overall crowd noises well and are often used for plays with large casts.

The type of microphone to use, in terms of both construction and directionality, depends on the purpose and need for which it is intended.[1]

Record Players

Most radio and TV facilities still have turntables that spin conventional records, although a relatively new invention called the compact disc player has been added to numerous stations.

turntables

The **turntable** has two functions: (1) to spin a record at the proper speed, and (2) to convert variations in the grooves of the record to electrical energy with the pickup arm. A typical broadcast-quality turntable consists of a heavy metal plate covered by a rubber or felt top. It will normally have a power switch to control the **motor,** a **gear shift** to select speeds, and an **equalizer** or

FIGURE 7.3

A broadcast-quality
record turntable.
*(Courtesy of QRK
Electronics Products)*

FIGURE 7.4

A compact disc player.
*(Courtesy of Sony
Corporation of America.
Sony® is a registered
trademark of Sony
Corporation.)*

filter to compensate for scratchy records. Next to the revolving table is a pickup
arm that houses the **cartridge** and **stylus** that actually pick up the signal from
the record. The cartridge receives the minute vibrations from the stylus and
converts them into variations in voltage. These variations are then sent through
wire to the control board. The stylus is a very small, highly compliant strip of
metal, the end of which is made of hard material, usually diamond.

Because turntables are used by radio stations primarily for playing musical selections, a control room will usually have two turntables so that a disc jockey can cue one record while another is airing.

The compact disc player operates on an entirely different principle. In fact, it even looks different because the "record," referred to as a **compact disc** or CD, is inserted into the machine rather than being placed on top of it. Instead of a cartridge and stylus, the compact disc player contains a **laser** beam to read material from the compact disc. This CD contains a great deal more music on a smaller disc. The sound, in addition to being compact, is recorded in digital rather than analogue form.

The use of a **digital** process improves sound quality and allows music to be transmitted and edited without loss of quality. The **analogue** method of encoding information is similar to creating a line graph to show a statistical analysis. In other words, there are no discrete points. All measurements are on a continuum, a line that curves up and down. Digital sound is encoded in a discrete way similar to looking at individual numbers in a statistical analysis and writing them down in a set order.

When the analogue continuous method is used for taking an electronic signal from one place to another (e.g. one tape recorder to another), the signal changes its shape and, thus, degrades the quality slightly. This is similar to someone trying to trace a curve on a graph. The reproduction will not be exactly as the original. When the digital method is used to transport the signal, this signal does not waiver or change. As it travels from one source to another, separate bits of information transmit with little loss of quality. This would compare to someone copying the numbers of the statistical study. The numbers would be copied accurately, so the results would not be distorted.

Digital technology is, of course, used for many audio and video components, but within the area of "record playing," it is rapidly replacing the analogue turntable, both in homes and in radio studios.[2]

Audio Tape Recorders

Audio tape recorders are devices that rearrange the iron particles on magnetic tape so that sound impulses can be stored on the tape and played back at a later date. This rearranging of particles is undertaken by stationary **heads.**

Quality recorders usually have three heads—one to erase, one to record, and one to play. The erase head is always positioned so that old material is erased an instant before new material is recorded. When the machine is in play, the erase and record heads are disengaged. Cheaper recorders have only two heads, one for erase and another for both play and record.

The three basic types of recorders are open reel, cassette, and cartridge. On an **open-reel** recorder all the tape is initially on one reel and must be threaded past the heads to a take-up reel. This kind of machine lends itself to

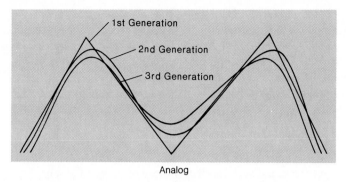

Analog

FIGURE 7.5

The analogue wave changes while the digital wave remains constant. *(From Gross, Lynne Schafer,* The New Television Technologies. © *1983 Wm. C. Brown Publishers, Dubuque, Iowa. All Rights Reserved. Reprinted by permission.)*

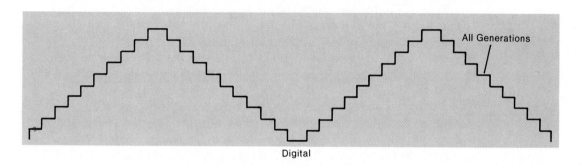

Digital

editing because the tape is readily accessible. Audio editing usually involves physically cutting tape and putting it back together with an adhesive-based tape. If the tape is accessible, this cutting procedure is fairly simple.

A **cassette** machine uses cassette tape, which is two reels already threaded inside a plastic container. The cassette is inserted so that the tape passes against the heads. Because of the small size, it is often used for taping material at locations away from the studio.

A **cartridge** machine is similar to a cassette machine in that the tape is slipped into a machine with no threading necessary, but the cartridge is constructed differently in that it is a container with a loop of tape. The tape is wound onto the center spool with the end next to the spool slightly raised. When the tape is cut from the master reel, it can be spliced to the raised end, thus making a continuous loop. Radio commercials are usually recorded on cartridges since they are brief and can profit from a cartridge recorder's cue tone, which automatically starts and stops the machine.

Most tape recorders can record at different speeds, the most common of which are seven and a half IPS (inches per second) and three and three-quarter IPS. This means that either seven and a half or three and three-quarter inches

cassette

cartridge

speeds

of tape passes the head per second. The more tape that goes past the head, the better the recording quality because more frequency responses can be placed on the tape. For this reason, there are also recorders that tape at fifteen IPS in order to achieve high-quality reproduction of musical sounds.

Digital has also entered the audio tape world in the form of **digital audiotape** (DAT), which offers the same high quality as CDs.[3]

Audio Control Boards

The audio control board or console is the primary consolidating piece of equipment for sound sources. It has three primary functions: (1) to enable the operator to select any one or combination of the various inputs (microphones, turntables, tape players, and so on) and mix them together; (2) to amplify the sound; and (3) to enable the operator to route the inputs to a number of outputs (monitor, transmitter, tape recorder, and so on).

The various inputs come to the board through wires after they have been converted from sound to electrical impulses. For example, the signal from a microphone enters the board at a connection for microphone input. This mike signal consists of a very small voltage that the board directs through a **preamplifier** so that it is increased enough to be sent to the **potentiometer,** or slider. Here the resistance to the sound is varied, thus increasing and decreasing the volume. At the output of the pot there is a **key.** When the key is in the "off" position, the signal is stopped. When the key is in the "on" position, the signal is sent to the program amplifier, the final amplification stage before being distributed. The signal is here amplified enough so that it can be sent through the line out to a tape recorder or transmitter.

At the "line out" position, the signal is also sent to the **volume unit indicator** (VU meter), a metering device that enables the operator to determine the level of sound going out the line. Here, too, sound is sent to a monitor amplifier and then to a speaker that enables the operator to hear the signal.

The audio control board as discussed so far has only one input—a microphone. Because the purpose of the board is to mix sound, most boards have several inputs so that several mikes, turntables, and tape recorders can be on at once. By using the various controls, one song can be faded into another or a disc jockey can talk over a record.

If a station is producing in stereo, both the inputs and outputs will be dual to manage left and right channels. [4]

(a)

FIGURE 7.6

Tape recorders. (*a*) An open-reel recorder. (*b*) A cartridge recorder. (*c*) A cassette recorder. (*d*) Tape recorder head arrangement. *(a,c courtesy of Teac Corporation of America; b courtesy of Spotmaster)*

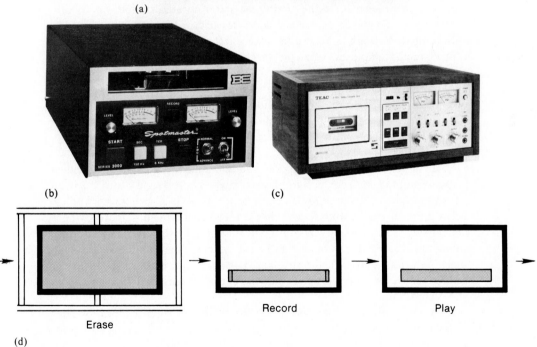

(b)

(c)

Erase

Record

Play

(d)

Equipment   187

FIGURE 7.7

This audio board can accept up to twelve inputs at once. *(Courtesy of Opamp Labs, Inc.)*

Radio Studios and Control Rooms

The equipment needed for the production end of a radio station is housed in a studio or studio and control room. Gone are the days of velvet-draped studios with complicated sound-effects gimmickry. In their place are all-in-one studio/control room complexes, often no larger than a department store display window. They house the control board, the mikes, the turntables, the tape recorders, the record and tape library, and, somewhere tucked in the middle, the disc jockey-engineer-salesperson-station manager-janitor.

small stations

Not all stations are this way, of course, but many small stations do operate primarily from one soundproof room, with perhaps a small outer reception area.

large stations

In larger stations the disc jockey and microphone are in one room called a studio, and the engineer, record players, tape recorders, and board are in an adjoining room called a control room. Usually there is a soundproof window between the engineer, who is spinning the records, and the disc jockey, who is announcing, so that they can communicate by hand signals. They can also communicate verbally through a talkback system when mikes are not being used on the air. A studio setup this large can usually also afford the luxury of genuine offices for the sales staff and the executives, as well as a newsroom and record library.

Many stations contain several studio complexes so that tapes for later airing can be produced in one while another is on the air or being set up for on-air. There are even a few large studios with adjoining control rooms that produce talk shows or live musical groups.

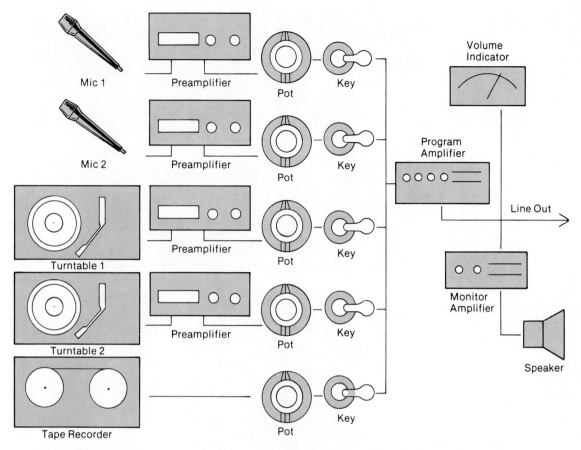

FIGURE 7.8

Diagram of how an
audio control board
operates.

Some radio stations have custom-built facilities, but many simply rent office space and soundproof as necessary.

Most small stations that originate all their programming in one or two studio complexes send the signals directly to the transmitter by phone line. However, where operation is more complicated, involving remote broadcasts or network feeds, a **master control** room is utilized. This control room contains power supplies, monitors to hear the various sounds, jack panels to route the sounds in many different ways, and numerous other buttons, switches, and signal lights.

master control

Some stations have very little in the way of either studios or control rooms because they use equipment that allows them to broadcast on an **automated** basis. These stations operate mainly with multiple tape decks, some of which contain reel-to-reel music and some of which contain cartridge commercials. Subaudio tones that cannot be heard over the air signal one tape recorder to

automation

Equipment   189

FIGURE 7.9

A radio studio that includes cassette machines, cart machines, CDs, turntables, reel-to-reel recorders, and other related equipment. *(Courtesy of KLOS-FM, Los Angeles)*

FIGURE 7.10

Automated radio equipment. The music is on the reels and the commercials and other announcements are on the carts. All the equipment is controlled by the computer off to the side. *(Courtesy of Harris Corporation)*

stop playing music and another to start playing a commercial, or vice versa. Taped station break announcements can also be inserted with specialized clock mechanisms. In this way, a station can program for a whole day without having either a disc jockey or an engineer; the sound is sent directly from the tape deck to the transmitter.

Now that programming can be transmitted by **satellite,** some stations pick up their programming from the dish and retransmit it. In this way, they do not need automated equipment, but they can be fully or partially automated. The satellite can send the bulk of the programming and then, through a special signal, turn on local carts that play station breaks and commercials.[5]

Portable Radio Setups

Radio station production can also be done outside the formal studio. Anyone with a microphone and a tape recorder can produce a radio program or part of a radio program. News reporters operate this way, taping and later editing interviews for actualities. Radio news crews also roam the city in automobiles or occasionally helicopters, either recording news or sending it back live to the studio by telephone or microwave.

ease

Cameras

Television audio operates in a manner similar to radio and also incorporates much of the same equipment—mikes, record players, tape recorders, and audio boards. These are used to create the sound portion of TV and to mix the sound in either stereo or mono. But because TV involves both picture and sound, it is a much more complicated process than radio.

The picture part of TV begins in the camera, which consists of a viewfinder, a lens system, and an electronic system. The **viewfinder** is simply a small TV monitor mounted on the camera to allow the camera operator to see the picture.

viewfinder

The **lens** system is similar to that of a photographic camera. With it, the camera operator selects and focuses the image and varies the light that strikes the electronics within the television camera. Selecting can be done by pointing the lens and zooming in or out until the proper amount of picture appears in the frame. Focusing is accomplished by adjusting a focus ring until the picture is sharp and clear. Light adjustment is accomplished by opening and closing an iris in the lens. If there is scant light in the room, the iris must be opened wide in order to take advantage of all available light; if there is profuse light, the iris can be partially closed.

lens

The camera itself contains the **electronic components** that convert the image focused by the lens into electrical impulses. For many years the only method for accomplishing this conversion was to use a camera **tube** that consisted of an **electron gun,** a **target,** and a **photosensitive surface.** A newer type of camera called a **CCD** (charge-coupled device) uses a chip to convert the image to electronics.

electronic system

In tube cameras, the picture to be televised goes through the lens to the photosensitive layer of chemicals arranged in microscopic, but individual, dots. Each dot emits an electrical charge that varies in intensity depending on the amount of light hitting it. For example, the dots of the photosensitive layer reacting to a white shirt would give off a stronger electrical charge than the dots reacting to dark hair.

tube cameras

Electrical charges emitted from the back of this photosensitive plate hit a target and reproduce an electronic equivalent of the image. Each little speck of information on the target is then picked up individually by an electron gun mounted at the back of the tube. This gun emits a steady stream of electrons that scan the target from top to bottom. In the present American system of

FIGURE 7.11

Parts of a television
camera.

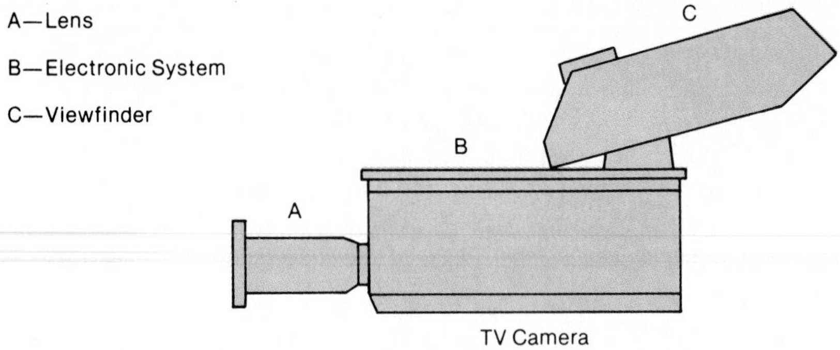

A—Lens

B—Electronic System

C—Viewfinder

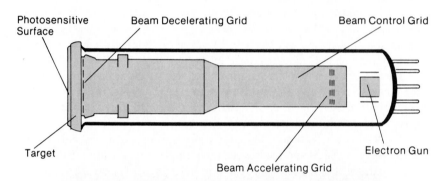

TV Camera

FIGURE 7.12

Television camera tube.

Photosensitive Surface | Beam Decelerating Grid | Beam Control Grid

Target

Beam Accelerating Grid

Electron Gun

FIGURE 7.13

A CCD camera.
(Copyrighted by Sony Corporation)

television, the electrons scan 525 horizontal lines from left to right each 1/30 of a second. This means that every 1/30 of a second, the tube generates one picture frame.

CCD cameras

In CCD cameras, the image from the lens goes to a chip (charged-coupled device) that contains a large number of "photo sites." Each site reacts to the light falling on it and liberates electrons in relation to the amount of light for 1/30 of a second. Again, more electrons will be liberated for a light scene than a dark scene. The charge accumulated at each photo site is then transferred into memory and the chip reacts to the light received by it for the next 1/30 of a second. No scanning is involved, but the system does generate 525 horizontal lines just as the tube does.

In the case of both the tube camera and the CCD camera, a new configuration of electrons is generated each 1/30 of a second. In the tube, these electrons wind up in the electron stream generated by the electron gun; in the CCD, they wind up in a memory unit. From there they are transmitted by wires to video tape recorders or other electronic components.

If the camera is black and white, it has only one tube or chip. If the camera is color, it usually has three tubes or chips working simultaneously; one for greens, one for blues, and one for reds. Some color cameras have only one tube and all three colors, in essence, "take turns."

At present, the CCD cameras do not have resolution that is as good as the tube cameras, however; they have many advantages. They are lighter in weight because the chips are much smaller than the tubes, and they operate better in very dark and very bright environments.[6]

One change that is taking place involving cameras, be they tube or CCD, involves **high-definition TV** (HDTV). This system incorporates 1125 horizontal lines instead of 525. The main advantage of HDTV is that the picture has much better resolution, which makes it look very similar to 35mm film. The HDTV system, which Sony and NHK of Japan have developed, also changes the aspect ratio of the TV picture from four units wide by three units long to a five by three aspect ratio. This means pictures would be wider than they presently are.

HDTV

HDTV cameras have been manufactured and used experimentally, especially for making movies, which are then transferred to film. However, in order for a high-definition signal to reach the viewer at home, all equipment must be able to accept and reproduce the 1125-line HDTV signal, including the consumer's home TV set. Another problem presently precluding the adoption of HDTV is that attempts are being made to develop an international standard so that all countries scan the same number of lines in the same amount of time and, therefore, can easily exchange programs. Some of the differences in scanning techniques of different countries have not been compromised, particularly the fact that European countries scan at 1/25 of a second rather than the 1/30 of the American system.[7]

Mounting Equipment

The camera unit (consisting of viewfinder, lens, and electronics) must be mounted on something so that it can be moved about. When cameras are taken to remote locations, this "something" is often a person's shoulder. Sometimes the person wears a **Steadi-Cam,** a harness device that attaches around the waist and keeps the camera from jiggling.

FIGURE 7.14

(*a*) A tripod. (*b*) A pedestal. (*c*) A crane.
(a, b courtesy of Quick-Set, Inc.; c courtesy of Chapman Studio Equipment)

(a)

(b)

(c)

tripod, pedestal, and crane

When a camera is in a studio, it is likely to be mounted on one of three types of basic mounting equipment: tripod, pedestal, or crane. A **tripod** is a three-legged stand. A **pedestal** is a large tube with which the camera moves up and down by a counterweight, hydraulic, or air-pressure system. A **crane** is a large machine that can raise the camera to a high level.

skycam

Occasionally unique devices are invented for holding cameras. One such device is the Skycam, which enables a camera to be suspended above a site such as a football stadium. It includes cables attached to rooftops and winches to move the camera holder about.[8]

Film Equipment

Some of the material seen on TV is produced using film cameras rather than TV cameras. Obviously, movies that were first shown in theaters and later run on TV utilize the film medium, but so do many dramas, comedies, commercials, and documentaries.

Films are shot using either 35mm or 16mm film. The 35mm film has higher quality because each frame is larger, but sometimes 16mm is good enough quality for material shown only on TV.

All film cameras have a lens, similar to a TV camera lens, that focuses and lets in light. Film is loaded on one reel and run to a take-up reel. In the process, it passes behind the lens opening and records the picture on photosensitive chemicals that are used to store the latent images upon which the lens has focused. After the film is removed from the camera, it is sent to the lab to be processed. basic process

If sound is needed for the film, it is usually recorded on a special tape recorder located near the film camera. The sound is later synced with the picture. sound

Editing film can be accomplished by using film-editing equipment, which enables the editor to view the film and physically cut out unwanted frames and then tape or cement the film back together. Often, however, material shot on film is transferred to video and edited with video equipment.[9] editing

If material is shown on film and edited on film, it must still be converted to electronic material before it can be sent through the airwaves. This is done by projecting it through a **film chain** (also called a telecine), which consists of a film projector, a slide projector, a set of mirrors, and a TV camera. film chain

The film projector looks like any regular film projector, but it is specially designed to adapt film, which is usually shot at twenty-four frames per second, to TV, which operates at thirty frames per second. The projector compensates for the difference so a flutter will not appear on the screen.

The slide projector is used for 35mm slides and usually has a dual lamp assembly that allows slides to be changed rapidly without the blank screen phenomenon that occurs with most carousel projectors.

The mirrors are usually contained in a unit called a **multiplexer.** The purpose of the mirrors is to direct the various film chain inputs to the TV camera. This is accomplised by mechanically rotating or moving the mirrors so that they are in position to deflect the proper pictures.

The camera part of the film chain operates on the same principle as the studio TV camera. It gathers the picture coming from the film or slide projector via the multiplexer and converts it to electrical impulses.[10]

Another method called the **flying spot scanner** is sometimes used to transfer film images to tape. It reads the information more directly than the film chain in that it does not utilize mirrors; therefore, its quality is better. flying spot scanner

New technological developments in the film area have been the cause of controversy. In the mid-1980s, several computerized methods were invented for adding color to black-and-white films. One of these methods involves analyzing films scene by scene by people called **colorists** who freeze frames at the beginning, middle, and end of a scene and then assign colors to the various shades of grey. The computer then "colors" the scene. The controversy has arisen because directors, actors, and other creative people do not want their art, which they created for black and white, changed through coloring.[11] colorizing

FIGURE 7.15

A film chain in operation. *(Courtesy of General Telephone Company)*

Character Generators and Computer Graphics

previous graphics

In the early 1970s, **character generators** (c.g.s) were introduced as a method of creating graphics digitally. Prior to this, graphics, such as opening and closing credits, charts, graphs, and diagrams, were created by graphic artists. To create words, such as titles or credits, the artists would print white letters on black paper. Then one camera would focus on the letters and another would focus on a scene from the TV program being produced. The two pictures would be combined, the black of the paper would disappear, and what would appear on screen would be white letters over a scene from the TV program.

character generators

With the advent of c.g.s, these cards were no longer needed. A character generator consists of a typewriter keyboard that electronically types words onto the TV screen. Usually these c.g.s contain a variety of fonts so that different letter styles can be typed. They also have provisions for changing letter size, position, color, background, and method of display. In addition, some of them have limited ability to draw pictures or charts.

computer graphics

The ability to create complicated pictures digitally, however, came with the advent of **computer graphics,** which experienced large-scale development in the 1980s. Computer-generated imagery sometimes encompasses the ability to type words and numbers and, therefore, includes a keyboard. However, its main function is to create, display, and manipulate pictorial information.

For this reason, the computer graphics devices usually include something to draw with, such as a light pen that creates lines wherever it touches the TV screen, or a **"mouse,"** which can be moved on a separate pad that translates the movements into lines on a TV screen.

FIGURE 7.16

A character generator.
(Courtesy of Chyron Corporation)

FIGURE 7.17

An example of computer graphics.
(Courtesy of Aurora Systems)

As the field of computer graphics has become more sophisticated, it has added a concept known as **"paint boxes."** Artists can select colors they wish to use and mix these colors together, or they can select the size of "brush" they want and lines will appear in different thicknesses. Systems have also been designed that can utilize pictures or TV images that have already been created. For example, a photograph of a car can be fed into the computer and can then be outlined, recolored, or manipulated in other ways. Similarly, one frame of a TV picture can be frozen and operated on graphically. Some computer graphics systems can also create animation, as well as figures that look three-dimensional.

Computer graphics have become very popular, particularly in the creation of commercials. They represent one step in the video process that is fully digital.[12]

Switchers and Digital Effects Generators

Most TV programs produced in a studio or edited in a postproduction configuration incorporate inputs from a number of sources. For example, a studio program might start with a series of slides from the film chain over which c.g. program titles are shown, then change to a long shot of the host on camera 1,

inputs

followed by a close-up of a guest on camera 2. A commercial being edited might start with computerized graphics in the top half of a screen and a pre-taped moving automobile in the bottom half of the screen. In order to accomplish effects like these, a switcher/digital effects generator must be utilized.

functions

A switcher, in its simplest form, consists of buttons and levers that place the proper picture on the air or on tape. For example, if the director wants the picture on camera 1, the person operating the switcher would punch the button labeled "camera 1." When the director wants camera 2, the "camera 2" button would be pushed. This is called a **"take"**—a quick change from one picture to another.

takes

Theoretically, a switcher could consist of two buttons if the only inputs available were two cameras and the only switching desired was "takes." A third button for "black" (nothing) could be added if it is desirable to have a blank screen before the program or at the end of it.

dissolves

If a **"dissolve"** (a slow change in which one picture gradually replaces another) is desired, then three more buttons and a lever can be added to enable a slow transition from one picture to another or from a picture to black.

digital effects

Many switchers are attached to digital effects generators that allow complicated effects such as wipes, turns, inserts, squeezing, stretching, and flipping. These are all possible because a digital effects generator can grab, at any time, any video frame from any video source, change it into digital information, manipulate it in a variety of ways, store it, and retrieve it on command. The digitized signals are inputted to the switcher where they can be mixed with other inputs or sent on to the next stage.[13]

Some of what a digital effects generator does is very similar to what can be accomplished with computer graphics. In fact, sometimes it is impossible to tell where, in the overall TV system, a particular effect was created.

Monitors

The pictures from the various video sources are displayed near the switcher on monitors so that the director can see them and choose the picture he or she wants to go over the air or onto tape.

type

A monitor looks just like a TV set but, unlike a home TV set, it has special electronics so that it can display a picture coming directly from a camera, and generally it has no tuner for selecting different channels. It also has no audio because the sound is heard from audio monitors connected to the audio board.

labels

The monitors are labeled according to their output: camera 1, camera 2, film chain 1, VTR 1, and so on. Usually there will be two monitors that are larger than the others, one labeled "program" and one labeled "preview." The program monitor shows the picture chosen to be on-air or taped at that moment, and the preview monitor allows for setup of the next shot or some particularly difficult special effect.[14]

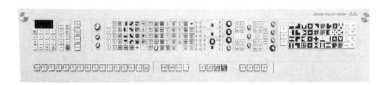

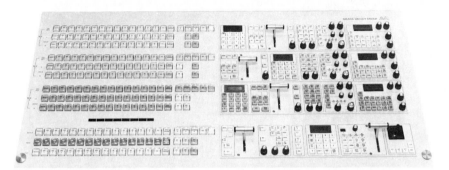

FIGURE 7.18

A switcher. *(Courtesy of the Grass Valley Group)*

FIGURE 7.19

A bank of monitors in a TV control room. *(Courtesy of KHJ-TV, Los Angeles)*

Video Tape Recorders

Video tape recorders (VTRs) are used for a myriad of purposes: taping studio productions, rolling pretaped excerpts into a studio production, taping material at remote locations, executing instant replay, editing program material, airing programs or commercials, or accepting a network feed for delayed broadcast.

FIGURE 7.20

The helical recording
process.

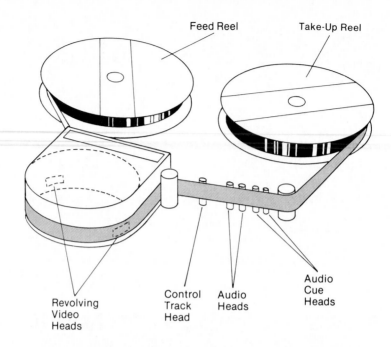

Feed Reel   Take-Up Reel

Audio
Cue
Heads

Revolving
Video
Heads

Control
Track
Head

Audio
Heads

The first VTRs developed were used primarily for taping studio productions that were performed from beginning to end with no stops. That was
because these VTRs were not capable of starting cleanly once they had been
stopped.

Unlike film, the "pictures" stored on videotape are not pictures that the
eye can see, but rather are rearranged iron particles. In this way, videotaping
is similar to audiotaping, but there are some significant differences stemming
mainly from the fact that much more needs to be recorded for a video signal
than for an audio signal—nearly one hundred times more.

If video information were recorded by stationary heads as in audiotape
recording, the tape would have to move at fifty-five feet per second, or nearly
thirty-six miles per hour. A one-hour videotaped program would require approximately 198,000 feet of tape. Obviously, sending tape past stationary heads
will not do.

The first method engineers designed for video tape recording involved
two-inch tape and rotating video heads, putting the information on the tape
vertically instead of horizontally as audio tape recorders do.

Later, engineers developed a method of putting information on diagonally or slant-track; that is, the method used by most modern day tape recorders usually referred to as **helical** recorders. On some models, tape from a
raised supply reel is wrapped around a drum at a slant and then run onto a
take-up reel at a lower level. The drum contains two or four heads that spin
at a rapid speed and place the video impulses on the tape. On other models
the head mechanism is slanted and the tape is straight. Audio information is
laid down horizontally as is control track information, which keeps the picture
stable.

(a)

(b)

(c)

Different helical recorders use different size tape; one, three-quarter, one-half, and one-fourth inch. Obviously, the different sizes are not compatible (i.e., material recorded on a one-inch cannot be played back on a one-half-inch machine). But even within the different formats there is incompatibility. For instance, tapes recorded on a one-half-inch recorder manufactured by Sony cannot playback on a one-half-inch recorder made by Matsushita. This is because the method of tape winding and other parameters are not the same. In general, one-inch recorders are open reel and all the rest use cassettes and are referred to as **videocassette recorders** (VCRs).[15]

incompatibility

FIGURE 7.22

A Sony camera and
half-inch recorder unit.
*(Courtesy of Sony
Corporation of America.
Sony® is a registered
trademark of the Sony
Corporation.)*

During much of the 1970s, most stations used one-inch tape recorders for programs that they taped in the studio or aired and used three-quarter inch out in the field for remote productions, primarily news inserts. One-inch had several different standards, but all three-quarter-inch machines were compatible with each other.

formats

Then in the early 1980s, along came both one-half and one-fourth-inch, but unfortunately both had multiple standards—Sony and Matsushita in one-half-inch and Bosch and Hitachi in one-fourth-inch. In fact, Sony and Matsushita each had multiple one-half-inch formats, some of which were not compatible with each other. Networks, stations, and cable systems liked the smaller formats because of their ease of handling and their quality, which surpassed three-quarter inch, but they hesitated buying because of the differing formats. Finally, CBS, ABC, and CNN chose Sony Betacam and NBC chose Matsushita M-II.

camcorders

These smaller formats were first chosen to be used in the field with **cam-corders**—single units that contain both a camera and a one-half-inch video tape recorder. This type of unit was first introduced by RCA in 1980 and had a quality advantage over the more traditional units because the signal could go directly from the camera to tape without having to go through a wire connecting the two. Some users of the one-half-inch format are so pleased with it that they are considering converting their one-inch studio taping and airing facilities to one-half inch.[16]

Since the advent of videotape, a great many features have been added to the recorders. Tapes can be edited, pictures can be shown in slow motion or still frame, and instant replay is possible.

Instant replay can be accomplished a number of different ways. Sometimes a specially designated VTR will record important elements of an event off the air and then be rewound to play back those elements. Other times one or more cameras and VCR units will be designated for instant replay and the outputs of those units will constitute replays from various angles. In some setups, video disc units, rather than VCRs, are used for instant replay. They have an advantage because on a disc, any part of a recording can be accessed without the delay involved in rewinding tape.

A special form of instant replay called **super slo mo** has also been developed. It uses rapid scanning technology developed for HDTV in order to record three times as much material as a conventional camera-recording system. The one-inch VTRs used in the system record the material at three times normal speed and then play it back at normal speed. The result is stable, smear-free pictures.[17]

Editing has become ultrasophisticated during the past decade, with many programs being taped in bits and pieces and put together later in the editing room. In fact, many programs are now videotaped by means of "film techniques." In other words, one or two cameras will be used to record a program, but each camera will be connected directly to a tape recorder instead of going through a switcher, in much the same way that a film camera is connected to its reels of film. One camera may take continuous close-ups and another continuous long shots. The two tapes will then be edited with the final result being an intermingling of close-ups and long shots.

Often master tapes are dubbed to other tapes, which are then used as work tapes for editing. Both the master tapes and work tapes are imbedded with SMPTE time code, a code that keeps track of each frame. For example, a frame marked 10:32:12 would be into a tape ten minutes, thirty-two seconds, and twelve frames. This time code can be seen on the screen during editing. All editing is planned and executed on the work tapes and programmed through a computer that remembers all the editing points. Then the master tapes are placed in video tape recorders and the computer executes the editing on the masters. All of the back-and-forth maneuvering necessary to determine the exact edit points is accomplished on the work tapes, thereby saving wear and tear on the masters.

Many elements can be utilized in the editing process, including character generators, computer graphics, and digital effects. Tapes can also be edited in a simple cuts-only fashion by going from a source tape recorder, where the original material is, to an editor tape recorder, where the finished program is built.

The next thing that may be on the horizon for video tape recorders is digital editing, which would be a great asset for production. VTRs are presently analogue and by the time the image has been dubbed three or four times, the picture is noticeably degraded. Digital editing would allow almost infinite generations with pictures that still looked like first generation.[18]

FIGURE 7.23

Time code seen over a
picture. *(Courtesy of
California State University,
Fullerton)*

FIGURE 7.24

A videotape editing
console. *(Courtesy of JVC
Industries Company)*

TV Studios and Control Rooms

The switcher, audio board, character generator, and monitors are usually located in a control room, while the cameras are located in the studio. Because
of their high noise level and because they are used for many purposes, the film
chain and VTRs are often located in a separate room away from the studio
and control room.

Usually the control room is situated near the studio but at a higher level control room so that cables from the camera and mike equipment can travel under the floor. In general, TV facilities have at least two studios and control rooms so that one show can be setting up and rehearsing while another is taping.

Most TV studios are a minimum of forty by fifty feet, with ceiling heights studios of about fifteen feet. This height is needed because lighting is used for TV operation and the lights must hang from a ceiling **grid.** Most studios have a curtain around the edge that can serve as a background, a floor that is smooth and extremely hard to permit smooth camera operation, and walls and ceilings that are acoustically treated to prevent outside noise interference. There should be no windows because outside light would make controlled lighting impossible, and there should be a large soundproof door for bringing in scenery. Air conditioning is needed to compensate for the heat of the lights.

In addition, power outlets and outlets for cameras and microphones are needed so that sets can be placed at various spots in the studio with easy access for microphones, cameras, and power equipment. Most studios contain a **loudspeaker system** that enables the director to speak from the control room to the studio before or after taping. During taping, however, a loudspeaker system would be picked up by the microphones and interfere with the audio of the program content. Because it is necessary for the director in the control room to give instructions during taping to the camera operators in the studio, an **intercom** system connects all members of the crew. This system consists of headphones and small microphones so that everyone can communicate with one another.

Master control of a TV station is usually a bustling place. This is often master control where the VTRs and film chains are contained, where program feeds are received, and where programming fare is sent to the transmitter for airing. [19]

Portable TV Setups

Like radio, TV can be produced from remote locations. Truly portable television equipment was developed in the 1970s. Equipment was transported from the studio to remote locations as early as the 1930s, but such equipment was usually contained in two trucks and was a far cry from today's miniaturized equipment.

During the 1950s and 1960s portable equipment could be contained in one truck. When the truck reached its destination, the cameras were wheeled early trucks to the area to be televised and the truck, which contained a switcher, monitors, video tape recorders, audio board, and other gear, was used as a control room. Many station remote trucks contained **generators** so that they could supply their own power. **Microwave dishes** mounted on the roof of the truck could send signals back to the studio. These trucks were used for all types of remote production but were particularly valuable for live or taped sports coverage and for taping events to be shown during news broadcasts.

In the 1970s small portable cameras attached to backpack three-quarter-inch video tape cassette recorders were developed to the degree that one person could carry both the camera and recorder. To make life a little easier, one

FIGURE 7.25

ENG and EFP

camcorders

trucks

person could shoulder the camera and another, walking in tandem, could carry and operate the VCR. A person or persons carrying this equipment had much easier access to fast-breaking news than a crew in a truck. This compact gear was initially used for news gathering and, hence, was called **electronic news gathering** ENG equipment. It was not long, however, before many uses were found for this equipment in various areas of informational and entertainment programming and the term **electronic field production** (EFP) became prominent. This was coupled with improvements in videotape editing that allowed bits and pieces taped with portable equipment to be edited into a unified whole.

Then in 1980 RCA introduced the Hawkeye, the first ENG system to include a camera and a half-inch videocassette recorder as a single unit. This smaller format began its inroad into ENG. The mobility and carrying ease of these camcorders greatly exceeded that of the camera and three-quarter-inch deck.

Remote trucks are still used, particularly for complicated shows in which switching must occur during the broadcast. It would be impossible, for example, to broadcast a live football game using only cameras connected directly to VCRs because some camera selection or switching must occur as the event

FIGURE 7.26

A remote truck.
(Courtesy of ESPN)

is taking place. In this case, studio-type cameras are located at fixed positions and the maneuverable EFP-type cameras roam the area, but all are connected to a central switching system.[20]

Conclusion

Because radio is a less complex medium than TV, it has less equipment. Microphones, turntables, CDs, tape recorders, and audio boards are the main pieces of equipment needed to produce radio programs. TV requires all the equipment needed for radio plus the equipment necessary to produce the video portion of the program—electronic cameras, film cameras, mounting equipment, film chains, character generators, computer graphics, switchers, digital effects generators, monitors, and video tape recorders.

Some of this equipment, such as microphones, turntables, and audio tape recorders, has been around for decades and has undergone changes in quality but not function. Other equipment, such as cameras and video tape recorders, has undergone so many changes that its function has changed primarily in terms of portability. Others, such as digital effects generators, character generators, and computer graphics, are recent developments that are still being explored in terms of ultimate function and features.

Audio boards and switchers have elements in common because they both select from available inputs, mix the inputs together, and send them on. An audio operator who fades one song into another is executing an effect similar to a video dissolve.

Electronic cameras and film cameras have similarities in that both have lenses to focus images and to adjust light. However, film cameras record these images on a chemical base that is then developed into pictures, while video cameras convert the images into electronic impulses—either through a tube-scanning process or through a chip—and then record them on videotape.

Tape recorders, both audio and video, maintain sounds and images by rearranging iron particles on magnetic tape. For audio signals, the tape moves horizontally across a stationary head, but for video, because of the greater amount of information, the tape threads past moving heads that record diagonally. Audiotape is edited by cutting and videotape is edited through a complex dubbing process that often involves identifying frames and using work tapes. Video editing is often assisted by a computer. Computers also play a large role in the operation of character generators and computerized graphics.

Both radio and TV equipment are housed in studios, control rooms, or master control areas and both are used in locations away from the studio. Portable audio equipment is very small and compact. Portable TV equipment continues to shrink in size and weight, although some portable truck setups contain all the elements of a studio.

Thought Questions

1. Do you think that TV equipment will entirely replace film equipment in the future? Defend your answer.
2. In the future, will more TV programs be produced remotely and fewer in the studio? Explain.
3. Do you think CDs will replace turntables? Why or why not?

Distribution

8

Overview

Once a program is produced, it must be distributed in some manner to the viewer. Within the telecommunications field there are a variety of distribution methods, some of which are used in conjunction with one another. These methods range from the very simple pick-up-and-carry bicycle method, whereby a person literally carries a tape to and from locations, to more elaborate methods that involve wires, satellites, microwave dishes, and broadcast transmitters.

These methods are covered by a discussion of the following major topics:

The electromagnetic spectrum, with special emphasis on the portions of it that are important to program distribution

Differences between AM and FM in terms of modulation methods and placement on the spectrum

Types of AM and FM stations

Radio reception

Transmission methods for regular broadcast TV, LPTV, MDS, STV, and teletext

Positioning of VHF and UHF stations

Broadcast TV reception, including some of the newer possibilities such as HDTV, digital TV, flat-screen TV, large-screen TV, and 3D

Transmission by telephone wires and coaxial cable

The wiring setup of cable TV systems

Characteristics of fiber optics

Various uses of microwave relay

The technical functions of satellites and ground stations

The bicycle method of distribution, particularly as it relates to videocassettes and video discs

The messages wirelessed ten years ago have not yet reached some of the nearest stars.

Guglielmo Marconi
inventor

The Spectrum

frequencies

Most systems of telecommunications distribution involve the use of the electromagnetic spectrum. This is a continuing spectrum of energies at different frequencies that encompasses much more than telecommunications distribution. The frequencies can be compared to the different frequencies involved with sound. The human ear is capable of hearing sounds between about sixteen cycles per second for low bass noises and sixteen thousand cycles per second for high treble noises. This means that a very low bass noise makes a vibration that goes up and down at a rate of sixteen times per second. These rates are

hertz

usually measured in **hertz** in honor of the early radio pioneer Heinrich Hertz. One hertz (Hz) is one cycle per second. A low bass note would therefore be at the frequency or rate of 16 Hz, a higher note would be at 100 Hz, and a very high note would be at 16,000 Hz.[1]

As the numbers become larger, prefixes are added to "hertz" so that the zeros do not become unmanageable. One thousand hertz is referred to as one kilohertz (KHz), one million hertz is one megahertz (MHz), and a billion hertz is one gigahertz (GHz). Therefore, the 16,000 hertz high note could also be said to have a frequency of 16 KHz.

radio waves

Radio waves, which are part of the electromagnetic spectrum, also have frequency and are measured in hertz. However, their frequencies are higher than sound waves and they can be neither seen nor heard, but are capable of carrying sound. It is with these radio waves that most of the telecommunications distribution takes place.

Above radio waves on the electromagnetic spectrum are infrared rays and then light waves, with each color occupying a different frequency range. After visible light come ultraviolet rays, X rays, gamma rays, and cosmic rays.

The radio wave portion of the spectrum is divided into eight sections, each comprised of a band of frequencies measured in Hz. Because most of the telecommunications applications are in the megahertz range, these frequencies can best be discussed in terms of MHz.

radio wave divisions

.003 to .03 MHz are very low frequencies

.03 to .3 MHz are low frequencies

.3 to 3 MHz are medium frequencies

3 to 30 MHz are high frequencies

30 to 300 MHz are very high frequencies or VHF

300 to 3000 MHz are ultrahigh frequencies or UHF

3000 to 30,000 MHz are super high frequencies

30,000 to 300,000 MHz are extremely high frequencies

wave length

Sometimes the low frequencies will be referred to as **long waves,** the medium frequencies as **medium waves,** and the high frequencies as **short waves.** These terms refer to the length of the wave rather than to its frequency, although the two are related. When a wave cycles slowly, it can stretch out further than

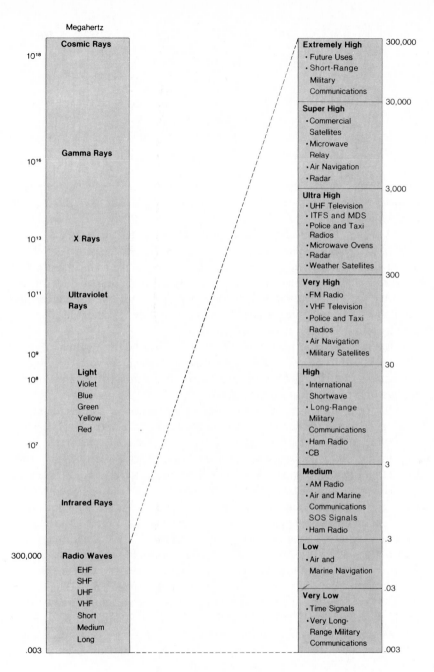

Megahertz

10^{18}	Cosmic Rays	
10^{16}	Gamma Rays	
10^{13}	X Rays	
10^{11}	Ultraviolet Rays	
10^9		
10^8	Light / Violet / Blue / Green / Yellow / Red	
10^7		
	Infrared Rays	
300,000	Radio Waves / EHF / SHF / UHF / VHF / Short / Medium / Long	
.003		

Extremely High — 300,000
- Future Uses
- Short-Range Military Communications

Super High — 30,000
- Commercial Satellites
- Microwave Relay
- Air Navigation
- Radar

Ultra High — 3,000
- UHF Television
- ITFS and MDS
- Police and Taxi Radios
- Microwave Ovens
- Radar
- Weather Satellites

Very High — 300
- FM Radio
- VHF Television
- Police and Taxi Radios
- Air Navigation
- Military Satellites

High — 30
- International Shortwave
- Long-Range Military Communications
- Ham Radio
- CB

Medium — 3
- AM Radio
- Air and Marine Communications SOS Signals
- Ham Radio

Low — .3
- Air and Marine Navigation

Very Low — .03
- Time Signals
- Very Long-Range Military Communications

— .003

FIGURE 8.1

The electromagnetic spectrum.

FIGURE 8.2

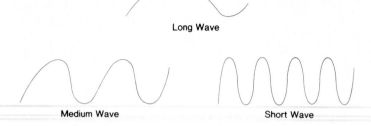

In the same distance, the short wave has a greater frequency than the long wave.

when it cycles quickly because it has more time between cycles. Therefore, the lower the frequency, the longer the wave and the higher the frequency, the shorter the wave.

Although all radio frequencies in the electromagnetic spectrum are capable of carrying sound, they are not all the same. The frequencies that are toward the lower end of the spectrum behave more like sound than do the frequencies that are at the higher end. These higher frequencies, in turn, behave more like light. For example, the lower frequencies can go around corners better than the higher frequencies in the same way that you can hear people talking around a corner but cannot see them.

characteristics of radio waves

Radio waves also have many uses other than the distribution of entertainment and information programming usually associated with the telecommunications industry. Such uses range from the opening of garage doors to highly secret reconnaissance functions.

uses of radio waves

The spot on the radio portion of the spectrum at which a particular service is placed depends somewhat on the needs of the service. The lower frequencies have longer ranges in the earth's atmosphere. For example, long-range military communication is transmitted on the very low frequencies, while short-range communication is transmitted on the extremely high frequencies.

However, many placements are an accident of history. The lower frequencies were understood and developed earlier than the higher frequencies. In fact, not too long ago people were not even aware that the higher frequencies existed and today the extremely high frequencies are still not used to any great extent. AM radio was developed earlier than FM, so the former was placed on the part of the frequency that people then knew and understood. The ultrahigh frequencies were "discovered" during World War II and led to the FCC's reallocating television frequencies after the war into two categories, very high frequency (VHF) and ultrahigh frequency (UHF).

The portions of the electromagnetic spectrum that are of greatest importance to the distribution of entertainment and information programming connected with electronic media are the following:

placement of services

.535 MHz to 1.605 MHz: 107 AM radio channels (this band of frequencies is usually referred to as 535 to 1605 KHz)

54 MHz to 72 MHz: TV channels 2, 3, and 4

76 MHz to 88 MHz: TV channels 5 and 6

88 MHz to 108 MHz: 100 FM radio channels
174 MHz to 216 MHz: TV channels 7 to 13
470 MHz to 890 MHz: TV channels 14 to 83

In addition, portions of the spectrum between 800 MHz and 13,000 MHz (.8 to 13 gigahertz) are used for microwave and satellite communication.

Other portions of the spectrum are obviously important to broadcasting-related operations, too. The area around 2,500 MHz is used for instructional television fixed service (ITFS) and multi-channel multipoint distribution service (MMDS). A large portion of the high-frequency band is used for short-wave broadcasting, primarily to foreign countries.[2]

Radio Broadcast

AM radio broadcasting emanates from the medium frequencies and FM comes from VHF. The position in the spectrum, however, has nothing to do with AM (amplitude modulation) and FM (frequency modulation).

Sound, when it leaves the radio station studio in the form of variations of electrical energy, travels to the station's **transmitter** and then to the **antenna.** At the transmitter it is **modulated,** which means that the electrical energy is superimposed onto the **carrier wave** that represents that particular radio station's frequency. The sound energy cannot go through the air itself because it does not move fast enough. It must be carried on a radio wave that is of much higher frequency.

modulation

The transmitter generates this carrier wave and places the sound wave on it using the process of modulation. Modulation can occur as either amplitude modulation or frequency modulation, regardless of where the carrier wave is located on the spectrum. In amplitude modulation, the amplitude (or height) of the carrier wave is varied to fit the characteristics of the sound wave. In frequency modulation, the frequency of the carrier wave is changed instead.[3]

amplitude modulation
frequency modulation

AM and FM have different characteristics that are caused by the differing modulation methods. For example, FM is static-free while AM is subject to static. This is because static appears at the top and bottom of the wave cycle. Because FM is dependent on varying the frequency of the wave, the top and bottom can be eliminated without distorting the signal. AM, however, is dependent upon height, so the static regions must remain with the wave.

modulation differences

AM and FM also have differences because of their placement on the spectrum. AM signals can travel great distances around the earth while FM signals are more or less line of sight and often cannot be heard if a building or hill comes between the transmitter and the radio that is attempting to receive the station. This is because FM is higher up on the spectrum and closer to light than is AM. Light waves do not travel through buildings or hills, and FM signals are similarly obliterated.

spectrum differences

AM, at the lower end of the spectrum, is not so affected, but these lower frequencies are affected by a nighttime condition of the ionosphere that, when hit by radio waves, bounces the wave back to earth. In this way, AM waves

FIGURE 8.3

Diagram of AM wave.

Consider this to be an electrical wave representing the original sound.

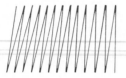

Consider this to be the carrier wave of a particular radio station. Notice it is of much higher frequency than the electrical wave.

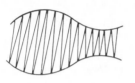

This would be the modulated carrier wave taking the sound signal. Note that the sound signal makes an image of itself and that the amplitude, or height, of the carrier wave is changed—hence, amplitude modulation.

FIGURE 8.4

Diagram of FM wave.

Consider this to be the electrical wave representing the original sound wave.

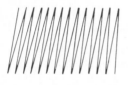

Consider this to be the carrier wave.

This would be the modulated carrier wave. The frequency is increased where the sound wave is highest (positive) and the frequency is decreased where the sound wave is lowest (negative). The amplitude does not change.

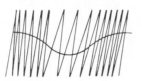

Perhaps this can be better seen by superimposing the sound wave over the carrier wave.

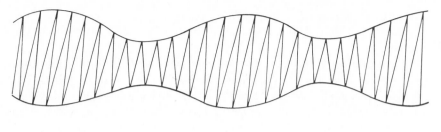

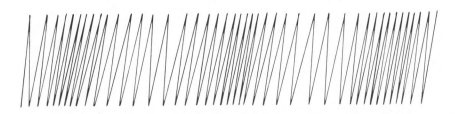

FIGURE 8.5

Comparison of AM
and FM waves.

can be bounced great distances around the earth's surface—from New York to London, for example. Because of this phenomenon, some AM stations are authorized to broadcast only during daylight hours so that they will not interfere with other radio station signals that are traveling long distances because they are bouncing. FM is not affected by this layer because of its position in the spectrum, not because of its manner of modulation. Theoretically, if FM waves were transmitted on the lower frequencies, such as 535 kilohertz, they could bounce in a manner similar to AM.

Another difference between AM and FM is that the **bandwidth** is greater for FM stations. Although each station is given a specific frequency, such as 550 kilohertz or 88.5 megahertz, the spectrum actually used covers a wider area. For an AM radio station this width (bandwidth) is 10 kilohertz and for FM the bandwidth per station is 200 kilohertz. An AM station at 550 kilohertz actually operates at from 545 to 555 so that it can have room to modulate the necessary information and prevent interference from adjacent channels. Because FM stations have a broader bandwidth, they can produce higher fidelity and contain more information than can an AM station.

This makes FM more adaptable to **stereo** broadcasting. An FM wave has sufficient bandwidth to carry more than one program at a time. Therefore, part of the wave can carry sound as it would be heard in the left-hand side of a concert hall and another part of the wave can carry sound as it would be heard in the right-hand side; the result is a stereo transmission. Similarly, quadraphonic sound can be transmitted if four signals are carried on one channel.

AM stereo is also in operation now. Although there were initially several different methods for producing AM stereo, the one that is being adopted most widely is the Motorola system called C-Quam (Compatible Quadrature Amplitude Modulation). For this system, two separate carrier waves of the same

bandwidth

FM stereo

AM stereo

frequency are modulated with separate left and right channels. These channels are then combined for transmission and then later separated in the radio receiver.[4]

SCA

The broader bandwidth of FM enables independent signals other than those needed for stereo and quadraphonic sound to be multiplexed on an FM radio station signal. This is referred to as **subcarrier authorization,** or SCA. The information is carried along on the FM signal, but can only be received by those with special receivers. Under the original FCC SCA rules, commercial FM stations had to offer broadcastlike services, such as music piped to doctors' offices. Noncommercial FMs could offer only educational services, such as reading for the blind, on a nonprofit basis. However, a 1983 rule changed that and now both commercial and noncommercial FMs can carry all kinds of communication services on a for-profit basis. This opens the door for stations to offer such services as paging, electronic mail, and dispatch.[5]

antennas

After the sound waves have been modulated (either AM or FM) onto the proper frequency carrier wave at the transmitter, they are radiated into the air through the radio station antenna at the assigned frequency and power. As noted earlier, because FM frequencies are line of sight, FM antennas are usually placed on high places where they can "see" a large area. AM antennae can operate effectively at relatively low heights.

allocations

The FCC has established a complicated chart of frequency and power allocations to allow a maximum number of stations around the country to broadcast without interference. Between 535 and 1,605 kilohertz there is room for 107 stations of 10 kilohertz each, but there are actually about 4,900 AM stations nationwide. This is possible because many stations throughout the country broadcast from their antennas on the same frequency, but in a controlled way that takes into consideration geographic location and power.

AM categories

The allocation chart divides AM stations into four categories; **class I** (clear), **class II** (secondary), **class III** (regional), and **class IV** (local). Clear channels usually operate with 50,000 watts and serve large centers of population and remote areas. There are only one or two class I stations per frequency and there are forty-five frequencies on which these channels transmit.

Class II stations used to be clear channels, each having its own frequency. But as pressure for more stations increased, the FCC gradually allowed greater use of these frequencies. Now these stations operate from 250 to 50,000 watts serving population centers and adjoining rural areas. There are twenty-nine secondary frequencies that operate in such a way that they do not interfere with the clears.

Class III stations use 500 to 5000 watts and serve cities and rural areas. There are forty-one regional frequencies and more than two thousand stations.

Class IV stations serve local areas and have a maximum power of 1000 watts during the day and 250 watts at night. There are six frequencies for these local channels, each occupied by 150 or more stations.[6]

directional antennas

In order to avoid interference, many AM stations are required to have **directional antennas.** When a signal is sent out from an antenna, it will travel equally in all directions, causing a circular coverage pattern. In order to prevent the signal from going in a direction where it might interfere with another

station, directional antenna systems are set up. Such systems require two towers—one normal tower and another to cancel out the signal coming from the normal tower. By varying the signal sent to each tower, the direction of the station signal can be limited.[7]

FM stations are divided into **class A, class B,** and **class C** stations. Class A stations use power from 100 to 3,000 watts and generally serve a very limited area. Class B stations serve a larger area and use power from 5,000 to 50,000 watts. Class C stations, the most powerful, can use up to 100,000 watts. There used to be class D noncommercial stations operating at 10 watts, but many of these have upgraded to class A because of the 1978 FCC rulings that required the class D stations to upgrade or move to less desirable frequencies. Class D allocations are no longer given out by the FCC.[8]

FM categories

The waves sent out from a radio station antenna are always in the air, but they do not succeed in distributing program material unless someone turns on a radio receiver. The purpose of the receiver is to pick up and amplify radio waves, separate the information from the carrier wave, and reproduce this information in the form of sound.

receivers

The first stage in radio reception involves the antenna, which picks up the signal being sent out by the radio station antenna. From here the signal goes to the **radio frequency amplifier,** which has two purposes: (1) to amplify the signal (carrier wave plus information); and (2) to select the frequency. This is accomplished by a device within the amplifier called a tuner, which makes it possible to tune in the signal of one particular radio station and exclude all other stations. If the tuner were not used, all stations broadcasting in the area of the receiver would be heard simultaneously.

Only the station tuned in gets to the end of the radio frequency amplifier. From here it goes to the **detector,** where the carrier wave is separated from the information it is carrying. This detector then discards the carrier wave, which has already done its job, and feeds the information in the form of electrical energy to the next piece of equipment, the **audio frequency amplifier.** Here the impulses are strengthened and sent to the speaker, where the electrical energy is converted back into sound waves that can be heard.[9]

When radio was first invented, a huge console, which usually sat in the living room, was used to house all the required equipment. Now, of course, it can all be housed in the smallest of receivers that can be worn in the ear, on the wrist, or in a pocket.

cable radio

A home radio receiver need not rely on its own antenna to receive transmitted radio signals. They can be picked up and distributed to subscribers by a cable TV system.

AM radio stations are the oldest form of telecommunications program distribution, but the methods devised many years ago both for modulation and transmission are still quite effective. If the bandwidth were wider, AM radio would be more flexible, but as it stands, it is a useful medium. FM radio, with its higher fidelity, has made great strides in recent years, but it is hampered somewhat by its shorter range and line-of-sight characteristic.

Television Broadcast

modulation

As with radio, the transmitter is the device with which television information is superimposed or modulated onto a carrier wave. With television, audio and video are sent to the transmitter separately—the video signal is amplitude modulated and the audio signal is frequency modulated. The two are then joined and broadcast from the antenna is either monaural or stereo.

scrambling

For subscription TV, which is sent out over the airwaves, the signal is **scrambled** at the transmitter so that it can only be received by TV sets that have descrambling boxes (also called decoder boxes). Although all the bits of information from the television picture are modulated onto the carrier wave, they are placed in a slightly different way from that needed to receive the picture in its proper form without a decoder. A person viewing a scrambled signal without a box will see picture information, but it will have no cohesion.

bandwidth

Again, as with radio, each TV station has a frequency band on which it broadcasts. However, a television station uses a great deal more bandwidth than a radio station. While AM stations span 10 KHz and FM stations 200 KHz, the bandwidth for a TV station is 6,000 KHz or 6 MHz. This means that a TV station takes six hundred times as much room as an AM station and that all the AM and FM stations together occupy less spectrum space than four TV stations. The reason for this, of course, is that video information is much more complex than audio information and, therefore, needs more room to modulate and to protect the modulated information from interference.

placement of channels

TV channels are placed at various points in the spectrum. There is a small break between VHF channel 4 and channel 5 and a larger break between channel 6 and channel 7 that encompasses, among other services, FM radio. In general, in any particular area, two adjacent channels cannot be used because they would interfere with each other. For example, Detroit could not have operating stations on both channel 2 and channel 3. However, because of the break in spectrum between 4 and 5 and between 6 and 7, communities can use those adjacent channels. For example, New York has stations on both channel 4 and channel 5.

Channels 14 through 83 are well above channels 2 through 13 in the UHF range. Because these channels have not been highly utilized, the FCC has given some of the space set aside for channels 71 to 83 to land mobile radio. Even higher yet in the UHF range are the channels used by MMDS (multichannel multipoint distribution service), ITFS (instructional television fixed service) and OFS (operations fixed service). These cannot be received on a regular TV set even though they are broadcast in a manner similar to VHF and UHF stations.

Both UHF and VHF waves follow a direct, line-of-sight path, so both are most effective if located at a high point. UHF signals, because they are closer to light, are more easily cut off by buildings and hills and are more rapidly absorbed by the atmosphere. Therefore, they require higher power at the transmitter to make up for these losses.

UHF and VHF

Although wattage varies considerably, generally VHF stations operate between 10 and 400 kilowatts and UHF between 50 and 4000 kilowatts. Low-power TV, of course, transmits with lower power—10 watts for VHF and 1000 watts for UHF.

wattage

Multiplexing on a TV station signal is not as easy as multiplexing on an FM station signal because of the greater complications and requirements of TV, but this process is still undertaken. For example, **teletext** and **closed-captioning** for the deaf are now broadcast along with the regular station signal. The words, numbers, and figures of these are much simpler than moving video pictures and can be carried in a much smaller space, namely the retrace of scanning. After the electron gun in a camera scans a frame from top to bottom, it turns off momentarily to return to the top of the picture and begin scanning again. The teletext information or captioning is placed in this space where the gun is turned off.

multiplexing

Once a TV signal is sent from an antenna, it must be received in order for the distribution process to take effect. Signals that are transmitted are intended primarily for individual homes, but like radio, they can be picked up by cable TV systems for retransmission. Sometimes they are also picked up by low-power unattended transmitters or **translators,** which then rebroadcast the signal into an area that has poor reception.

distribution

Regardless of whether the signals are retransmitted or not, however, they wind up on a home TV set. The process of displaying the picture on the face of the set is essentially the reverse of its creation. The modulated impulses are received by the home antenna, demodulated, and sent to the TV picture tube and speaker. TV audio is, of course, produced very similarly to audio in a radio receiver. Video is produced by a scanning method similar to that used in cameras.

reception

This method includes electron guns located in the rear of the **cathode ray,** or picture, tube. If the set is black and white, it holds a single **gun.** If the set is color, it contains three guns; one for reds, one for blues, and one for greens. The incoming video signal causes this gun (or guns) to scan the **phosphor screen** in the same manner as the camera gun—left to right, top to bottom, odd lines, then even lines—so that it creates what the eye perceives as a moving picture. When the beam strikes the phosphor layer, it causes the layer to glow according to the intensity of the signal, thus creating blacks, grays, whites, and various shades and combinations of color. A television picture is actually an array of small glowing dots blinking very rapidly in various degrees of brilliance.[10]

HDTV

Many improvements are on the horizon for reception. One of these is a high-definition TV (HDTV) system developed primarily by the Japanese. This system produces a much sharper picture than that on present TV sets because it scans 1,125 lines instead of the 525 lines currently scanned in the United States.

The drawbacks to this HDTV system include: (1) it needs much more spectrum space than the present system; (2) both TV program producers and consumers would have to buy new television equipment to utilize this new system; and (3) an international standard has yet to be agreed upon. CBS proposed a solution to the first problem that would have involved setting aside the direct broadcast satellite spectrum space to establish high-definition TV. In this way, the development of high-definition TV would not interfere with conventional broadcasting. However, this idea was not approved by the FCC, but it has surfaced again in some quarters.

The second problem, that of everyone needing to buy a new TV set, is being worked on by CBS. It is developing a system that will make possible the reception of high-definition signals on conventional TV sets. If this is successful, the changeover of both TV cameras and receivers could be gradual.[11]

The third problem, that of international standardization, is a thorny one. Some people propose that the U. S. and Japan, who have been the primary developers of HDTV electronics based on the parameters of their broadcasting systems, go their merry way and produce and distribute HDTV. However, most people feel that genuine effort should be made to come up with a world-wide standard so that programs can be distributed internationally.[12]

Another system called advanced compatible television (ACTV) has been proposed by NBC. It consists of 1050 lines and theoretically is compatible with present TV and can be broadcast on a regular channel. HDTV has also been demonstrated using two adjacent TV UHF channels. Cable TV systems, videocassettes, and video discs would be ideally suited to distribute HDTV because, through wires, they have the spectrum capacity to handle it.[13]

digital

People who have seen HDTV feel that consumers will be willing to buy the improved quality TV set, especially if HDTV is combined with several other developing technologies. One such technology is **digital TV,** which encodes a regular video signal into one that is less susceptible to distortion and noise. Like HDTV, it uses more bandwidth than the conventional signal, which is presently encoded in an analogue format. Digital technology can be used with high-definition TV, but it does not of itself increase resolution. It simply makes the best signal of what is already there, be it 525 lines or 1,125 lines.[14]

(a)

(b)

FIGURE 8.7

(*a*) A flat-screen TV.
(*b*) A large, screen-
projected TV.
*(a coyrighted by Sony
Corporation; b courtesy of
Valley Cable TV)*

Another related technology is **flat-screen TV.** Much of the technological development in this area involves replacing the bulky TV picture tube with some other source of light, such as the liquid-crystal displays in digital watches. Several configurations of this form of flat-screen TV are now available to consumers in both black and white and color. Other experiments involve reducing the size of the tube and placing it below the screen rather than behind it. Flat-screen TVs can be made very small so that they can be battery operated and carried in a handbag like a paperback book. They can also be made very large and hung on the wall like a painting. In this configuration, they could utilize the improved technical quality of HDTV and digital.[15]

Large-screen TVs are now available, but they operate on a projection principal. They have lenses that project the image from a modified TV onto a large screen so that the image can be seen effectively by a large group of people. Some of these projected TV images are placed on the screen from the front and some are projected from the rear. Some units come in two pieces (the projector and the screen), and others are a single unit. The quality of large-screen TV projection has improved immensely since its introduction in the 1970s, but it, too, could improve with high-density and digital TV.

Three-dimensional television is also a possibility. Experiments have been conducted with 3-D TV for many years, but most companies have placed this type of development on the back burner because the need does not appear

flat screen

large screen

3-D TV

great. When a program is produced for 3-D, two different cameras are placed side by side, each taking a slightly different picture. This is to approximate the eyes, which are able to perceive depth because they are placed slightly apart. When the program material is played back, two images appear. In order for viewers to see only one image, they must wear special glasses, each lens of which filters out one of the images. Sometimes these glasses are different colors, and sometimes they are polarized. Watching a 3-D program without special glasses is very annoying because of the two images. Therefore, some methods that utilize only one image have been tried. With one, the person viewing 3-D wears glasses that delay the picture in one eye, letting it get to the brain later and therefore giving a 3-D effect. Other methods use six different lenses in production and reproduce the picture in a layered manner similar to 3-D postcards.[16]

other advances

Other devices to enhance viewing are in various stages of development. For example, by pushing a button, a picture being telecast can be freeze-framed until the button is pushed again, returning the viewer to the original program. Similarly, short instant replays can be programmed by the set owner, and several programs can be viewed at once by sets that insert different signals in different corners of the tube.[17]

The future looks exciting for TV reception. The signals sent from the transmitter, modulated onto the station carrier waves, and then sent out by the antenna may soon be received with greater clarity and variety than ever before.

Wire Transmission

Broadcasting that utilizes a transmitter and antenna is fine for disseminating programming in a local area; however, if the programming is to go a great distance, some other form of distribution must be used.

special telephone wires

One such form of distribution is copper wires. Telephones are the main users of wires, and now that data services are available, both voice and data are large customers for telephone wires. Special telephone wires that give higher fidelity than regular telephone wires can also be used for audio transmission. For many years, radio network programming was sent across the country on such wires, but now much of it has gone to satellite. A wire system is augmented by amplifiers so that the signal remains strong. AT&T owns most of the telephone wires that cross the country, so networks or other entities that wish to use the wires lease time on them from AT&T.

Wires can also be used in the production process to distribute audio signals. When a radio station wishes to broadcast a sports remote, it rents phone lines so that the game announcing can travel from the stadium back to the station. On an even simpler level, microphones and turntables are connected to audio boards by wires.

If a station is not located at the same place as its transmitter, the program can be sent over wire from the station to the transmitter. Videocassette recorders and video disc players are connected to TV sets by wires.

Phone wires are not adequate for television signals because they cannot carry the amount of information needed for video. A special type of wire, **coaxial cable,** has been developed for television. Coaxial cable, amplified periodically, can transport television programs great distances and can also carry programs from a station to a transmitter or from a remote location to the station. It is also coaxial cable that carries the TV signal to various apartments in SMATV setups.

One of the most common uses of cable, however, is in cable TV systems. A cable system consists of a headend where all of the inputs, such as local TV stations, local radio stations, pay-cable services, and basic cable services, are received. At this headend, all the services are put on wire that is either buried underground or hung on telephone poles. This wire can then be taken into a subscriber's home and connected to the TV set.

Because the wire is physically in the home, signals can be sent back up the wire to the headend, enabling cable to be two-way or interactive. Although coaxial cable must be shielded, signals on the cable are not as subject to interference as signals traveling through the airwaves. Therefore, it is possible to put a large number of signals in a fairly small wire and use all the channel numbers. In other words, both channels 2 and 3 can be used for cable TV, though not for broadcast TV.

The signals of the various services must be processed at the headend so that they can be assigned to a particular channel. For example, if the local cable system decides to put Home Box Office on channel 6, the Home Box Office signal must be encoded so that it is recognized by the TV when the knob is turned to channel 6. It would certainly be possible for a cable TV system to change a city's station from one channel to another. In other words, a station that is broadcast as channel 4 could be on channel 37 of the cable connection. Recently, some cable channels have been moving independent stations to the very high, little-watched numbers and putting cable network services on the lower channels. This repositioning has angered the independent stations.

Most cable TV systems have channels far in excess of what is available on the TV set. In order for all of these channels to "talk to" the TV set, the customer must be given a **converter box** with all the channel numbers so that he or she can call up the desired channel.

The engineering involved in designing cable, converters, and encoding systems for the modern cable TV system is complex, but the challenges are generally being met effectively.[18]

Looming on the horizon are **fiber optics,** which could completely revolutionize wire transmission. With this invention, information in the form of light can be sent through an optic strand that is less than a hundredth of an inch in diameter. Because this strand is made of glass, it is relatively inexpensive, lightweight, strong, and flexible. A fiber optic cable that is less than an inch in diameter can carry 400,000 phone calls simultaneously—ten times the amount that can be carried on conventional wire. It can also carry audio and video signals very effectively. Using light makes transmission less susceptible to electrical interference than coaxial cable or telephone lines.[19]

FIGURE 8.8

A fiber optic. *(Courtesy of Corning Glass Works)*

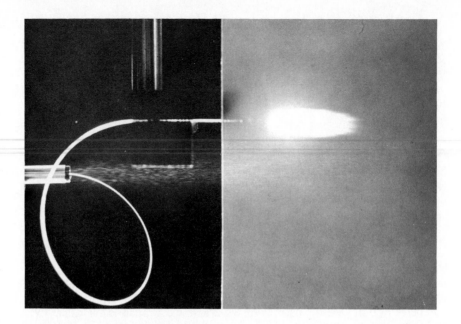

Fiber optics have already proven themselves in a number of applications. ABC used them for the 1980 Winter Olympics and the 1984 Summer Olympics; they link the various pavilions at the Walt Disney EPCOT Center in Florida; cable TV systems have installed fiber rather than coaxial cable in a few places; the Canadian phone company has a link that is over 3000 kilometers long; and American phone companies are beginning to replace old cable with fiber optics.[20] As of 1985, fiber optics accounted for 6 percent of intercity telephone transmission, but that number was expected to increase rapidly.[21] Plans were also being made to lay fiber across the Atlantic Ocean.[22] Although phone lines for audio and data and coaxial cable for video are well established as devices for distribution, they could eventually be replaced by fiber optics. In addition, satellites are losing some of their business to fiber.[23]

Microwave

Stringing cable is not always the most effective way to accomplish long-distance distribution of programming material. Installing cable in mountainous, uninhabited, or inclement terrain is not efficient in terms of initial installation or maintenance.

relays

The most common method for distributing television network signals across the country for many years was through a microwave relay link, but this, too, has swtiched to satellite. Microwaves are very short waves higher up in the spectrum than are radio or TV station allocations. They are line of sight, so relay stations must be in sight of each other and not more than about thirty miles apart. Each microwave station across the nation is mounted on a tower or tall building that occupies a relatively high place. The first station in the

(a)

(b)

FIGURE 8.9

(*a*) A microwave tower. (*b*) A cable TV headend with a microwave tower and satellite dishes.
(a courtesy of AT&T Long Lines; b courtesy of Valley Cable TV)

chain sends the signal to the next station, which receives the signal, amplifies it, and sends it on to the next station. The last station need only receive the signal.[24]

Along with the telephone lines, AT&T owns the microwave links and the users of the microwave links lease time from AT&T. For this reason, no user relays material indefinitely on any particular microwave frequency. AT&T decides which frequencies are available for its various customers. In general, microwave relays are common in rural areas and coaxial cable distribution is common in urban areas.

Microwave can be used for purposes other than cross-country distribution. Stations often place a microwave dish on the top of a remote truck to send the signal back to the station. Cable systems often import distant signals by way of microwave.

remotes

Cable systems have yet another use for microwave. If a cable system covers a large area, it will often microwave relay its signals to various points within its franchise area called hubs and then wire from these hubs. In other

hubs

Distribution   225

FIGURE 8.10

This hypothetical franchise area shows how the headend and hubs relate to each other.

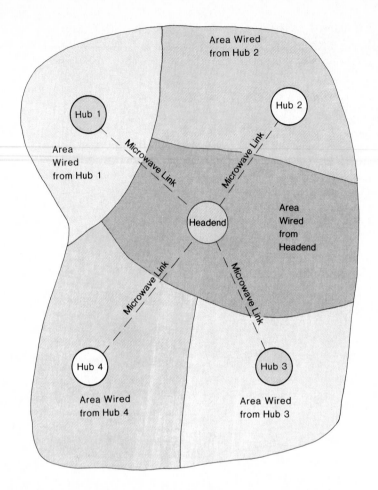

words, the cable system will receive the various signals from satellites, transmitters, microwave stations, wires, and the like at its headend and then retransmit those signals in two different ways. It will send them by wire to the homes located near the headend and will send them by microwave to hubs in areas distant from the headend. At the hub, a microwave receiver will pick up the signals that will then be sent by wire to homes in that location.

Satellites

Satellite distribution is essentially glorified microwave distribution. Instead of sending signals from one point on the earth's surface to another by microwave, signals are sent by microwave from the earth to a satellite in outer space and back to the earth again. The satellite can be powered by energy that it gathers from the sun.

compared to microwave

Satellite distribution has many advantages over microwave relay, one of which is that the cost of satellite distribution is less than the cost of conventional microwave distribution. Also, when waves travel through outer space,

FIGURE 8.11

ESPN's satellite
ground station dishes.
(Courtesy of © ESPN)

they do not encounter all the interference and inhospitable weather conditions that waves traveling through the earth's atmosphere encounter. Once a wave leaves the earth's atmosphere, it is virtually home free for the rest of its trip to the satellite and back again to the earth's atmosphere. Furthermore, with terrestrial microwave, a signal can travel only from one microwave relay station to another. Once a signal has reached a satellite, it can be received by any earth facility that is equipped to receive that signal. In other words, microwave is point to point and satellite is point to multipoint.

Being equipped to receive the signal means owning a **ground station** (also known as an earth station) satellite dish positioned in such a way that it lines up with the signal being sent from the satellite. Signals are sent to the satellite by large ground station dishes and are received on the satellite by individual **transponders.** Each satellite contains about twenty-four transponders, which are roughly analogous to channels. A satellite with twenty-four transponders can receive twenty-four different program signals at one time.

These transponders then transmit the program signals back to earth, where they are picked up by the ground station satellite dishes. The dishes that send information to a satellite are more expensive than the dishes that only receive, so reception points, such as cable TV headends, usually buy receive-only dishes.

The satellites are positioned 22,300 miles above the equator. These **synchronous satellites** travel in an orbit that is synchronized with the speed of the rotation of the earth, thus appearing to hang motionless in space. In this way they can continually receive and send signals to the same points on earth.[25]

The satellites that transmit to the United States are positioned along the equator between 55 and 140 longitude. From there they can cast a signal, called a footprint, over the entire United States.[26] Satellites positioned at other longitudes have **footprints** over other sections of the world. Most of the surface of the globe can be covered by three strategically placed satellites. In this way,

ground station

transponders

orbits

footprints

instant worldwide communication is possible through a worldwide satellite network and, of course, pictures can be beamed from anywhere in the world to the United States.

The first three companies to launch satellites were Comsat, Western Union, and RCA, which named their satellites Comstars, Westars, and Satcoms respectively. Later other companies, such as Hughes and GTE, entered the satellite launch business.[27]

uses

Satellites are, of course, used for many purposes other than communications, including military and weather-gathering applications. The early communication uses of satellites included the transmission of such noteworthy international events as Pope Paul's visit to the United States, splashdowns of U.S. space missions, track meets in Russia, Olympic games, and international talk shows. The broadcast networks used this material either as special programs or as inserts for news broadcasts.

network remotes

Gradually, other uses were found for satellite interconnects. Networks began using them to send remote programming, such as a football game in Missouri, back to the network headquarters in New York, where it could then be sent to affiliate stations through the network's microwave link leased from AT&T. In the same manner the Johnny Carson show was sent by satellite from Burbank, California, where it was produced, back to New York, where it was sent by microwave back to KNBC in Burbank and, of course, to NBC affiliates in other cities.[28]

In the late 1970s the Public Broadcasting Service and National Public Radio became the first networks to distribute all their programs to their affiliates by satellite.[29]

The Robert Wold Company leased large blocks of satellite time and then, acting as a broker, sold the time to companies who wanted to use satellites on a one-time or limited basis. For example, U.S. Tobacco, which specializes in the sale of snuff and chewing tobacco, identified a number of cities that it felt were prime potential sales areas. Through Wold, which rented the tobacco company not only satellite time but also sending and receiving dishes, it joined together stations in those cities for the sponsored telecast of the Fort Worth National Rodeo Championship.[30]

However, it was HBO's placement of its pay service on RCA's Satcom I in 1976 that opened the floodgates. Once the cable industry took to satellite, the demand for transponders well outstripped the supply.[31] When Satcom III, launched by RCA in December 1979, was lost, the cable industry companies that were planning to place their services on that satellite complained so bitterly that RCA leased time on other companies' satellites and re-leased it to the cable companies.[32]

Syndicators of such programs as "The Merv Griffin Show" and "Hour Magazine," who previously had mailed or hand carried tapes to stations, began simultaneously distributing by satellite and convincing the stations who carried the program to buy satellite receiving dishes.[33]

Some individuals bought satellite receiving dishes for their backyards or rooftops and were then able to receive most of the signals that were being transmitted by cable services, syndicators, networks, and others.

Several new radio networks sprang up and delivered their programming exclusively by satellite.[34] Established networks, both radio and TV, converted from phone lines or microwave relays to satellite in order to deliver to their affiliates.[35] Local stations got in on the act by using satellite feeds for local news. A new term, SNG (**satellite news gathering**), was bandied about the newsrooms because local stations could now easily access national and international events and give them a local twist. For example, if a plane that crashed in Chicago had survivors from Atlanta, an Atlanta TV station could send a reporter to Chicago to beam back a story with local interest live via satellite.[36]

All of this further heightened the demand to the extent that the transponders on the first Hughes satellite, Galaxy 1, scheduled to be launched in 1983, were all spoken for by 1981.[37] Eventually, enough satellites were launched that supply equaled demand and the rush for satellite space subsided.

In fact, the satellite business began to turn a bit soft. The people who needed transponders had them and didn't need any more. DBS became stalled and did not need satellite technology. NASA's launch disasters in the late 1980s greatly increased the cost of launching a bird because companies had to use private launching contractors who charged twice what the government-subsidized NASA had charged. In addition, fiber optics began to erode some of the satellite business that was point to point as opposed to point to multipoint.[38]

public broadcasting

brokering

cable TV transmission

syndicators

backyard dishes

radio

SNG

high demand

lower demand

During their short history, satellites have become an important method of program distribution. They have been placed in synchronous orbits with specific footprints and their transponders can both receive and transmit to earth stations. Uses for satellites have gone from occasional special pickups to twenty-four-hour program services. Although they have had problems of late, they will no doubt remain an important element in the distribution process.

Pick Up and Carry

From the ethereal to the mundane, many forms of program distribution involve the physical transporting of a program from one place to another. Usually this means that one or more human beings on foot or in some sort of vehicle actually pick up a tape and carry it. One term used to describe this type of distribution is **bicycling** because, at one time, messengers on bicycles did carry material from one point to another.

educational TV

The old educational television network that mailed copies of tapes from one station to another was using bicycling. Some cable TV access programming currently involves bicycling in that the community people who produce the programs hand carry the tapes to several different cable TV systems, where they are placed on video tape recorders and cablecast on access channels.

cable TV

videocassettes

One of the best examples of pick up and carry is the distribution pattern for videocassettes and, to the extent that they exist, videodiscs. The program material is duplicated onto many cassettes or discs, and these are then shipped to wholesale houses or retail stores. Consumers, on foot, in a car, or perhaps even on a bicycle, go to the store, buy or rent the program, and take it home and put it in their machine.

Although the pick-up-and-carry method of distribution is not as glamorous as satellites, microwave, coaxial cable, optical fibers, or broadcast transmission, it is often the most effective and least expensive means of distribution.

Conclusion

Many methods of distribution of program material are currently in use. Most of them make use of the electromagnetic spectrum that encompasses a continuum of frequencies. These frequencies have varying characteristics and uses that involve terms such as hertz, VHF, UHF, short waves, AM, FM, modulate, line of sight, bandwidth, stereo, multiplexing, clear channels, and carrier waves.

Often several methods of distribution will be used in conjunction with one another before the programming gets from the producing agency to the consumer. For example, a radio station may send a program from the station to the transmitter by wire and then send it out the antenna through the airwaves where it can be received by radio sets in homes and cars. This same

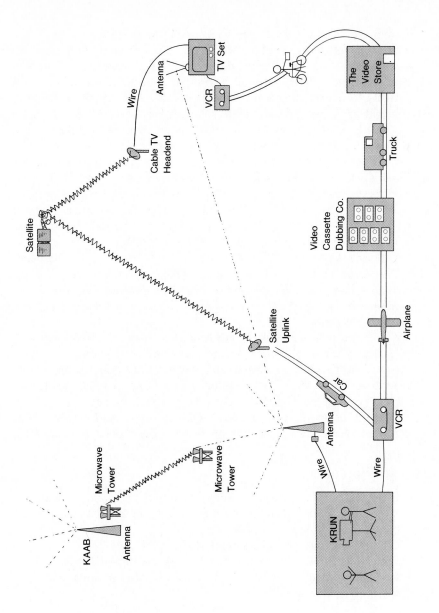

FIGURE 8.13

A distribution pattern.

TV Set

Antenna

Wire

VCR

The Video Store

Cable TV Headend

Truck

Satellite

Video Cassette Dubbing Co.

Satellite Uplink

Airplane

Car

VCR

Microwave Tower

Microwave Tower

Antenna

Wire

Wire

KAAB Antenna

KRUN

program may be sent to a satellite and received by other stations with satellite receivers; these other stations then send it out their transmitters to radio receivers in their areas. The program might even be duplicated onto an audio cassette and copies sold in stores by means of bicycle distribution.

Cable TV reception and distribution probably illustrate the ultimate in combining distribution systems. At the cable headend, local AM and FM radio station signals are received by an antenna as they are transmitted through the air. Sometimes a cable system will also pick up a multiplexed signal of an FM station, particularly if this signal is in the public interest, such as readings for the blind. Local TV stations will also be received by an antenna. If the community has subscription TV or teletext, this, too, may be received by antenna.

UPI cable news and various other video data information can be received through wires that also come to the headend. One or two distant stations may be received by a microwave link involving two or three microwave relay stations that bring the signal from the distant station to the cable system headend.

The headend will also have at least one satellite dish to receive pay-cable, superstations, and other cable network services. The headend may also include video tape recorders to which members of the public carry their access programs for cablecasting.

All of these inputs come to the headend where they are converted in such a way that they can be placed on the various cable system channels.

Now the inputs become outputs and they are sent through coaxial cable to subscribers' homes. If the cable system covers a large area, the signals may be microwaved to a hub and then sent through cable to homes.

With all the distribution processes in mind, try to describe what is happening to the exercise program being produced in the drawing in figure 8.15.

Thought Questions

1. If frequency modulated radio were placed in the medium frequencies rather than VHF, would the technical quality of radio be better than the present AM or present FM? Explain.
2. What will be the future for satellites? For fiber optics?
3. Which of the reception improvements—HDTV, digital TV, flat-screen, projected TV, or 3D—do you feel will be most important to consumers? Why?

Programming

The product of the electronic media is programming. It is the antics and actions of people and objects that constitute the reason for existence. Ranging from the ridiculous to the sublime, this programming is executed by human beings with ordinary intelligence and ordinary prejudices. A select few program for the many, but the many, by a simple flick of a switch, affect the decisions and even the careers of the few. Programs inform, teach, entertain, or merely occupy time, but they are often the basis for conversation, interaction, or even decision making.

PART

IV

Entertainment Programming

9

Overview

Entertainment programming is the predominant form of programming in both radio and television. And yet, it is often difficult to distinguish between entertainment and information. A talk show contains information, but people watch it mainly for the enjoyment of seeing celebrities. Sports programming shows real events as they are happening, and yet many people watch to be entertained or exasperated. Children's programming can be entertaining, informational, or both.

The program forms covered in this chapter are those primarily intended to be entertaining. In that regard, the following topics are covered:

The role of music on radio

The functions of music licensing organizations

The role of music on TV, including the phenomenon of music videos

Controversies surrounding classical music and rock lyrics

The elements of broadcast drama

Controversies surrounding sex and violence on TV

The format and changes of comedy shows

Problems of producing variety shows

The use of specials

The various stages of movies on TV

Talk shows through the years

The evolution of quiz shows and game shows

Changes and similarities in soap operas through the decades

Changes in children's programming

Controversies revolving around children and TV

Watching television is like making love—not a reasoning activity.

Television Quarterly

Music

Music is the mainstay of radio, with disc jockeys' chatter and platters filling the airwaves. A listener who is patient enough can uncover just about every type of music imaginable.

Because of the impact that radio airings have on record and CD sales, large stations are usually deluged with promotional copies and do not need to pay for them. Smaller stations do buy records and CDs, but often at reduced rates.

All stations, however, do have to pay for the right to air the music through arrangements with the American Society of Composers, Authors, and Publishers (**ASCAP**), Broadcast Music, Inc. (**BMI**), and the Society of European Stage Artists and Composers (**SESAC**). These are music licensing organizations that collect fees from stations in one of two ways.

ASCAP, BMI, SESAC

One way is called **blanket licensing**—for one yearly fee, a station can play whatever music it wants from the license organization without having to negotiate for each piece of music. The other way is called a **per-program fee** whereby the station pays a set amount for each program that utilized music from the licensing organization.

blanket and per-program fees

All of these fees have been historically controversial, both in terms of amount and how they are determined. Radio has been happier with the situation than television. Because they air so much music, radio stations generally opt for the blanket license, which is based on a percentage of the station's revenue. Having paid that, the station can air whatever music it wants that is controlled by the music licensing company without further negotiations. The only controversial part is the percentage of revenue the stations must pay. This fee is frequently renegotiated between the music licensing organizations and a group called the All-Industry Radio Music License Committee, which represents radio stations. Presently the rate is about 1.4 percent.[1]

radio

The per-program rate has a more complicated base of pay determination. Generally a station has to pay about fifty times its highest one-minute advertising rate plus a percentage of the revenue it receives for the particular program on which the music is to be aired. This amounts to about three times what the blanket license fee would be for the same period of time. However, for stations such as all-news stations, which use very little music, it is less expensive than a blanket license. Generally the radio stations that choose per-program licenses use so little music that they do not find the fee offensive.

Such is not the case with television. TV stations do not like blanket licenses because they use very little music in the local programs they produce. They see their fees for music constantly increasing, even though their use of music does not. This is because the fees are tied to the ever-increasing revenues of stations. In addition, most of the music heard on TV stations is on programs produced by production companies who have already paid a one-time, **"needle-drop"** fee, which enables them to distribute and exhibit the programming with the music in it. The stations feel they have, in one way or another, paid for the programming and should not need to pay again for the music contained in it.

TV

At one point the TV stations took ASCAP and BMI to court over the legality of blanket licensing, but in 1985 the Supreme Court ruled that blanket licensing was legal.[2] Since that time the music licensing organizations, the courts, and the All-Industry Television Station Music License Committee have been trying to decide on an equitable method by which TV stations should pay for the use of music. In the meantime, stations are paying at the 1980 rates.[3]

other entities

Other entities besides commercial radio and TV stations pay money to music licensing organizations. These include bars, restaurants, concert halls, public TV stations, and some of the newer media. Usually the licensing organizations do not bother a new entity while it is in the formative stages. Public access, for example, does not need to pay music fees. Some cable systems that have healthy local origination pay fees and BMI has started to collect from low-power TV stations.[4]

distribution of money

ASCAP, BMI, and SESAC distribute the money they collect to composers and publishers in accordance with the number of times the music has been aired. The top hits, naturally, gain the largest percentage of income. In order to determine which pieces are aired most frequently, each licensing organization periodically surveys a representative sample of stations and asks them for their **play lists.** The information from these lists of records and the number of times they were broadcast is then fed into a computer, which determines the pay rate for each piece.

Obviously, one licensing agency rather than three could handle this chore, but three have evolved, mainly to ensure proper competitive practices. In the early days of radio, only ASCAP existed. When it raised its fees to an extent that radio stations considered exorbitant, the broadcasters countered by

history of licensing organizations

forming BMI. The idea behind BMI was that stations would play only music by composers and publishers represented by BMI, circumventing the need for ASCAP. However, ASCAP ceased its high rate demands and most stations now play music represented by both ASCAP and BMI and pay licensing fees to both. ASCAP and BMI try to woo successful composers through special financially rewarding contract provisions, so musicians profit from having the two competitive organizations. SESAC primarily represents foreign and religious music; therefore, many stations do not bother to contract with it. In recent years, SESAC has captured a few hits and it is now seen more frequently on station expense records.

company structure

ASCAP, BMI, and SESAC pay only composers and publishers, not performers and record companies. Exposure on radio is assumed to increase record demand, from which both performers and record companies profit. However, record companies and performers generally do not feel this way and sometimes performers compose their own music in order to reap the benefit of the licensing payments. Licensing agencies are nonprofit by nature, so after paying expenses connected with surveys, computer calculations, and personnel, they distribute the rest of the money. The largest number of employees are field representatives who handle problems related to collections and monitor nonsubscribing entities to make sure they are not playing music represented by their licensing organization.

One of the controversies surrounding music on radio involves the lyrics of rock music. In the mid-1980s, a group of Washington parents became particularly concerned about the sexual and violent nature of the lyrics. Some stations became somewhat careful about the lyrics they aired, but most said they did not broadcast the offensive lyrics and that attempts to police what aired amounted to censorship. This continues to be a controversy.[5]

rock lyrics

Historically, music as an entity in itself has had a more minor role in TV than in radio. Most thirty-minute or hour shows by prominent musicians had to become variety shows in order to maintain the attention of viewers. Lawrence Welk was one of the few musicians who was able to survive for a long period with a primary product of music. "Your Hit Parade," performances of the top ten tunes of the week, lasted from 1950 to 1956 but was eliminated by rock 'n' roll, music that did not fit the program's talent or format. Dick Clark's "American Bandstand" is a glorified disc jockey show on which teenagers dance to the current hits.

music on early TV

Public broadcasting regularly airs concerts of classical music from various concert halls around the country. Leonard Bernstein was successful with his "Young People's Concerts," which were a combination of music performance and low-key instruction about music. "Voice of Firestone," a classical music program, maintained a small but appreciative audience from 1949 to 1963, when its cancellation caused a great deal of controversy. Firestone Tire Company wanted to continue sponsoring the program and was willing to pay all bills, but the network canceled the show anyway because it could not deliver a large enough audience to serve as a lead-in to other network programs.

The minuscule role of classical music on both radio and TV is decried by broadcasting critics. Only about a dozen classical music stations are left on the AM band; most of the rest blare with the current Top 40, twang with country-western, or reminisce with yesteryear's Top 40. On the FM band, classical music is found primarily in the public radio portion of the spectrum.

classical music

In several instances, a proposed change in classical music format has brought citizen unrest. In the late 1960s the sole classical music station in Atlanta, Georgia, announced that it was going to be sold to new owners who planned to change its format. Shortly thereafter, WEFM, a Chicago station that had been playing classical music since 1940 made a similar announcement. In both instances, citizens protested the transfers and the issue wound its way through the FCC and the courts, picking up a similar case involving WNCN in New York along the way. Conflicting judgments were issued at various judicial levels concerning whether or not format change should be considered in licensing decisions. In 1981 the U.S. Supreme Court ruled that the FCC need not consider the uniqueness of a radio station's format before granting a license renewal or transfer; it should let the marketplace determine format decisions. This was a setback for cultural organizations and classical music lovers.[6]

The music concept that has had the most success on recent TV is **music videos.** These were begun in 1981 by MTV, Warner Amex's twenty-four-hour-a-day, advertiser-supported cable network. These three-to-four-minute mini-films made to accompany rock music quickly became a big hit with teenagers,

music videos

Entertainment Programming   237

FIGURE 9.1

MTV's taping of the British rock group, Squeeze. *(Courtesy of Warner-AMEX Satellite Entertainment Company)*

enabling MTV to become one of the few cable services making a profit. With MTV's success, rock music videos began to appear everywhere—on networks, stations, other cable channels, videocassettes, and at dance clubs. They were criticized, however, for their sameness, sexism, and violence. The music video business also spread to other forms of music—The Nashville Network initiated country-western music videos and several companies started making music videos for oldies.[7] None of these ever reached the fervor-pitched success of MTV and the popularity of rock videos began to wane by the late 1980s as the novelty wore off.[8]

Drama

Dramatic programs have changed greatly over the decades. Radio was replete with them until TV took over, at which time radio drama essentially disappeared. From time to time various radio programs initiate a revival in an attempt to reawaken public interest and acceptance.

1950s anthology drama

The TV **anthology dramas** of the 1950s, *Marty* and *Requiem for a Heavyweight,* probed character and motivation and emphasized the complexity of life. Although these plays were popular with the public, they became less acceptable to the advertisers, who were trying to sell instant solutions to problems through a new pill, toothpaste, deodorant, or coffee. The sometimes depressing, drawn-out relationships and problems of the dramas were inconsistent with advertiser philosophy and largely led to their demise by the 1960s.

episodic drama

What replaced these anthology dramas were **episodic serialized dramas** with set characters and problems that could be solved within sixty minutes. With series such as "Gunsmoke," "Route 66," "Marcus Welby, M.D.," and "Miami Vice," plot dominated character and adventure, excitement, tension, and resolution became key factors. Westerns, detective stories, mysteries, science fiction thrillers, and medical shows all tended to have good guys and bad guys. Although the main characters were the same week after week, they rarely

FIGURE 9.2

A typical dramatic program expression from Susannah York on PBS. *(Courtesy Mobil Masterpiece Theatre, a BBC Production)*

seemed to profit from lessons learned on previous programs and were as pristine at the end of the episode as at the beginning. Problems of individual episodes could be solved, but never the overall motivation for the series because that would mean the series itself would have to end.

Various forms of dramatic programs show cycles of popularity, with doctor shows being big one year, police shows dominating the next, and lawyers holding the limelight a year later. Most of these forms had precursors in style and content within novels, films, and radio, where danger, panic, pursuit, and climax had held sway for years. However, television called for changes in concept because the small screen demanded intimacy rather than spectacle, few characters rather than many, and reliance on close shots rather than the long shots of movies or the imagination-induced shots of radio.

various dramatic forms

The more probing type of drama has surfaced occasionally with network single presentations, such as *Death of a Salesman* and *The Glass Menagerie,* and public broadcasting series, such as "Hollywood Television Theater" and "Visions." Dramas of longer duration, known as **miniseries,** became popular with the 1975 serialization of Irwin Shaw's *Rich Man, Poor Man* and the 1977 docudrama *Roots,* Alex Haley's saga of his slave ancestors, which was aired eight straight nights to the largest TV audience up to that time. Since then, miniseries have continued to be a dramatic form, but they do not air on TV as often or as successfully as in the past.

Most TV drama is designed for escapism rather than for thought, and it contains predictable chase scenes and predictable plot ideas designed for high ratings. However, there are shows that attempt to deal with social and humanistic problems.[9]

content

One of the main problems encountered by TV writers and networks revolves around the fact that, of all forms of entertainment, drama has the greatest capacity to evoke strong and even disturbing emotional responses in its audience. For this reason, TV drama is particularly susceptible to criticism and censorship within and outside of the industry. Examples of network or sponsor management deleting or rejecting controversial content are numerous, but probably even more numerous are outcries from pressure groups, government agencies, and the public at large. Although presentations regarding politics, bigotry, religion, and other controversial topics have come under fire, the subjects of sex and violence in TV drama conjure up the hottest arguments.

sex

Sexual permissiveness, both in society and on the TV screen, has increased over the years. The low-cut dresses that raised eyebrows in the 1950s are considered modest by today's standards. TV drama has broached sexually sensitive subjects, such as homosexuality, premarital intercourse, and incest, usually to the almost instant outcry of critics. In the end, the heat passes and the networks go on to conquer another sexual taboo.

The newer media are now receiving most of the criticism regarding sexual programming. R- and X-rated movies are available over cable TV and on cassettes. The argument is made by the producers of this material that the programs are not available to anyone who happens to tune in but are distributed in a restricted way to those who wish to pay for them. Opponents argue that such pornographic material is tasteless and should not be available because it leads to greater promiscuity, particularly among the young. They also feel it desensitizes those who watch it to many acts considered socially unacceptable. The battle lines have been drawn and future developments will be interesting.[10]

violence cycles

Violence, even more than sex, appears to recycle in a predictably unpredictable manner. A hue and cry will emerge from various segments of society followed by a TV impoundment of guns, crashing cars, knives, and fists. But to some viewers, nonviolent programming seems bland, and it does not draw the audience that its more violent counterpart does. So gradually the guns and knives reemerge until they are so prevalent that a hue and cry once again arises.

As far back as 1950 Senator Estes Kefauver asked the U.S. Senate if there was too much violence on TV. The first major outcry arose in 1963 after the assassination of President Kennedy. Claims were made that all the violence on TV had led to the possibility of assassination. In 1967 an antiviolence crusade led to an investigation by the Senate communications subcommittee chaired by John O. Pastore.

In 1972 the Surgeon General issued a report that stated the causal relationship between violent TV and antisocial behavior is sufficient enough to merit immediate attention.[11] Hardly anyone paid immediate attention, but a growing protest against violence reached a crescendo in 1976 and 1977 and led to network program changes that led to diminished protest. Part of the 1977 outburst against violence was the result of a court case in which it was alleged that a fifteen-year-old boy killed his elderly neighbor because he watched too much violent TV.[12]

In the early 1980s the subject of violence, coupled with sex and profanity, was resurrected by Reverend Jerry Falwell, head of the Moral Majority, and Reverend Donald Wildmon, who organized the Coalition for Better Television. They took credit for the fact that several advertisers canceled sponsorship of violent programs.[13] More Congressional hearings took place in the mid-1980s, and Senator Paul Simon introduced several bills to curb violence on TV.[14]

Both quantitative and qualitative problems plague the violence debate. Measuring violence is not like measuring cups of sugar. Is pushing someone in front of a runaway cactus the same violent act as pushing someone in front of a car? Should the humorous "pie-in-the-face" slapstick comedy be considered violent? Is it violent for one cartoon character to push another off a cliff when the one pushed soars through the air and arrives at the bottom with nothing injured but his pride? Should a heated argument be treated the same as a murder? Is it worse to sock a poor old lady than a young virile man? Should a gunfight be considered one act of violence, or should each shot of the gun be counted? Should a bona fide news item about a kidnapping be considered?

Despite all these measurement pitfalls, indices abound in an attempt to tell whether violence on TV is increasing or decreasing. One of the oldest violence-measuring systems was developed by Professor George Gerbner of the University of Pennsylvania. For over ten years he has had trained observers watching one week of TV fare a year to count acts of violence according to his complicated formula. Because his count includes just about everything remotely violent, the networks take offense at his calculations, and CBS and ABC have developed their own violence indices.[15] The National Citizens' Committee for Broadcasting, somewhat with tongue in cheek, developed a "violence index" that calculated how many years each network would have to spend in jail if convicted of all the crimes it portrayed in one week—the range was from 1,063 to 1,485 years.[16]

Measurement is not the only pitfall connected with violence. The effect of TV violence on society is also debated and hard to determine. Many research projects and surveys have been conducted, the findings of which have generally been severely (perhaps even violently) challenged by both friends and foes of TV fare. Some of the results that have surfaced are enlightening. (1) People who watch killings and woundings on TV show a greater immediate tendency toward aggressive behavior than do those who watch chase scenes and arguments. (2) People who watch violence on a large screen show greater tendencies toward aggressive behavior than do those who watch on a small screen. (3) The inclusion of humor in a program dampens the tendency toward aggressive behavior on the part of the viewer. (4) There is little difference between news programs that contain only nonviolent news and ones that contain both violent and nonviolent items in terms of increased inclination toward aggression. (5) The more children identify with violent characters in a program, the greater is their inclination toward aggression. (6) There is little correlation between what children consider to be violent acts and what mothers consider to be violent acts. (7) People think there are more bloody scenes on

violence indices

violence effects

Entertainment Programming   241

TV than there actually are. (8) Four out of ten people feel that violence is harmful to the general public and to children in particular. (9) Four out of ten people say that they avoid watching violent shows. (10) Hardly anyone believes that watching violence hurts him or her personally.[17]

Innumerable organizations have joined the battle to curb violence. The PTA held hearings on the issue in eight cities in 1977. The American Medical Association wrote a letter to advertisers urging them to refrain from advertising on violent programs. The National Citizens' Committee for Broadcasting (NCCB) distributed a list of the advertisers who most frequently advertise on violent shows. Consumers' groups have boycotted products advertised on violent programs. Even interindustry groups, such as the National Association of Broadcasters and the Screen Actors Guild, have at times called for a halt to violence.

The entire violence issue tends to generate heat and may remain unresolved in future generations.[18]

Situation Comedy

Situation comedy shows are perhaps the purest form of entertainment in that their aim is to make people laugh. This is not an easy task. It takes strong-penned writers and strong-willed actors and actresses to crank out humorous lines and actions week after week.

The grande dame of situation comedy is Lucille Ball, whose antics will probably live forever in reruns. Others who have made their mark in this form of programming are Henry Winkler as the Fonz in "Happy Days," Robert Young in "Father Knows Best," Bob Newhart on "Newhart," Alan Alda of "M*A*S*H," and Dick Van Dyke and Mary Tyler Moore, first together and then on separate shows.

format

The general successful format for a comedy show is the development of characters who are placed in a situation that has infinite plot possibilities, the creation of complication, the reign of confusion, and the alleviation of the confusion. The problems encountered are usually the result of misunderstanding rather than evil, and the audience can relax because it knows the problem will be solved.

early comedies

The early situation comedies made an attempt to be believable, but the necessity to crank out programs accelerated a trend toward paper-thin characters and canned laughter. One of the mainstays became the idiotic father ruling over his patient and understanding wife and children. The advent of "The Beverly Hillbillies," a family of nouveau riche hillbillies who moved to Beverly Hills where their uncultured life-style clashed with the accepted standards, was cited as evidence of the decadence of TV and perhaps the depth of silly exaggerated situations, slapstick corny plots, and unbelievable characters. But these early comedies live on because reruns of them, and later ones, fill enormous blocks of time on independent stations and some cable networks.

(a)

Drama is a predominate fare on network prime time. In any given week, over twenty different continuing drama series are programmed by NBC, CBS, and ABC between 8:00 and 11:00 P.M. Although they are often criticized for their sameness, the ones that survive for any length of time often develop at least one unique trait—in some cases something that the originators did not plan, but that came about through audience perception. For example, (a) "Moonlighting" developed a sexual tension between Maddie (Cybill Shepherd) and David (Bruce Willis) that kept viewers tuning in. (b) "Dallas" was the first of the truly successful nighttime soap operas and made a "cult villain" out of J. R. Ewing (Larry Hagman). When J. R.'s fate was left hanging over the summer of 1980, "Who Shot J. R.?" became a burning question both nationally and internationally. (c) "L.A. Law" took an irreverant look at the legal profession, which previously had been lauded in drama. (d) "Murder She Wrote" starred a dynamic mystery writer/solver (Angela Lansbury), who was older than such leads had been traditionally. (e) "Cagney and Lacey," starring Sharon Gless and Tyne Daly as two police officers, gave impetus to strong-willed roles for women. (f) "Miami Vice" took the detective series to a new level of glitz—in part due to art design and music. (g) "thirtysomething" became the substitute for a psychiatrist's couch for the "baby boomers" who watched and said, "That's us."

(b)

(c)

(d)

(e)

(f)

(g)

(a)

Informational programming comes in many varieties and receives both praise and criticism. For example, (a) News programs, such as those presented by Dan Rather of CBS, keep the public informed and are a major source of communication during disasters. These same programs have been criticized for being biased, sensationalized, capsulized, tasteless, and trite. (b) Magazine documentaries such as "60 Minutes" present palatable news stories and features, but are few in number and are often criticized for the allegedly biased views that evolve through editing. (c) Sports programs, such as coverage of the Olympics, allow people all over the world to share in the glory of victory, but also overcommercialize and alter sports. (d) Religious programming, such as that conducted for many years by Jim and Tammy Bakker, bring religion to those who are housebound. However, such programming has been criticized for its overly evangelistic nature and for the sex, drug, and financial scandals of the late 1980's. (e) Public affairs programs such as "Meet the Press" present worthwhile, up-to-date material, but are often dull and aired at undesirable times. (f) Political broadcasts, such as candidate debates, inform voters but cause fairness problems in terms of equality for all candidates. There is also the question of how far the press should go when reporting on the personal lives of the candidates. Special interest programs, such as those on the (g) Weather Channel and the various (h) home shopping services, can fill the needs of particular audiences, but also can become repetitive and almost addictive.

(b)

(c)

(d)

(e)

(f)

(g)

(h)

FIGURE 9.3

Bill Cosby, one of the masters of comedy.
(Photo/Viacom, Hlwd.)

A breakthrough in comedy series occurred in the 1970s with the debut of Norman Lear's "All in the Family," whose bigot lead, Archie Bunker, harbored a long list of prejudices. This series, unlike any previous comedy series, dealt with contemporary, relevant social problems and even with politics, heretofore taboo for comedy series. This program and subsequent similar ones raised the status of situation comedy in the eyes of critics and the public alike.

social relevance

In the 1980s two comedy programs stole the show, mainly because they defied stereotypes. "The Cosby Show," starring Bill Cosby, dealt with a black family, but provided situations that anyone could relate to. The other program was "The Golden Girls," which dealt in a very wholesome energetic way with three older women who shared a house in Florida.

antistereotype

Situation comedy still comes in for its share of criticism, however, because of the emphasis on sex, the improper portrayal of members of minorities, too slavish obedience to ratings, and the outlandish financial demands of the stars. Situation comedy, like its dramatic counterpart, is criticized for representing American life incorrectly and for making it appear that all problems can be solved in thirty minutes. This concept was particularly pursued during the 1960s when the first television-weaned generation began demanding wholesale reforms. There were those who felt these young people had been exposed to so much TV that their view of reality was a simplistic one in which all problems could be solved easily and quickly, not allowing for real-world complexity.

criticisms

But the producers of situation comedy shows are still concerned with making people happy. As Norman Lear has said, "I would hope—at least that's the intention—that people turn off these shows and feel a little better for having seen them."[19]

Variety

Variety shows are a hard act to follow—especially for the variety performers themselves. In the days of vaudeville, a stand-up comedian, juggler, or musical group could survive for years by keeping on the move and performing for new audiences in each town. Not so with television. In one prime-time hour, a comedian will have exhausted his or her supply of jokes before an unseen audience of twenty million. What staves off unemployment? Because variety shows absorb jokes faster than writers can write them, juggling acts faster than performers can learn new ones, and musical numbers faster than singers and orchestra can rehearse them, few variety shows have enjoyed longevity.

material exhaustion

The longest running variety show was Ed Sullivan's "Toast of the Town," which was seen on CBS every Sunday evening at 8:00 P.M. for sixteen years. Different talent appeared each week, with Mr. Sullivan giving straight-laced and straight-faced introductions to each. No one ever accused him of being a stand-up comedian, so he did not run out of funny material, but he did know how to put together a first-rate weekly show with wide audience appeal that included such coups as The Beatles and the Singing Nun.

representative shows

One of the most controversial variety shows was hosted by the Smothers Brothers, who rose from obscurity in the 1960s to become a top-rated team in their first season. In a period of political division caused by the war in Vietnam, the brothers leaned heavily on political satire. Constant battles raged over the censorship that the two brothers felt CBS used to stifle their creativity. Amid arguments over edited segments of programs and lawsuits, the brothers were relieved of their show.

The list of entertainers who were on prime-time network variety shows includes Jackie Gleason, Sid Caesar, Dinah Shore, Andy Williams, Carol Burnett, and Donny and Marie Osmond.

Nowadays, most of the variety shows have disappeared from prime time. A late-1980s attempt at a revival of variety programming with a show starring Dolly Parton was unsuccessful. Late night is the bastion for what remains, particularly in "Saturday Night Live," which has its own unique form.

few in number

pay-TV

Early pay-cable adopted a version of the variety show by taping stand-up comics appearing at clubs or engaging in a friendly competition to determine who was the best comic. This was fare that could be taped even though union agreements had not been signed. Later pay-cable continued with stand-up comics, some taped especially for the services.

production

Variety shows, when they are produced, generally use studio time in the networks. Most of the drama and situation comedy shows are produced by independent production companies and sold to the networks, but variety shows are usually a product of the network itself. Part of the reason for this is that these programs lend themselves more easily to multicamera videotape production methods than to the film techniques often employed for drama and situation comedies. The spontaneity of the acts can be best maintained if it can be captured by a bevy of cameras simultaneously, and rarely is there a need for a car chase or outdoor scene that cannot be taped within the confines

FIGURE 9.4

A 1987–88 attempt at a full-scale musical-variety show was ABC's "Dolly" starring Dolly Parton. *(Courtesy of Sandollar Television, Inc.)*

of a studio. Some variety programs are among the more expensive shows to produce because of the lavish costumes and sets, as well as the highly paid talent.

Criticism of variety shows includes their few number and the sameness; they generally revolve around singers and comedy skits, some of which are overly risqué. However, most variety shows provide wholesome family entertainment and they do offer variety within the television diet.[20]

criticism

Specials

"Special" is a word used to designate a program that is not within the regular network schedule. Frequently, variety shows start and/or end as specials. A performer may be brought on to do a special and be such a hit that he or she is given a weekly berth. Or a star that has been performing for a weekly variety show and has run out of material and/or energy may "retire" to an occasional special. Stars who are primarily popular in other entertainment forms, such as radio, records, film, or the stage, often opt for specials.

types of programs

Some of the major names associated with outstanding specials are Barbra Streisand, Frank Sinatra, Bob Hope, Mary Martin, Ethel Merman, and Fred Astaire. Broadway plays restaged for TV are occasional specials, as in the case of *Peter Pan, Annie Get Your Gun,* and *Kiss Me Kate.* There are also old standby specials that have lasted over the years, such as beauty pageants, holiday parades, and entertainment awards shows. Hallmark is a company name associated with high-quality specials. "Live Aid," the rock musical to raise money for the starving in Africa, was its own form of special extravaganza because it employed satellite-fed production from various points around the world.

reasons for airing

Specials are often aired by a network in order to boost sagging ratings. The regular show scheduled to be aired at that time is canceled and the star-spangled special replaces it, usually walking away with the ratings race if the

FIGURE 9.5

A scene from a Bob Hope Christmas special. Bob, Peter Leeds (*center*), and Steve McQueen (*right*) televised this from the Air Force Academy in Colorado. *(Courtesy of Peter Leeds)*

other two networks stay with their regular programming. What often happens, however, is networks will schedule competing specials at the same time, particularly if it happens to be an important ratings week. This often leaves the viewer irritated by feast and famine.

cost

Specials, like variety shows, are usually produced at network facilities and are, if anything, more expensive to produce because sets and props can only be used once. However, the cost usually seems justified to the networks, for they feel they are getting more holler for the dollar.[21]

Movies

Movies on TV is another evolving area in which the emphasis is shifting from local TV stations to networks to pay-cable to videocassettes. Many of the regular drama and comedy series are produced on film, and as such could be considered movies for TV, but they are not generally placed in this category. All movies that are first shown in the theater and then released to TV fall into the movie category, as do films without continuing characters that are made specifically for TV.

local

In the early days of television, theatrical films were the mainstay of local independent stations. With a twenty-year backlog of films just sitting on the shelf, the film studios were happy for this new source of revenue. No union contract had envisioned this bonanza, so at first there were no **residuals** to be paid. But as the use of movies on TV became popular, both the guilds and unions negotiated contracts with producers calling for the payment of residuals. With costs thus greatly increased, the producers turned to the networks,

networks

which obviously had larger pockets than independent TV stations. The phenomenon of movies on TV caught hold in a big way and by 1968 there were movies on at least one network each night of the week. Soon the twenty-year backlog of movies was depleted.

The networks, led by ABC, then began contracting for movies made especially for TV. These are still being produced in fair abundance. Many of these made-for-TV movies are low-budget, quickly produced, grade-B movies, but occasionally one emerges that is good enough to achieve critical acclaim or to make the rounds at movie theaters after its TV debut.

made-fors

Of course, big box office movies still find their way onto TV, but the cost per film to a network is generally in the millions. Such film contracts usually stipulate that the films cannot be shown on TV for a specific period of time after their release—to try to ensure that the TV showings will not divert revenue from the movie theaters. This period of time is referred to as a **window.** A film with a two-year window cannot be shown on TV until two years after it is released to theaters.

At present, three categories of movies are seen on local and network broadcast TV: the oldies that are making the rounds for the umpteenth time, the made-for-TV movies, and the recent releases that are being shown over the airwaves for the first time. Today movies do not account for as large a percentage of broadcast time as they did during the 1960s. Nevertheless, they still make their presence known.

In the 1970s several new outlets for movies arose in the form of pay-TV, subscription TV, videocassettes, and videodiscs. The early local subscription TV services featured movies, but usually were not able to afford the big blockbusters. When Home Box Office's national satellite service began to reap success, HBO began negotiating for the more glamorous films. Once again, union contracts had not envisioned a booming pay-TV market, and the fledgling pay-cable and subscription TV services were able to obtain movie product fairly inexpensively during their initial years.

pay-TV

They were also able to negotiate a shorter window than for commercial TV. As a result, films were sometimes seen on pay-TV within months rather than years of their theatrical release. The home video services—cassettes and discs—eager to distribute movies, were able to negotiate an even shorter window. The general release pattern became movie theaters, then cassettes and discs, then pay-per-view, then subscription and pay-TV, then commercial network TV, then local stations. As the success of movies on pay-TV increased and as union negotiations solved payment and residual problems, the major pay-TV services began ordering made-for-pay-TV movies.

windows

Controversies have arisen concerning the content and presentation of movies on both pay- and commercial TV. One recent controversy involves the **colorization** of films. Old black-and-white films are being made into color films through computerized processes. The distributors ordering the colorizing (led by Ted Turner) are doing so because they believe color films have more audience appeal than black-and-white films. Directors, talent, set designers, and

controversies

FIGURE 9.6

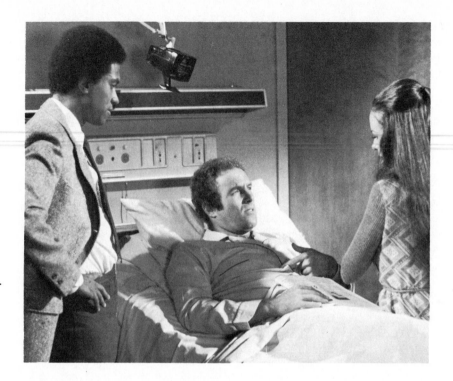

Billy Dee Williams (*left*), James Caan, and Shelly Fabares in the made-for-TV movie, *Brian's Song.* This 1971 movie dealt with the deep friendship of football players Gale Sayers and Brian Piccolo, ending with Brian's fatal illness. It was such a success that it was shown in theaters after being shown on TV—the first such made-for-TV movie to achieve this distinction. *(Courtesy of Columbia Pictures Television)*

other creative people disagree, saying they created the films for black and white and do not want their artistic integrity tampered with by color.

The commercial element of broadcast TV irritates many viewers because the movies are interrupted so frequently by commercials. The pay-TV and cassette businesses attribute a substantial portion of their success to the fact that viewers are so upset by commercial interruptions that they are willing and even eager to pay to see movies without commercials.

Another complaint directed against commercial TV is that the content of movies is so heavily censored to eliminate sex and violence that the films are edited beyond recognition or sensibility. On the other hand, there are complaints that the violence and sexual material present in the movies is much too explicit and that films dealing with homosexuality, rape, prostitution, and similar subjects should be kept off both commercial and pay-TV.

The quality of many made-for-TV, thin-on-plot, heavy-on-action films has also received criticism. This criticism generally does not apply to the home video market, which distributes primarily unedited theatrical films without commercials. Because they are rented or purchased, the content of films in this form do not raise as much ire as when they travel through open airwaves. Occasional cries are heard from theater owners who say that movies shown on any other distribution form hurt their business. In an "if you can't lick them, join them" style, some theaters have set up videotape movie rental facilities in their lobbies. But movies, in some form or another, will no doubt continue to be a staple of television.[22]

FIGURE 9.7

Orah Winfrey interacting with her audience. *(Courtesy of Harpo, Inc./Paul Natkin, photographer)*

Talk Shows

Most TV talk shows capitalize on the average person's desire to know what makes celebrities tick. The shows constantly parade celebrities past hosts or hostesses who attempt to bring out the unusual or peculiar in the guest. The late-night talk shows are probably the best known, with Johnny Carson being the dean of talk show hosts. David Letterman is another of the most popular hosts, primarily because of his unique brand of humor. Of late, an afternoon talk show rivalry has heated up between Phil Donahue and Oprah Winfrey. Their shows utilize not only guests, but reactions and questions from members of the audience.

Radio talk shows often involve celebrities, too, but many are much more audience oriented. Some are devoted entirely to members of the public who call to express opinions or ideas on specific or general subjects. Others feature experts on various subjects who then answer listeners' questions.

The cost of talk shows depends primarily on the quality and demand of the guests and host. Some talk shows are virtually free, for they are beset with requests from aspiring authors, dog acts, one-man bands, and the like, who wish to appear on the program for the free publicity. Other shows pay top price to obtain "hot properties" of the show business and political worlds.

Public access cable TV has also become a fertile area for talk shows. Hosts and hostesses interview friends, acquaintances, business associates, and sometimes celebrities or would-be celebrities in an effort to further their causes or their careers.

Naturally, not all guests on talk shows turn out to have scintillating personalities, so the shows are occasionally criticized for being boring. Some hosts capitalize on abrasiveness in order to get a rise out of guests, and this too is

on TV

on radio

cost

on public access

criticism

Entertainment Programming　　　249

criticized. Some of the subject matter discussed on talk shows seems out of the bounds of propriety to some elements of the public.

Talk shows run the gamut of network produced, station produced, and independently produced. Some radio talk shows draw top ratings in particular markets and TV talk shows, although they are not the largest of drawing cards, tend to draw consistent audiences.[23]

Audience Participation Shows

Audience participation shows have always been popular. Early radio had its quiz shows for both children and adults, and today quiz and game shows abound on TV.

early shows

TV took over the quiz-game program idea early in its history with a 1942 simulcast on radio and TV of "Truth or Consequences," a program on which contestants who could not answer questions had to participate in generally silly activities. Other notables through the early years were "Beat the Clock," "What's My Line?" and "Name That Tune."

quiz scandals

Most of these early shows had modest prizes for the winning contestants, but during the mid-1950s, the stakes began to increase as such programs as "The $64,000 Question," "The $100,000 Surprise," and "The $64,000 Challenge" made their debut. Of course, the 1958 **quiz scandal** gave the quiz-game show area a temporary blow. For a while, no chance-oriented shows dared touch the airwaves, but gradually additional low-stakes programs, referred to as game shows, emerged during daytime hours. The prize for "The Dating Game" was an expense-paid date for the contestant and the person he or she selected. "The Newlywed Game" featured household items for recently married couples who agreed on answers to questions about each other.

present-day shows

Gradually money and expensive prizes crept back in and the game shows dallied into the evening, particularly the early evening hours when local stations have control of the programming fare. The biggest hit in game shows became "Wheel of Fortune," which is the highest rated syndicated show in television history. It brought fame to its card turner, Vanna White.

The gamut of opinion regarding game-quiz shows ranges. Some people think the games are educational because of the information contained in the questions. Others think the games feed on avarice and gambling instincts, and still others believe the games make fools of all the contestants who participate and waste the time of those who watch.

cost

Game shows are among the least expensive to produce. All talent except for the host is free, the set can be used over and over, and the prizes are donated by companies in exchange for mention on the show.

criticisms

The degree of commercialization inherent in these programs is often questioned. Some programs appear to be one long commercial as the merits of the various prizes are revealed. The games themselves are criticized for being inane and childish and for a sameness that seems to permeate most of them. However, many viewers compete or empathize with both winners and losers, and there is never a lack of people lined up to try their luck or skill on big-time TV.[24]

FIGURE 9.8

Hostess Vanna White and host Pat Sajak show a winning contestant his new car on "Wheel of Fortune." *(Courtesy of Wheel of Fortune)*

Soap Operas

Soap operas arose during the heyday of radio and succeeded in dominating the afternoon hours with stories dealing mostly with the housewife struggling against overwhelming adversity—sick and dying children, ne'er-do-well relatives, weak husbands.

Television adopted the soaps, often referred to as "daytime TV," at about the same time other programs switched from radio to the new medium. Many of the original traits were retained: each program is serialized in such a way that it entices the viewer to "tune in tomorrow"; the plot lines trail on for weeks; music is used to designate transition; very little humor is included in the dialogue, as adversity is the common thread. Soap opera characters, unlike their evening dramatic and comedy counterparts, live with their mistakes and are constantly affected by events that happened on previous programs. They also grow old and have children who grow older.

traits

What has changed from the old radio soap opera days is the program content. Although there are still some housewives struggling against overwhelming adversity, the emphasis is now much more on male-female sexual relationships. Infidelity, premarital sex, artificial insemination, mate swapping, impotence, incest, venereal disease, frigidity, and abortion have been added to nervous breakdowns, sudden surgery, and missing wills. For a period of time, subject matter was often tried first on soap operas to determine if it would be fit for evening drama and movies. Soap operas then became part of evening programming with such successful programs as "Dallas" and "Dynasty."[25]

changing content

Sometimes soaps are produced by the networks and sometimes they are produced by independent companies. The daytime soaps are among the most profitable TV ventures, for production is cheap and ads are plentiful. The same paper-thin scenery is used day after day, and because soaps are a world of words and close-ups with very little action, hardly anything is consumed or destroyed. In recent years, some soaps have taped on location, but most are still studio bound. The nighttime soaps have budgets and sets more akin to prime-time drama.

Daytime soap opera stars are paid much less than prime-time talent, a fact that the former often decry because they work at a much more hectic pace. While the nighttime stars are working to crank out one program a week, the talent of the soaps must produce one program a day. Understandably, this leads to some production sloppiness where blown lines are left intact in the aired product. Such incidents are remarkably rare, however, if one considers the time pressures under which the actors are performing.

Soap opera regulars, if they can take the pace, can be fairly sure of long-term employment, for many soaps have survived while there have been dozens of turnovers in the prime-time area. Some of the longest running soaps include "Search for Tomorrow," "Days of Our Lives," "General Hospital," and "The Young and the Restless."

For the most part, daytime soaps are put in a second-class stepchild position by critics and the broadcasting industry alike—mainly because of their air time, cheap production, and maudlin story lines. However, there are those who feel that from a literary point of view, soaps are superior to nighttime dramas and comedies. Relieved of the chore of solving all problems in thirty minutes, the soap writers can explore character and probe motivation in a way that provides viewers with more realistic, albeit exaggerated, situations.

It was assumed for many years that only middle-class housewives and shut-ins made up the audience for soap operas. But in recent years many "closet-case fans" have emerged, including baseball players, nighttime TV stars, college students, politicians, and many men and women who work nights. In fact, a small weekly magazine, *Soap Opera Digest,* which prints capsule plots of each soap on the air that week, has been very successful marketing its product to those who must miss an episode of their favorite soap, but do not want to fall behind the story line.[26]

So, despite the fact that Heather has been jilted at the altar by John, who has discovered that his father is impotent and he is the love-child of an affair between his mother and Dr. Winton, thus making him a first cousin of Sharon, who is in love with Tom, the husband of Tricia, who has just had an abortion in order to cover up her affair with Richard while Tom was out of the country searching for his child of a previous affair who had been put up for adoption—soap operas will no doubt continue.

FIGURE 9.9

A wedding scene from "General Hospital." (© 1987 American Broadcasting Companies, Inc.)

Children's Programs

Never could Sky King, The Lone Ranger, The Green Hornet, Howdy Doody, Kukla, or Mickey Mouse envision the furor that has arisen over children's programming.

The air time for children's programming has not changed since the 1930s. Saturday morning and after-school hours were the domain of the young in early radio days and still are with TV today. However, radio, by virtue of its aural nature, emphasized imagination and sound effects, whereas TV emphasizes sight and action. Modern-day children's radio programming is essentially nonexistent except for a few programs on public radio, most notably "Kids America," a call-in show aired on American Public Radio.

TV networks started children's programming early with an emphasis on puppets, such as Howdy Doody and his real-life friends Clarabell the Clown and Buffalo Bob, and Kukla and Ollie with their real-life friend Fran Allison. The longest running kiddie show on network TV was "Captain Kangaroo," starring Bob Keeshan. Beginning in 1955, it ran Monday through Friday until 1982. In that year, CBS changed the air time to Saturday and Sunday in order to add an hour to the "CBS Early Morning News," which aired on weekday mornings. By 1985 the Captain had left the air altogether.

early programs

Children's programming was important on early local TV stations, too. Most programs consisted of a host or hostess whose main job was to introduce cartoons and sell commercial products. During the 1960s, networks overwhelmingly adopted the likes of "Felix the Cat," "The Roadrunner," "The Flintstones," "Popeye," and "Tom and Jerry"—and therein began the controversy.

For many years these cartoons dominated Saturday morning TV, making for one of the most profitable areas of network programming. The cartoons were relatively inexpensive to produce, and advertisers had learned that children can be very persuasive in convincing their parents to buy certain cereals, candies, and toys. The result was profits in the neighborhood of $16 million per network just from Saturday morning TV.[27]

But gradually the situation changed. Parents who managed to awaken for a cup of coffee by 7:00 A.M. Saturday noticed the boom-bang violent non-educative content of the shows along with the obviously cheap mouth-open/mouth-close animation techniques. A group of Boston parents became upset enough to form an organization called **Action for Children's Television** (ACT), which began demanding changes in children's programs and commercials. Scholastic Aptitude Test scores started going down as the first television generation took the tests. Researchers realized that children under five were watching 23.5 hours of TV a week and that by the time they graduated from high school, they would have spent 15,000 hours in front of the tube. Government agencies also discovered that those nutritious cereals weren't so nutritious after all. The Children's Television Workshop developed "Sesame Street," and its successful airing on public TV proved that education and entertainment could mix.

All of this led to a long, hard look at children's TV. Various independent researchers, most of them at universities, undertook studies to determine the effects of TV on children. In 1969 the U.S. Surgeon General appointed a committee of twelve prestigious researchers to investigate the effects of television crime and violence on children. These twelve worked for two and a half years on the project and came to the conclusion that a modest relationship exists between viewing violence on TV and aggressive tendencies.[28]

The overall body of research on children and TV yielded some conflicting results, but overall seemed to indicate that changes did need to be made in children's TV. Studies have shown that children do learn reading and vocabulary from TV, but the children who watch TV the most are the ones who do poorly in reading in school. Watching TV generally cuts down on book reading, but certain TV programs that refer to books actually increase book reading. Children three and under understand very little of what they watch on TV, and yet they will sit mesmerized before the set. Nine out of ten children between the ages of seven and eleven understand social messages when these are present in programs. Some studies show that children predisposed to violence are more apt to increase violent behavior after seeing it on TV than are so-called normal children; other studies show exactly the opposite. It has been determined that watching TV is an activity involving mainly the brain's right hemisphere, which contains nonverbal, nonlogical, visual, and spatial components of thought. From this research it is theorized that watching TV may hamper development of verbal and logical abilities. One study conducted on highly creative children found that their creativity dropped significantly after three weeks of intensive TV viewing.[29]

Led and cajoled by ACT, a number of other organizations began demanding reforms in children's TV. They took note of the fact that there were some fine children's programs on TV—"The Wonderful World of Disney," "Lassie," and "Mr. Rogers' Neighborhood,"—but they were after changes in the cartoons, slapstick comedy, and deceptive commercials. In 1974 the FCC issued guidelines for children's television. It stated, among other things, that stations would be expected to present a reasonable number of children's programs to educate and inform, not simply entertain. It also stated that broadcasters should use imaginative and exciting ways to further a child's understanding of such areas as history, science, literature, the environment, drama, music, fine arts, human relations, other cultures and languages, and basic skills such as reading and mathematics. ACT was also largely responsible for getting the National Association of Broadcasters to include in its 1973 code an amendment reducing the number of commercial minutes in children's programs from sixteen per hour to twelve; this was later further reduced to nine and a half minutes.

program changes

By the mid-1970s most stations and networks had acquiesced, at least in part, to the reform demands. Programs with names such as "Kids' News Conference," "What's It All About?," "Let's Get Growing," and "Villa Allegre" hit the airwaves. Many of these shows attempted to teach both information and social values, and they contained advertising that was lower keyed than it had been in the past. These were more expensive to produce than cartoons, so the amount networks and stations spent on children's programs increased.

ABC developed after-school specials that dealt dramatically with socially significant problems faced by children, such as divorce and the death of a friend. In the area of cable TV, Warner-Amex established a satellite service called Nickelodeon that consisted solely of nonviolent children's programming. Unfortunately, most of the socially relevant programs did not receive high ratings.

With the 1980s, the tide turned again and the FCC, in the spirit of deregulation, dismissed the idea of maintaining or adopting standards for children's programming. Regulations regarding the amount of commercial time allowed in children's TV shows and the types of programs TV should air were abandoned.[30]

deregulation

The networks, eyeing their sinking children's TV profits with fear, canceled many of the expensive education-oriented programs and returned primarily to the world of cartoons. Some of these cartoons were based on toys and became known as **toy-based programming.** The toy was developed with the idea in mind that a TV series would be developed around it. Some people viewed this programming as a thirty-minute commercial for the toy.[31]

FIGURE 9.10

A scene from "Fraggle
Rock." This animated
series, which airs on
network TV, originated
from the live action
"Fraggle Rock"
created by Jim Henson
and shown on cable
TV. *(Courtesy of Henson
Association, Inc., and Marvel
Productions, Ltd.)*

Citizens groups and Congress complained about all of this,[32] but broadcasters pointed out that children, like adults, need entertainment as well as education and that, given a choice, they will still choose "The Flintstones" over "Young People's News" in the same way that adults choose to watch an old movie rather than a sterling documentary. This phenomenon, they say, is a problem beyond their realm. Parents are the ones who must control the set, and as long as parents use it as a cheap baby-sitter, children's viewing hours will not be curtailed and their habits will not be changed.

By the late 1980s, it looked like the pendulum might swing back. The courts told the FCC it should not have abandoned the rules regarding the number of commercial minutes allowed in children's programs. And, for some unknown reason, the small fry were beginning to abandon the TV set.[33]

The conflict over children's television is likely to continue. Parents and citizens groups will argue that broadcasters should provide quality children's programming, even if it is not profitable because such programming meets the nation's needs. Those in the television industry will defend their programming and their bottom line.[34]

Conclusion

Entertainment programming has both friends and foes. The music available on radio is often applauded because stations, overall, provide variety to meet the tastes of everyone. Music on TV has experienced high points, such as the Bernstein concerts and the Firestone Hour. MTV has been a financial success.

Dramatic programming is sometimes applauded because it provides the escapism that many people need in their lives. There have been probing, humanistic high points in the drama area, such as *Roots* and *Rich Man, Poor Man*.

Situation comedies are healthy in that they make people laugh and have the ability to highlight social problems in a humorous manner.

Variety shows generally provide wholesome family entertainment and lend a change of pace to the television fare.

In a similar manner, specials are complimented because they provide good entertainment and are well produced.

Audience participation shows can be educational, and the networks like them because they are inexpensive to produce.

The characters in soap operas can be well developed because the story line lasts almost indefinitely. People on these programs are affected by their mistakes, making this genre somewhat realistic. Soap operas also give networks a chance to try out ideas to see if they are potentially acceptable for nighttime TV.

Children's programming provides both entertainment and baby-sitting for children. Some of this programming has been of an excellent nature, such as "Captain Kangaroo," "Sesame Street," and after-school specials.

Criticisms expressed toward entertainment programming are more abundant than the applause, but it is always easier to criticize than it is to create.

Music programming is criticized because both radio and TV have essentially abandoned classical music despite court cases to attempt to retain it. Music lyrics are criticized for the sex and violence they encompass. Within the industry, performers and record companies are dissatisfied with the method of music licensing payment.

The glow of nostalgia affects drama in that people criticize current dramatic presentations because they are not as creative or "deep" as the radio and TV drama of the 1950s. The recurring waves of sex and violence are deplored by many organized groups and individuals.

Situation comedy also comes under attack for allowing too much sexual permissiveness. Other criticisms revolve around misportrayal of minorities, oversimplification of solutions to problems, exaggerated situations, and slapstick approaches for getting a laugh.

Freshness is hard to maintain on variety shows and soon they succumb to a sameness that many people abhor.

Networks frequently schedule specials opposite each other, to the chagrin of the viewers.

Movies are often relatively old by the time they are broadcast because of the window used to protect theater owners. Pay services and home video have access to movies sooner, but at a cost to the consumer. The constant interruption of movies for commercials on broadcast TV irritates many viewers, and the battle over too much or too little censorship becomes heated on occasion. The colorization of movies has created an internal battle within the movie industry.

Talk programs are criticized for being dull and sometimes abrasive.

To many, audience participation shows are an inane, childish waste of time that capitalizes on the gambling instincts of the human race.

Soap operas, like many other forms of programming, are criticized for their high degree of sexual content and their often morose nature. The work needed to produce soaps on a daily basis is often hard on the actors and others involved in their production.

Slapstick, violent children's programming is considered not only a waste of children's time but a detriment to their growth and development. Both organized groups and experimental studies provide continual pressure for the upgrading of children's programming.

Entertainment programming will continue to evolve as society itself evolves. Ironically, radio, which pioneered most of the forms of programming that now appear on TV, uses very few of them. Subject matter once taboo on soap operas is now handled in children's programming. Organizations influential in entertainment, such as ASCAP, BMI, ACT, and the numerous production companies, will continue to interact. The forms and quantity of entertainment programming available to the public will no doubt multiply, but the public will still determine success or failure as it sits in judgment.

Thought Questions

1. Does violence on TV adversely affect society? Support your answer.
2. Should performers and record companies receive money from ASCAP and BMI? Explain.
3. Should networks present more specials and fewer series?
4. How can children be encouraged to watch quality programs?

News and Information Programming

Overview

As more and more people turn to television and radio as their major source of information about what is happening in the world, the programming that generally falls within the news and public affairs area develops more social significance. However, it is often an uphill fight for those committed to news and information programming to see their program ideas reach fruition. Not only is such material likely to envelope a station or network in controversy, it is also generally expensive to produce.

News and information programming is not watched by nearly as many people as entertainment programming and, hence, cannot sell high-priced ads—in fact, sometimes it cannot sell ads at all. Because public service programming generally costs three times as much as entertainment programming per rating point, it must depend on the profits of its more glamorous sisters to support it. Yet this area is of great importance in the realm of social significance.

In view of this fact, the following information and controversies are discussed in this chapter:

News-gathering processes

Selection and presentation of news

Compliments and controversies surrounding the news

Problems connected with producing and airing documentaries

The content and controversies of editorials

The interrelationships between sports and the electronic media in areas such as rights fees, overexposure, structural set-ups, athlete ads, and blackouts

The complexities of producing sports programming

The rise of magazine shows

Educational programming and the lack thereof on radio, TV, and cable TV

The history and controversies surrounding religious programming

Pros and cons of public affairs programs

Basic provisions related to programs concerning political candidates

The overall role of broadcasting in the political process

The proliferation of special interest programming

10

Television has learned to amuse well; to inform up to a point; to instruct up to a nearer point; to inspire rarely. The great literature, the great art, the great thoughts of past and present make only guest appearances. This can change.

*Eric Sevareid,
longtime CBS commentator*

News

speed of news

Americans declared their independence on July 4, 1776, but it wasn't until many months later that the British learned of the declaration. Likewise, in the War of 1812, the Battle of New Orleans was fought weeks after the war was actually over, for word had not gotten to New Orleans. Today the slightest little rift between countries can be reported, analyzed, and even blown out of proportion within a matter of minutes. More people today are aware of what is happening in the world than ever before, and basically radio and television can take credit for this. The greatly increased worldwide communication of the past few decades is also painless to the viewer or listener. A mere flick of a dial can bring one up to date on current events or at least ensure that no great disaster has occurred.

In times of disaster it is the electronic media that become the main source of help and information—directing victims to sheltered areas, seeking help from outside sources, communicating vital health and safety information, and calming jangled nerves. The advent of portable equipment and satellite transmission for both radio and TV allows news to be reported rapidly and allows for firsthand reports through actuality interviews— question-and-answer sessions conducted with those involved in the news story.

Gathering news is generally a complex process, with stations and networks depending on a variety of sources for news. The news **wire services,** Associated Press (AP) and United Press International (UPI), have been a traditional source of national and international news from the early days of newspapers. They have specialized services for both radio and TV that include copy written in electronic media style and visuals to accompany the stories. Wire service news is collected by a bevy of reporters stationed at strategic points around the world. These reporters have regular beats, such as government offices and police stations, which they cover to gather news.

wire services

The written stories they gather are sent by satellite, wire, or phone to central AP or UPI offices where they are assembled and often rewritten. Then the stories are typed into a special machine and sent by teletype to stations and networks, where they are printed out on a machine specially made to receive the signals. Stations and networks, along with newspapers and magazines, pay the wire services subscription fees for both machine and news. Visual material is sent by microwave or satellite.

Stations in major markets have their own wire services to help them keep track of local news events. The use made of this wire service news is largely up to the organization that receives it. Some radio stations simply rip the news from the wire service machines each half hour or hour and read the writing on the paper verbatim over the air. Other stations select news items and rewrite them in ways they feel will appeal to their audiences.

In addition to wire services, there are many other news agencies that provide news to stations and networks. Independent News Network supplies stories to TV stations not affiliated with ABC, CBS, or NBC. A number of agencies cover news in large cities, such as Washington, New York, and Los Angeles, and offer this news to stations in other cities.[1]

Then, of course, the major networks—ABC, CBS, and NBC—have their own national and international reporters who cover stories for their network news. Cable News Network also has an elaborate news-gathering structure that makes news available to others besides CNN. All of these agencies operate similarly in that correspondents around the country and around the world gather and report the news and then transmit it, usually by satellite, to its destination.[2] network correspondents

The gathering of local news has also become quite complex. Local stations and local cable TV origination channels have their own news staffs responsible for covering the local scene. In some instances, such as in small music-oriented radio stations, this staff may consist of one person who makes phone calls to local officials. Other stations have numerous crews of people in vans and helicopters covering city council meetings, protests, fires, robberies, and other local news events.[3] local news

The advent of **satellite news gathering** (SNG) has led to many changes in local news coverage. This concept was pioneered by Conus, a Hubbard Broadcasting-owned company, in 1984. Initially it gathered news with trucks equipped with satellite uplinks and sold this to specific stations equipped with downlinks. Stations that wished to could send one reporter to a news site to get the local slant on a story and Conus would transmit this back to the local station. For example, if a plane crash occurred in Chicago, a station in Atlanta could send one reporter to Chicago to report on surviving passengers from Atlanta. Conus encouraged stations to buy uplinks and make the news stories they covered available to other stations through the Conus structure. SNG

Other entities set up regional or national cooperative arrangements. For example, stations in the state of Florida established an SNG network called Florida News Network. With uplinks in five different cities, it provides coverage of the state primarily for stations throughout the state.

As more and more stations purchased **satellite news vehicles** (SNVs), they were able to interchange more news and depend less on the national news-gathering organizations, including the major networks. Stations were also able to go further afield to gather news. The older microwave trucks that stations used to gather news could only transmit about thirty miles, so stations could not cover stories further away than that. Satellite signals can travel throughout the country, so station crews can cover stories in more remote or distant areas.[4]

While correspondents and reporters are gathering worldwide, nationwide, and local news, news writers and producers remain at network, station, and cable facilities reading wire service and newspaper stories, viewing satellite-delivered stories, seeing that footage sent from the field is edited, drafting opening and closing copy, and rewriting stories. writers and producers

Many newsrooms are now computerized so that stories and ideas can be typed into a computer by a writer and recalled from the computer onto a screen in another location by the producer, who can then rewrite or in other ways act upon the story. Computers also aid in providing story ideas and background information, in helping to avoid overlooked details, in keeping track of crew status and location, in distributing messages to the staff, and in filing material used in the past.[5] computers

All of these functions are of great help to writers and producers. The producer uses this input plus experience to make decisions regarding the value and importance of various news items so that the top stories reach the audience. There are almost always more stories being fed in than can be used during broadcast time.

With most radio stations, this selection process is ongoing because news is broadcast many times throughout the day. Radio networks generally feed stories to their stations as they occur and also provide several news programs throughout the day. Thus, the news producer of a station affiliated with an ABC network might decide that the 8:00 A.M. news should consist of five minutes of ABC news as broadcast by the network; a story on a fire in Delhi that is on the UPI wire but ABC did not cover; an update of the condition of a hospitalized city official that the producer obtained by phoning the hospital; the report on a liquor store robbery gathered by one of the station's reporters; a report of a murder received from the local wire service; and the weather as received over the phone from the weather bureau. For the 9:00 A.M. news, the stories of the Delhi fire and liquor store robbery may be included again, slightly rewritten and accompanied by two stories excerpted from the 8:00 A.M. ABC news broadcast, a report phoned in live by a station reporter covering a school board meeting, and updates on the hospitalized official and the weather.

There are a fair number of all-news radio stations that broadcast news continuously throughout the day. They have a large number of crews, a variety of news services, and a large staff of people working at the station. Needless to say, an all-news format is very expensive.

Most television news has a more defined countdown because the major effort is devoted to the evening news. There are exceptions, of course, including late night and early morning newscasts incorporated from time to time in both network and station schedules. Cable News Network provides twenty-four-hour news, so has a procedure more closely akin to an all-news station.

News producers at most local stations and the major networks spend the day assessing the multitude of news items received to decide which twenty or thirty will be included on the evening news and in what order. Usually the producers of the various newscasts meet early in the morning to assess what they feel will be major stories. Crews are then assigned to cover these events.

The crew may consist of one person recording shots to be included in a story delivered totally by the anchorperson, or the crew may be two or three people including a reporter and equipment operators. Once the story is covered, the reporter may return to the home base, write the story, and work with an editor to cut the footage. Or the reporter may turn everything over to a writer and go out on another story. Or the footage may be microwaved or sent by satellite and the reporter and crew move on to another story while writers and editors complete the first story. Sometimes the reporter does a standupper live from the location during the newscast.

Changing events of the day often alter how the news is gathered. Communication transpires frequently between producers and reporters in the field—"There's a hurricane warning in Florida. Should I cover it?" "Yes, forget the Georgia peanut contest and get there right away."/ "Forget the vice-president's luncheon and cover the French embassy picketing. Try to get an interview with the ambassador."/ "The prime minister just resigned and I lined up an interview with his son. Can you give me five minutes of the show?" "Not five, but maybe two."/ "Got the interview with the ambassador." "Send it on the satellite. We'll use it."/ "The Senate just confirmed the president's nominee for the FAA. Should I try to interview him?" "Don't bother, we have too much other news. Get over to the secretary of state's news conference."/ "Rewrite this story on the earthquake prediction so it only takes thirty seconds. We have to make time for the ambassador's interview."

This type of decision exchange goes on at TV networks and local stations until close to air time when the "final" stories are collected, written, and timed. The anchorpeople get into position, the director and crew prepare for the broadcast, the edited tapes to be rolled into the newscast are put in order, the teleprompter copy is readied, the computer graphics are polished, and the newscast begins.

the newscast

Last-minute changes can still occur, for the news is sometimes changed even as it is being aired—"Just got a satellite damage report on the Florida hurricane. Substitute that for the earthquake prediction."[6]

Presentation of news is another important area. For radio, the days of the dulcet-toned announcers are gone; news is generally read by disc jockeys or reporters. However, stations do vary both content and presentation of news broadcasts in relation to their audience. The ABC radio network has four different services to suit the format and style of different affiliates. Because TV news is actually a money-maker for local stations, they are particularly anxious to lead the ratings; therefore, they attempt to find newscasters who will appeal to viewers. The networks hope to establish trustworthy, congenial newscasters who will maintain a loyal audience.[7]

presentation

Special elements, in addition to the daily news, fall under the jurisdiction of the news department. Some of these—such as presidential news conferences, congressional hearings, and astronaut launchings—can be predicted in advance and adequate preparations can be made. Others—such as riots, earthquakes, and assassinations—must be handled as well as possible by reporters and crews who happen to be close at hand covering other assignments. Often stations interrupt their regular programming to broadcast these special events.

special news elements

Broadcast journalism is proud of its service function. As early as 1932 New York area stations conducted around-the-clock operations to cover the Lindbergh kidnapping case. In 1937 radio provided the main communications for the flood-stricken Ohio and Mississippi Valleys, and in 1938 it did likewise for southern California. Many times since then radio has acted as the main communication conduit during disaster. Radio and TV have brought the world the stories of the Hindenberg explosion, the declaration of World War II, the invasion of Normandy, the surrender of Germany, presidential news conferences, deaths of presidents, political conventions, the visit by Soviet Premier Khrushchev, space orbits, riots, President Nixon's trip to China, Watergate

service function

hearings, Iran-Contra hearings, and most other major events. For most of this coverage, broadcasting has been praised for its decorum and service to the nation.

And yet broadcast journalism is one of the most criticized institutions in our country today. It is blasted by government officials, liberals, conservatives, middle Americans, and members of its own fraternity.

The criticism centers primarily on what broadcast news presents and how it presents it. Generally conceded is the fact that radio and TV bring important events to their audiences, but many critics claim that those media have not yet learned how to make the prime issues of our time understandable. The broadcast media are good at covering wars and fires, but inadequate in dealing with subjects such as inflation, unemployment, and economics.

visual emphasis

Television, particularly, is obsessed with visual stories and sometimes may downplay an important story simply on the grounds that there are no exciting visuals to accompany it. Sometimes stories are so highly visual that the facts become obscured. The weather, for example, can contain so many computer graphics, satellite feeds, and digital effects that viewers are left wondering whether or not it will rain.

capsulized news

Both media must provide capsulized news. The thirty-minute evening network news programs, when all commercials have been added, boil down to twenty-two minutes of news—approximately the equivalent of three newspaper columns. Obviously this cannot be the day's news in depth, explained and analyzed. And yet the news that is chosen is often trivial. Greater coverage is given to the president eating a taco than to his views on the nation's economy. If a station or network has achieved a scoop, it will dwell on that story even though the story itself is relatively unimportant.

creating news

Radio and TV stations are accused of creating or altering news by their very presence. It was certainly true, especially in the 1960s, that "spontaneous" protests began when the TV remote truck arrived and stopped as it pulled away. Many people feel that violence in the news is overplayed and sometimes appears on the news as violence for violence's sake.

publicity

Similarly, TV is sometimes held responsible for the evil effects of publicity given to violent lawbreakers. Murderers, kidnappers, skyjackers, and the like receive so much news coverage that they are boosted almost to a celebrity status. At times broadcasters have been encouraged simply to ignore such crimes in the hope that the perpetrators will stop when they see that no one is paying attention to them.

tasteless coverage

Tasteless coverage of the victims of violent acts is also a problem. There are reporters who attempt and often succeed in interviewing people who have just seen a close relative killed or have just lost all their possessions in a disastrous flood. The interview answers given by victims under these circumstances are often not rational and further sensationalize the news event. In their sensationalizing, news media are often accused of judging the guilt of a suspect before he or she has had a chance to receive a court trial. "Trial by the press" is a frequently used ploy of lawyers who think that their clients do not have a chance because of adverse radio and TV publicity.

In many types of stories, broadcast journalists accentuate the negative—unemployment going up is reported more frequently and with more fervor than unemployment going down. Because of the speed of radio in particular, news is sometimes reported inaccurately just so it can be first. With the advent of minicams and satellites, pictures of events can be shown before the reporter has had time to gather the facts.

Another objection that some individuals have raised, particularly against network broadcast journalism, is that it is a biased product of the "liberal eastern establishment." The criticisms range from charges of outright purposeful distortion of facts to editorializing by exclusion. Politicians and businesspeople, who are often targets of news darts, are particularly prone to cry bias. They bemoan not only what is reported, but how it is reported and what is left out of what is reported.

Some criticisms of the news process come from those working within the news industry itself. They are constantly pleading for more air time for news so that they can correct the flaws brought about by the short, capsulized treatments. One-hour evening newscasts rather than half-hour newscasts are proposed frequently, but these proposals have been largely unheeded—mainly because entertainment programming supplies more dollars.

As networks, stations, and cable systems tighten their belts, many of the functions and people associated with the news departments are disappearing. Foreign bureaus are being cut and less reporters are being sent to cover a single story. The total number of stories being covered has also been cut down because stories covered and not aired are money down the drain. The people within the news departments decry this and claim that the quality of news will suffer. They also worry that the gathering of the news will become more centralized and fewer interpretations will surface.

Tension also exists between local stations and networks, mainly because of SNG. With the satellite possibilities, the importance of the network is diminished. This could portend major changes in the status of network evening news.

Criticism is bound to be a permanent irritant to broadcast journalism. For one thing, most of the critics are members of the upper-middle class who do not get most of their information from television and are, therefore, not part of the mass audience to which broadcasting is attempting to relate. Much of the criticism of broadcast news is closely tied to news events themselves; however, broadcasters can in no way control world affairs. Perhaps most important, broadcast news is caught squarely between the two purposes of the broadcast industry. The first, as stated in the 1934 Communications Act, is to serve the public interest, convenience, and necessity. The second, as stated by broadcasting's management and stockholders, is to make a profit.[8]

Documentaries

Documentaries are designed to give in-depth coverage of subjects that can be dealt with only superficially in news programs. They require extensive research and expensive production.

FIGURE 10.1

A television news
cameraman covering
an accident story.
*(Courtesy of KOCE,
Huntington Beach,
California)*

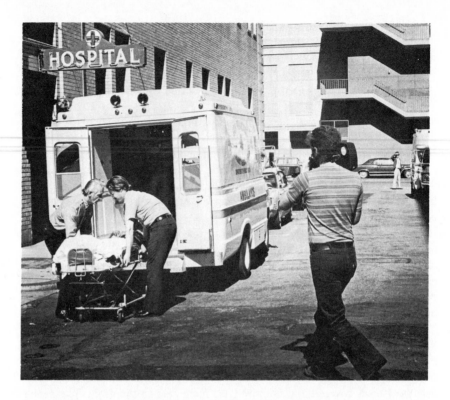

types of documentaries

Radio documentaries were not uncommon on early radio and occasionally are produced today by commercial and public radio networks or local stations. Local TV stations also produce documentaries of local issues. Sometimes they run these as **"mini-docs"** within the news. They will air three to five minutes about a certain subject each day of the week, and later they may edit the segments into one unified documentary to air as a program by itself.

changes in documentaries

However, the best known and most controversial documentaries of today are produced by the TV networks. Ed Murrow and Fred Friendly invented the TV news documentary in the early 1950s with "See It Now," which presented bold, strong programs on controversial issues. The series was canceled in 1958 because it lost its sponsor, its production costs increased, the network was tired of fighting the problems it caused, and Murrow and Friendly got tired of fighting the network. Since then documentaries have gone through phases of varying emphasis, depending largely on the degree to which the networks are pressured into presenting public interest programming. Sometimes documentaries appear only during the Sunday afternoon "intellectual ghetto" hours, and sometimes they enter prime time. The CBS weekly series "60 Minutes," a compilation of several documentary subjects, has often been tops in the weekly prime-time ratings.

hard and soft

Documentaries are usually divided into **"hard"** and **"soft"** on the basis of their subject matter. The "hard" documentaries, by far the more controversial, are usually the result of investigative research of current topics. Examples include: "The Uncounted Enemy: A Vietnam Deception," a program

FIGURE 10.2

A scene from the "Frontline" documentary dealing with the Nicaraguan Contras. *(Courtesy of a consortium of public television stations: KCTS Seattle, WGBH Boston, WNET New York, WPBT Miami, and WTVS Detroit/ Gustavo Sagastume, photographer)*

indicating that the number of enemy troops in Vietnam was purposefully underreported; "The Selling of the Pentagon," a program about the money allegedly spent by the Department of Defense for public relations efforts aimed at selling the American public on defense projects; "The Migrants, 1980," a documentary about working conditions of migrant workers; and "Boys and Girls Together," an exploration of teenage sexuality. "Soft" documentaries give in-depth information regarding less controversial subject matter, such as "The Louvre," "The White House Tour with Jacqueline Kennedy," and "Lyndon Johnson's Texas."

Documentaries cause innumerable problems for the networks with respect to both expense and content. If a dramatic program set in the 1800s employs words not in the vernacular of the time, it may be criticized, but with none of the severity that occurs when a public official is misquoted in an unfavorable light. Documentaries also have traditionally low ratings and barely recover their costs. Stations have been required by the fairness doctrine to give varying points of view for all controversial issues; therefore, if some individual or group could prove its point of view was not fairly represented, it could demand free time to present information. Obviously, this adds further cost.

A network frequently finds itself in a difficult situation after airing a controversial documentary. After CBS aired "The Uncounted Enemy," it found itself in a libel case against General William Westmoreland. When "The Selling of the Pentagon" was aired, CBS network executives were called to testify before congressional committees regarding the content of the show. Documentaries have been known to cause internal dissension within the network family. Producers argue with network executives who want to censor material, and stations within the affiliate family sometimes refuse to carry a program because of the subject matter.

problems created by documentaries

Documentaries are subject to some of the same criticisms leveled at news—a liberal bias, an emphasis on the negative, editorializing by exclusion, and unnecessary sensationalizing. In addition, they are criticized for appearing too infrequently on the program schedule and for being aired essentially all at one time. Many documentaries are aired during **"black weeks,"** the weeks when ratings are not taken. The only regularly scheduled prime-time documentary, "Frontline" appears on PBS. Network executives, of course, counter these arguments with dollars and cents and an appeal that perhaps what is needed in this country is less advocacy and less material that will divide the nation into fragmented groups.[9]

Editorials

Radio and TV stations are not mandated to editorialize, and many choose not to. This is due in part to fear of the controversies that may ensue and in part to the difficulty of complying with the present strings attached to editorializing. If a station is going to endorse a political candidate, it must notify all of the candidate's opponents. If it is going to say something negative about a person, it must give the person advance notice and an opportunity to reply. Small stations generally do not have the staff to handle these requirements and sometimes do not even have the staff to write the original editorials.

editorial ideas

Ideas for editorials are usually conceived by a station management team and then presented by a member of top management. Some editorial material comes from networks or syndicators in the form of commentaries. Occasionally these consist of a series of programs designed to present a spectrum of opinion on a particular subject and in that way cover all points of view that might be considered in the controversy. Commentaries are personal viewpoints, while editorials express the viewpoint of the station management.

pros and cons

Editorializing is itself a controversial subject. Some critics feel that because the number of frequencies is limited, radio and TV stations should not be allowed to editorialize at all because doing so gives these broadcasters an unfair advantage over others in the community. Other critics feel that editorializing is guaranteed by the first amendment and that, in fact, broadcasters who do not editorialize are shirking their public duty. Methods of editorializing are also debated. If editorials are presented within news programs, they may be mistaken as news, but if they are presented at other times, they can be a jarring interruption to program continuity.

blandness

Editorials that are aired are often criticized for their blandness. The subjects covered—such as public parks, automobile safety, and school crosswalks—are often so noncontroversial as to be hardly worth the status of editorial. Presentations are generally dry and nonvisual and are considered boring if they last very long. As a result, editorials are generally the least glamorous of radio and TV programming elements.[10]

Sports

Sports programming is a hybrid of information and entertainment programs. It involves material broadcast from the scene generally in real time and of real events, but most people watch it for entertainment purposes.

The most highly publicized of sports programming is that on the networks—CBS, NBC, and ABC. Cable has also made important inroads into sports, particularly with ESPN, which is virtually nonstop sports throughout the day and night and which has occasionally outwitted the networks in rights' battles. So much sports is consumed by television that independent networks, such as Hughes Sports Network and Television Sports Network, have prospered by selling commercial time nationally and paying independent stations to carry the shows. Throughout the country, local radio, TV, and cable systems have their own sports programs, usually in conjunction with a local college or university.

sports producers

Throughout their short history, sports and broadcasting have had an unusual symbiotic relationship. The Dempsey-Carpentier boxing match gave early radio its first big boost; early TV had its wrestling matches; ESPN was one of the first successful cable networks; subscription TV's biggest subscriber enrollment periods came right before major prize fights were shown; and many of the people who have installed backyard satellite dishes are sports junkies who delight in the thirty or more feeds they can watch at any time during the weekend.

interrelationship

As sports proved its value to broadcasting, the **rights fees** television had to pay to air the games skyrocketed. During the period between 1970 and 1986, National Football League rights climbed from $50 million to $430 million, the National Basketball Association went from $10 million to $42 million, and baseball rights rose from $18 million to $160 million.[11] The money was used to pay the ever-increasing sports stars' salaries and to build profit into the whole sports structure.

rights fees

For many years the networks, cablecasters, and stations broadcasting sports continued to make profits on them despite the rising rights fees. Advertisers were eager to access the affluent males regularly attracted to sports programming and, on occasion, paid over half a million dollars for a single spot in a sports event. In the late 1980s, the spiraling balloon began to lose air, however. Advertisers found cheaper ways to reach the audience they wanted and some of the ads in sports programming went begging. The glut of sports that was programmed on broadcast TV and cable overwhelmed even the most avid of fans and audience numbers per program declined. As a result, broadcasters reduced the avarice that went into outbidding each other for the rights and the fees leveled off—in a few instances, they even declined. This pinched the sports organizations somewhat because they had become totally dependent on television fees in order to remain solvent.[12]

Of course, this setback did not mean an end to the sports-broadcasting relationship, just a readjustment of terms. It was not the first time that the relationship had hit rocky times. In the early days of television, sports had been negatively affected by overexposure on TV. One of the first sports to suffer was boxing, which was one of the most popular events on early television, with fights virtually every night of the week. While everyone was watching boxing on TV, no one was supporting club boxing, so about 250 of the 300 small boxing clubs in the United States closed up shop between 1952 and 1959. The result was no fresh talent and a boxing lethargy that was not restored until the coming of Muhammad Ali. Baseball was affected too—while baseball club owners were greedily grabbing every golden nugget TV offered for the rights to their games, attendance at these games fell 32 percent between 1948 and 1953. Similarly, attendance at college football games dropped almost three million between 1949 and 1953.

Eventually, teams began restricting the number of telecast games, but it was a long hard pull to coax sports fans back into the stadium. Once there, however, the fans seemed to enjoy the media coverage. Sometimes the fact that a game is to be telecast seems to draw fans rather than drive them away.

One of the other sore points that arose between the sports world and the broadcasting world involved the degree to which sports were changed to accommodate television. For example, there were two kick-offs during the 1967 Super Bowl because NBC was in the middle of a commercial during the first kick-off. Likewise, the 1978 tennis competition time was changed at the last moment from 1:00 to 12:30 in order to accommodate taping requirements of TV. As a result, many paying fans missed several games of the match. These and other similar events caused indignation. But as the years passed, it was increasingly difficult for people in sports to register moral indignation because sports itself became largely show biz. The athletes seemed to adjust their adrenaline levels to fit the needs of the television world. Besides, it seemed a small price to pay when the result was that millions of spectators who otherwise would not have had the opportunity could enjoy the game.[13]

Another selling-of-the-soul aspect of the sports-broadcasting relationship involves the frequent commercials that feature top athletes. Part of the controversy revolves around whether sports stars should stoop to peddling, but the money earned for such work is certainly tempting. Large sports salaries plus money from commercials have turned many athletes into company presidents and entrepreneurs. Another aspect of the controversy revolves around the probable duping of the American public when they see sports idols extol the virtues of a product without, in most instances, any expertise or credentials in the area. In order to limit this phenomenon somewhat, athletes and others engaged in testimonials must now at least use the product they advertise.[14]

Even if athletes do not perform for commercials, TV still tends to require that they be actors as well as athletes. The number of interviews they must submit to is enough to make the shyest ones glib, and those who want to keep their private lives private find themselves hounded just like movie stars.

Another controversial aspect of sports and broadcasting involves **blackouts.** Sports owners realized that they were cutting their own throats when they televised all their games. As a result, a law was passed by Congress in

FIGURE 10.3

ESPN coverage of a soccer match. *(Courtesy of © ESPN/Thomas F. Maguire, Jr., photographer)*

1973 that allows football, baseball, basketball, and hockey games to be blacked out (not broadcast) up to ninety miles from the origination point unless the game is sold out seventy-two hours in advance of game time. The biggest critics of this policy are the fans who are deprived of seeing the game on TV.[15]

There are occasional outcries from the sports world against TV equipment that is so sophisticated it out-umpires umpires and out-referees referees. On the other hand, the sophistication of modern TV equipment often enhances the viewing with slow motion, split-screens of different parts of the game at the same time, and cameras placed at reverse angles to give a different point of reference to a play. Some stadiums now have large-screen TVs that show parts of the game in the stadium while it is in progress.

equipment sophistication

The sophistication of TV equipment is another factor that adds to the cost of sports. Aside from paying for the rights, broadcast and cable networks must pay production and transmission costs, making sports broadcasting one of the most expensive forms of programming that stations or networks can undertake. Obviously, the sport cannot be brought to the TV station's studio, so the studio must go to the sport. It is not uncommon for a network to gather twenty cameras, thirty microphones, two remote trucks, and one hundred technicians to cover a sporting event. Even local stations and cable TV systems find it expensive and complicated to cover the local high school football games. The 1984 Summer Olympics utilized two hundred and sixteen cameras, thirty-five hundred people, twenty-six mobile units, four helicopters, and three houseboats.[16]

costs

The actual broadcast of a game can produce ulcers for those involved. The announcer must attempt to be clever and articulate about plays while he or she listens through a headphone to instructions being barked by the producer and director—"After this play remember to do the promo for next week's

production

game and mention the sponsor's name. Tell the audience that Governor Flupadup is here because we want to get a shot of him"—and trying to comprehend messages being passed under his or her nose—"That was Schlocks's fourth time for hitting three triples in twelve games. Attendance is 27,982. Station break time." The director must choose the best picture from among the twenty or so displayed before him or her, usually with the help of assistant directors watching particular monitors—for example, one assistant director watching only the isolated instant replay cameras and another watching for interesting crowd scenes. Electronic chalkboards, super slo mo, reverse angle cameras, and a host of special effects add to production values. The camera operators must exercise common sense in following the action and in following the director.

Whatever the problems are with the sports-electronic media relationship, it is not likely that either party will initiate divorce proceedings. Broadcasters like the audience numbers that accompany sports events. So many sports organizations have built their entire budgets around television, that if television were to withdraw the money, the sports structure would collapse. The fans would not approve of that.

Magazine Shows

A magazine show is really defined more by its form than its content. It consists of short segments on a variety of subjects, similar to a printed magazine. Also, like its printed counterpart, its segments can cover virtually unrelated subjects or they can deal with material of a common theme.

Magazine shows first came to the fore in the late 1970s when the **prime-time access rule** was instituted. Under this FCC rule, networks were allowed to program only three hours of the 7:00–11:00 P.M. time period and local affiliates were to program the rest. Although most of the local stations opted for game shows and other inexpensive fare, Westinghouse tried a concept called "P. M. Magazine." The parent company produced a number of segments for the show that it distributed to all the stations it owned. These stations then added a few segments of their own local material. In some instances, the material produced locally was incorporated into the nationally distributed material.

P. M. Magazine

Westinghouse first put this program on its stations in 1977 and although it was not an instant hit, the company stayed with the concept and eventually it became very popular.

As often happens with successful shows, imitations began to appear. Other stations began producing magazine shows to fit into their prime-time access periods. Most of these dealt with events in the local community and featured a host and hostess who acted as host/interviewers.

spread

The concept spread beyond prime-time access and networks and syndicators began distributing magazine shows. These were varied in nature—some serious and some frivolous. For a time, a number of them dealt with unbelievable feats that people had accomplished, while another group of them dealt with lifestyles of the rich and famous.

FIGURE 10.4

Shooting a segment of "PM Magazine." This segment dealt with Kathy Swenson, a competitive dog musher who raced in the Yukon Quest. *(Courtesy of KFMB-TV, San Diego/Geary Buyaos, photographer)*

The word "magazine" has been connected to a number of different types of programs because it does deal with form rather than content. For example, "60 Minutes," "20/20," and similar concepts have been labeled as magazine shows, although some would argue that they should be called documentaries. The fact that they contain segments, some of which are rather light-hearted, makes them a hybrid.[17]

types of programs

Educational Programming

Educational programming is not an area America excels in. Most countries exceed the United States in both the quantity and quality of such programs. In many countries radio and television stations broadcast direct instructional material to schoolchildren as a matter of policy. This includes programs to supplement what is occurring in the classroom—such as science experiments or programs about foreign places that students might be studying—as well as course work in which the total instruction is from radio or TV. Although programs of this nature do appear on many American public TV stations, they are generally not utilized to the extent they are in many foreign countries.

in-school programming

In other countries, television is also used as a primary source of adult education. Some of the underdeveloped countries have full-scale TV courses to reduce illiteracy. In some totalitarian countries, watching certain television courses is one of the prerequisites for promotion. England has an open university that enables people to obtain college degrees through a combination of television and correspondence. In the United States, college credit and adult

adult learning

education courses are programmed in various sections of the country over public and commercial stations and through cable networks, but the extent and use of these is minimal compared to that of some other countries.

Other educational types of American programs, which again are handled with much less fervor here than in foreign countries, are about the activities of the schools and self-help, how-to programs. Networks sometimes publish study guides for regular programs so that teachers can assign viewing as homework and then follow up with educational activities in the classroom.

scheduling

In other countries, educational material is deemed worthy of prime time, but in the United States it is generally on commercial stations at the least popular time—6:00 A.M. on weekdays or on Sunday mornings. The reason for this is that educational programming is rarely sponsored—even if it were on at a better time, it would not draw a significant audience. Part of this is due to the subject matter of the programs, and part is due to the fact that these programs are generally low-budget and do not include all the production elements of more expensive entertainment shows.

In many instances, TV stations do not produce educational shows themselves, but purchase them from syndicators or allow local school districts to handle content and basic production using the facilities and crew of the TV station.

cable TV

Many cable TV franchises require that channels be set aside for education and some of these channels are being programmed by educational institutions.

radio

Very little is done educationally on radio, even though many of the concepts of educational TV programs could be presented quite adequately using audio only. Overall, the potential for educational material on the American electronic media is barely realized.[18]

Religious Programming

early programming

For many years, religious programming was another broadcast stepchild usually aired during the early morning or Sunday morning time slots on both commercial radio and TV stations. This form of programming was generally quite inexpensive for the stations because it was supplied, on film or tape, free of charge by the various religious sects.

The networks paid very little attention to religious programming, although one of the earliest "hits" on network TV was Bishop Fulton J. Sheen, who preached each Tuesday night opposite Milton Berle's "Texaco Star Theater." This caused Sheen to comment that he and Berle worked for the same sponsor—Sky Chief.

religious stations

The only religious programming to have much of an impact was that broadcast on radio or TV stations dedicated exclusively to religion. At first

FIGURE 10.5

Galileo, played by Aaron Fletcher, contemplates the law of inertia in this program for a college credit science series called "The Mechanical Universe." *(Courtesy of the Southern California Consortium)*

these were few in number and very local in nature. A great deal of the airtime was spent soliciting contributions from listeners, a procedure that was quite successful.

The advent of satellites and cable TV made religious television a much more powerful force. Programs that were local could now be distributed nationwide and receive contributions nationwide. The religions that used television the most tended to be evangelistic, leading to the coining of the term **"televangelism."** This televangelism became a billion dollar a year business and ministers such as Jimmy Swaggart, Robert Schuller, Pat Robertson, Jerry Falwell, and Jim and Tammy Bakker became known nationwide and launched political as well as religious movements.

national coverage

A major blow was dealt to TV evangelism in 1987 when Jim and Tammy Bakker of the PTL network were involved in a sex and hush-money scandal. The ensuing investigation led to a number of financial improprieties of the Bakkers and PTL, which had been subject to very little auditing as a nonprofit organization.

scandal

Congress held an investigation into the tax-exempt status of all televised evangelism, then matters seemed to simmer down and the religious forces went about trying to repair their tarnished image.

Religious programming of a more traditional nature still appears in its Sunday morning time slots, but the leaders of these religious groups are generally angered by televangelism, which they feel may be replacing the church.[19]

News and Information Programming   275

FIGURE 10.6

TV minister Jerry Falwell in the control room of his nationally telecast program "The Old Time Gospel Hour." *(Courtesy of The Old Time Gospel Hour/Les Schofer, photographer)*

Public Affairs

Public affairs programming is a general term for information programming that does not fall into other categories. Primarily it consists of interview, discussion, and on-site programs that deal with issues of concern to the citizenry. Some long-running network programs like "Meet the Press" and "Face the Nation" have featured prominent names in the news being interviewed (or grilled) by top journalists. The network morning shows and Ted Koppel's "Nightline" could also be considered public affairs, although some would put them in the news category.

types of programs

Most public affairs programs are local productions that deal with community problems. Some local radio and TV stations are very community oriented and program significant, timely, fairly frequent public affairs programs, but most try to skimp on both the quantity and quality of these programs because costs generally cannot be recouped through commercials.

Although stations are criticized by the public for their lack of good public affairs programming, they find that when they do air quality material, very few people watch, making that programming an unattractive buy for advertisers.

issues

These programs sometimes necessitate that a station give time and facilities to a group whose point of view was not represented when a particular controversial topic was discussed. This, of course, adds to the cost.

On the positive side, there are many instances in which public affairs radio or TV programs have been instrumental in helping correct injustices or problems within the community, thus constituting a worthwhile public service.

FIGURE 10.7

A typical setup for a public affairs talk show. *(Courtesy of KOCE-TV, Huntington Beach, California)*

A station whose personnel become involved in the community is usually held in higher esteem in that community than a station whose personnel remain aloof. Therefore, the community-involved station may receive overall higher ratings than the aloof competitor and can afford to put money back into public affairs programs that aid the community. In fact, all stations profit from the community they serve and, thereby, have economic justification for providing public affairs programs.

Finding adequate issues and program participants is generally not a problem faced by stations—all communities seem to have adequate spokespeople with opinions to express. Local cable TV operators have begun to capitalize on this phenomenon by encouraging "do-it-yourself" public affairs programming. Members of the community are trained in the rudiments of TV production and are able to use the cable TV facilities and equipment to produce their own public affairs programs, which are then cablecast over access channels.

The overall future of public affairs programming is uncertain. Economic and social fluctuations affect both the content and availability of these programs. When stations are economically healthy, they are more inclined to program public affairs. When social issues are of burning significance, both larger demand and larger audiences surface for public affairs programming.[20]

Politics

Sometimes all of a station manager's other problems seem dwarfed when he or she enters the area of political broadcasting. The biggest problems come during the years when there are major elections. It is then that political candidates fill the airwaves in their attempts to become elected or reelected. According to **Section 315** of the Communications Act, stations must give equal opportunity (commonly called **equal time**) to all political candidates who are running for the same office. If any candidates feel that they have not been given equal treatment, they can appeal to the FCC and the station then becomes involved in a hearing that has the potential of being appealed all the way to the U.S. Supreme Court. In some instances, the issues involved in these appeals can lead to Congress's amending or enacting new laws.

Although giving equal opportunity to all political candidates running for the same office may sound simple, it is actually a very complex process. Most major election years bring over forty FCC rulings concerning political broadcasting. They deal with questions such as When does a political campaign begin? Where is the precise line between a news interview with a political candidate (which is exempt from Section 315) and a political interview, (which is not exempt)? When is a so-called independent political committee that can spend unlimited funds on broadcast time really separate from a political candidate who has restrictions on political expenditures?

A few stations have attempted to avoid the pitfalls of Section 315 by broadcasting absolutely nothing political during election campaigns. This was discouraged by a 1972 amendment to the Communications Act stating that stations could have their licenses revoked for willful and repeated failure to allow reasonable access to candidates. Furthermore, this type of abstinence does not endear stations to the FCC, politicians, or their community because it represents a shirking of public service.

Each election season broadcasters try to do the best they can with Section 315, but its changing complexion and ambiguities make it difficult to administer.

There have been cries from broadcasters, from the FCC, and from some politicians to eliminate the equal time provisions altogether so that broadcasters would have the same freedom of choice as newspapers in supporting and reporting on candidates. This is staunchly opposed by the smaller minority parties and the nonincumbent candidates on the grounds that the incumbents could perpetuate themselves by getting an overabundance of coverage and that stations could influence elections to suit their biases.

strong role of television

Issues other than equal time also arise at election time. One is the overriding role of broadcasting in the election process. Television has become the most potent force a candidate has at his or her disposal, especially if the candidate is running for a national office. No whistle-stop campaign can get a face and views in front of as many people as can one television commercial. There are those who complain that for a candidate to win an election, he or she must project as a TV personality and that perhaps the country should not be run only by the glib and the glamorous. Candidates must rely more on their public relations firms and media advisors than on their political views and philosophies.

coverage of politics

Broadcasting's coverage of political events is also criticized for giving too much emphasis to frilly baby-kissing events and too little to the issues that divide the candidates. Most of the spots that stations sell to potential office holders are thirty seconds long—hardly time enough to tell how one would run major aspects of the government. There are also complaints that too much money is spent on broadcast advertising, making running for office a game for the rich. Devious methods used for obtaining campaign money have led to some public financing of presidential campaigns and other reforms, but it is still extremely expensive to seek office.

Coverage of political conventions is also a thorny topic because network coverage of the Republican and Democratic nominating conventions has changed the political structure. In some ways this has been an advantage—the delegates now tend more to business than to partying because they know they are under a watchful eye. But because the conventions are on TV, they have become show business, with major actions being programmed for prime time and with contrived suspense building so viewers won't flip the dial to a cop show. The actual cost to the networks of covering a convention is in the neighborhood of $3 million, which has led them to pool resources in order to cut costs.

conventions

Despite its imperfections, broadcasting brings to the people more information about candidates than they could gain by themselves, while at the same time it simplifies the campaigning process for the candidates.

Coverage of election results, whether they be national, statewide, or local, also provokes controversies. The fact that networks have reported election results in some parts of the country while people are still voting in other parts can distort election results. For example, Democrats in the West charged that the reporting of Reagan's obvious landslides in both the 1980 and 1984 presidential elections before the polls closed in their time zone meant that many Democrats did not bother to vote and that state and local Democrats suffered as a result. The fact that networks also predict winners early by asking voters who are leaving the polls how they voted has been decried. The networks have agreed to stop these **exit poll projections** in future elections. They are also trying to encourage Congress to remedy the time difference situation by instituting uniform poll closings throughout the country.

predicting results

Between election campaigns, politics is still evident in broadcasting as the media air press conferences and intramural political disputes that may arise between election years. The press tries not to come too close to the politicians it covers, but there have been complaints that broadcasters do have their political darlings and are in a position to foster the careers of those they like. Likewise, reporters have often been accused of unjustly causing the downfall of politicians through personal revelations that would not affect their ability to govern.[21]

Special Interest Programming

Special interest programming is primarily a product of the cable TV age. When the three networks dominated programming, they geared their material to the masses and had little room for programming that would appeal to only a small segment of the population. As the number of channels increased through the addition of many independent TV stations, as well as through cable and videocassettes, the concept of **narrowcasting** took hold and programming geared toward special audiences became fairly popular.

changing emphasis

News and Information Programming   279

FIGURE 10.8

Setup for a political broadcast. Local candidates are invited to the station to present views and answer questions posed by reporters. *(Courtesy of KOCE-TV, Huntington Beach, California)*

C-span

ethnic programming

home shopping

other shows

Within cable TV, whole networks are devoted to bringing people financial news, the weather, health information, and travel facts. These channels are on up to twenty-four hours a day and people watch them when they need specialized information.

C-Span is another type of special interest programming. The coverage of the House of Representatives and the Senate is watched from time to time by a fairly large audience, but most of the time the material is far from prime-time quality.

Ethnic programming is also of a special interest nature. Several cable TV networks feature it and it is also fairly common on independent TV stations. Many markets have one or more TV stations designated to showing programming, usually in a foreign language, of a dominant ethnic group. Some stations show several different types of ethnic programming at different times of the day.

The home shopping channels, stations, and programs that have sprung up are another example of a new type of special interest programming. People who like to shop watch for long periods of time; others tune in to try to find a good buy.

Some types of special interest programs have been on both radio and television for years, but they are more abundant now. These include shows about gardening, cooking, farming, and consumer education, as well as movie reviews and how-to programs. Some of the how-tos have been particularly successful on videocassette.[22]

Overall, radio and TV are dynamic information sources that have captured public attention in a short period of time. Because of their influence, they have great responsibility to provide accurate and thorough information.

FIGURE 10.9

Off-monitor shot of C-
SPAN's coverage of
the senate. *(Courtesy of
C-SPAN)*

Conclusion

Preproducing, producing, and airing information programming can be a complex process. The myriad of sources available for news—wire services, independent news agencies, network reporters, station reporters, and SNG services—must all be constantly checked to supply viewers and listeners with up-to-date newscasts on an ongoing basis, an hourly basis, or during the nightly newscast.

The research that must precede the airing of either a hard or soft documentary so that it is accurate is also enormous. Likewise, conceiving and researching editorials is a time-consuming process.

In the area of sports, a crucial preproduction activity is the negotiation of rights with sports organizations. The logistics of transporting equipment and people to the scene of the sporting event involves expense, personnel, and time.

Producing magazine shows involves gathering information on a variety of subjects and then tying it together, usually with some sort of loosely connected theme.

Preplanning and consultation must be a part of educational programming so that the material presented meets the intended educational goals.

Religious programming is generally underwritten and produced by the religion involved and tends to be of the evangelistic nature.

Public affairs programs are generally the cheapest and easiest to produce, but even at that they generally do not pay their way.

Those overseeing political broadcasting must study equal-time regulations carefully and, of course, the complicated logistics of setting up election-night coverage.

Special interest programs vary greatly in their level of complexity and have been brought about largely because of the greater number of channels available to the public.

Although the media are praised for the service they provide with their news and information programming, criticism of this type of programming also abounds. News is lambasted for being biased, sensationalized, capsulized, tasteless, and trite. It rarely undertakes complex issues and when it does, it seems to fail in making them comprehensible. More attention is paid to the looks and demeanor of the newscasters than to the news itself. The media can alter events by their presence and can bring glory to lawbreakers.

Documentaries share many of the criticisms lodged against the news. In addition, networks are criticized for the very limited quantity of documentaries they produce and the unfavorable times at which they usually schedule them.

Many stations that do not editorialize are criticized for shirking their public duty, and those that do editorialize encounter conflicts because their editorials are considered unfair or dull.

The hoopla and high financing that accompany sports events often alter the games. Athletes who become admired and revered are then criticized when they plug commercial products. Blackout rules bother fans but are part of the economic base of the sports-media connection.

The major criticisms of educational programming is the lack of quantity and, to some extent, the dullness.

Some people question the degree to which televangelism has replaced the church. Scandals have tended to lower the credibility of this type of broadcasting.

Public affairs programs suffer from dullness and poor airtime and can bring conflict and turmoil to stations.

Many of the problems with political broadcasting stem from the changing interpretations of Section 315 of the Communications Act. In addition, broadcasters are criticized for emphasizing the frills rather than the issues of political campaigns, for reporting election results before the polls are closed throughout the country, and for fostering the careers of politicians they favor.

News and information programming have become very important to the structure of our society. Regulations and customs that change this type of programming usually indicate major changes in society.

Thought Questions

1. Do you feel that broadcast news is biased and/or tasteless? Explain.
2. Should radio and TV stations be mandated to editorialize? Why or why not?
3. Is it proper to rearrange sports events in order to accommodate broadcasting? Explain.
4. Should political candidates be given free advertising time by stations? Why or why not?

Programming Decisions

Overview

Radio and television stations and networks are as individual as people in their handling of programming. However, for purposes of simplification, programming techniques will be broken down in this chapter into the processes undertaken by the different kinds of telecommunications entities: an independent radio station, a network-owned or network-affiliated radio station, a radio network, an independent TV station, a network-owned or network-affiliated TV station, a TV network, cable TV, and other media.

Among the topics included in this chapter are:

Overall individualization of programming decisions

Factors that influence independent radio station decisions regarding format and special feature programs

Similarities and differences in programming decisions of independent radio stations and owned and affiliated radio stations

The changing role of network radio decision making

Programming decisions in public radio

The various sources from which independent TV stations receive programming, including syndicators, film companies, and made-fors

Methods that independent TV stations use to develop local programs

The contractual process by which TV networks provide programs to their affiliated and owned stations

Advantages and disadvantages that affiliated program directors have in relation to independent program directors

The process by which TV network shows are chosen and produced

Issues in network programming, including cancellation, cost, counterprogramming, preemptions, and reruns

Public television programming policies

The patterns being set in cable TV programming

The evolving nature of program decision making in the newer media

11

Imitation is the sincerest form of television.

Fred Allen
performer

Independent Radio Stations

To understand how radio programming is developed, it is best to look at a small, independent station faced with the need for something to fill its airtime. Generally, a management team will decide what type of material the station will air. In some small stations this "team" may consist of one person. In larger stations it may include the general manager, program manager, sales manager, business manager, and chief engineer. Inputs from all are necessary to determine whether the costs of the programming decided upon can legitimately be borne by the revenue the station will generate.

Radio station programming decisions revolve around two main elements—format and special features. Some of the overall **formats** used by radio stations include contemporary hits, easy listening, country-western, oldies, progressive rock, classic rock, adult-oriented rock, soul music, classical music, jazz, all-news, talk, ethnic, religious, and agricultural programming.[1]

Some of the **special features** stations may use include business reports, comedy, drama, real estate reports, editorials, children's programs, local sports, live concerts, boating reports, farm reports, personality interviews, public affairs, health information, and science updates.[2]

Many factors affect these format and feature decisions. One consideration is the programming already available in the station's listening area. For example, an overall format of country-western music may be decided upon because no other station in town offers that type of music. However, if there are only a few country-western music fans in the local area, this format will probably not be adequate for drawing an audience. Therefore, the composition of the listening audience is another important factor. A rural audience will probably be more interested in frequent detailed farm reports than will a city audience. The interests of the community and the interests of the station management can also affect programming. If the town has a popular football team, the station might choose to broadcast football games. If the station manager is particularly interested in boating, he or she might promote boating reports.

Once the general format and features are decided upon, it becomes the program manager's responsibility to execute this programming. Often this is done with the aid of an hourly **"clock"** the program manager designs. This shows all the segments that appear within an hour's worth of programming. Generally each time segment within an hour resembles that same time at other hours. For example, if news comes on at 9:15, it will also come on at 10:15, 11:15, 12:15, and so on. In this way, the listener knows what to expect of a station at each segmented time of the hour.

Because most radio stations opt for some type of music format to constitute the bulk of their programming, the program manager must find disc jockeys whose talents fit the station's format. For example, rock 'n' roll music disc jockeys must be capable of fast, lively chatter and a classical music host must be more subdued and capable of pronouncing foreign works. For an audience call-in show, the on-the-air talent must have good rapport with people.

team decision

formats

special features

clock

disc jockeys

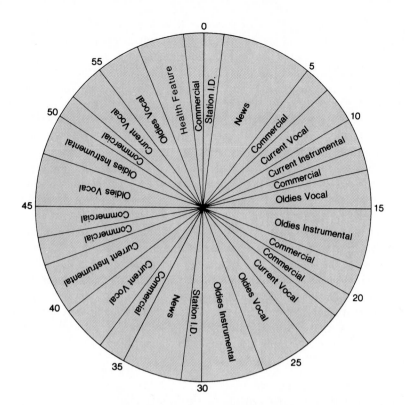

FIGURE 11.1

A typical radio station clock.

The program manager must also begin building a station record library. This is usually just a matter of making sure that the station is on the right mailing lists. Because record companies are eager to have their selections aired, the program manager sometimes finds that he or she must fend off salespeople wanting to give records they want the station to plug.

record library

Special features must be handled by the program manager as the need dictates. If local college news is to be aired, a communication system for obtaining the news must be set up. If public affairs discussions are to be held, participants must be contacted and coordinated. If a local fair is to be covered, the details for a remote coverage must be cleared. If an editorial is to be aired, people and organizations with opposing viewpoints must be allowed to present their points of view. If city council meetings are to be covered, arrangements must be made with the proper government officials.

local programs

Sometimes special features can be obtained from one of a vast array of companies that produce radio programming and **syndicate** it to interested stations either by satellite or on tape.[3] Sometimes stations pay cash for this material and then sell ad time within or around the features in order to recover the money they have spent. Other times they receive the features for free because the companies producing them have already placed ads within them and have recovered their costs that way. This is usually referred to as **barter.** Sometimes feature programs are sold on a **cash plus barter** basis. This means the

syndicated material

program producer has placed some ads within the program but has left some for the station to sell. The station pays the producing company a small amount of cash and can recover this by selling limited ads.

News demands particular attention at many radio stations. Those with an all-news format devote most of their energies to this area. Generally they will combine network news, wire service news, and news stories obtained by their own local reporters out in the field. They take care to write, rewrite, and update stories in order to present them in as interesting a form as possible.

A station with less emphasis on news might subscribe to a wire service and augment this with a small news department of a few people. These people will search out stories as best they can given their limited numbers and resources. Once they have gathered local information, they will write the stories and probably also rewrite or edit wire service stories. In a station where news has minimal emphasis, the entire news "department" may consist of a UPI or AP machine from which the announcer rips and reads.

Most of the programming of a small independent radio station is produced locally. There are music suppliers who produce taped music used mainly by automated stations, and there are syndicators and wire services to supply news and features. Generally, though, music is selected and introduced by a disc jockey operating from a radio station. News, sports, features, editorials, and public service programs originate from the studio or a local remote location.

Network-Owned or -Affiliated Radio Stations

Many of the country's ten thousand radio stations are **independent;** that is, they do not have formal association with NBC, CBS, ABC, or any other network. Some commercial stations are **owned** by a network, which means that a network organization has financial ownership of the station and supplies some of its programming. Because no organization can own more than twelve AM and twelve FM stations, the number of network-owned stations is very limited.

Other stations are **affiliated** with networks. In this case, the station receives programming from the network, but the network has no financial control over the station.

There is very little programming difference between independent commercial radio stations and network-affiliated or network-owned commercial stations. The owned and affiliated stations are often larger, more powerful, more influential, and, hence, have more employees. The general manager and the program manager's jobs are subdivided into such positions as news producer, sports producer, and public affairs producer.

However, like the independents, the stations owned by or affiliated with networks must select a format and produce most of their own programs. The networks provide them with news, short features, specials, and perhaps a few regular programs, but the quantity of network programming is not enough to significantly alter the local responsibilities of programming. The stations affiliated with music networks such as Satellite Music Network and Transtar Radio Network have more material supplied to them, but they still have the obligation to program some local material.[4]

Radio Networks

Radio networks have changed greatly over the years. During the 1930s and 1940s they supplied most of the popular programming of the time—soap operas, children's programs, drama, and comedy. When this programming moved over to television, the radio networks became basically news services. Gradually radio networks increased their services to include features, total programs, and a few began supplying music, but a great deal of radio network energy still goes into news.

Each network has reporters throughout the world and supplies its affiliates with up-to-date news. The networks broadcast news periodically throughout the day, usually by satellite, and stations merely tap into the network feed and retransmit the news that the network is broadcasting.

Other network programming is produced by network staffs and by affiliated stations. For example, a network affiliate in Los Angeles may produce a talk show that is distributed by the network to affiliated stations in other cities.

The method of payment for network programming varies from service to service. Sometimes the network secures ads for its programs and then pays affiliates a token amount for airing the news or features. Sometimes the stations pay the networks and then insert local ads. Sometimes the network sells some of the ad time and the station sells the rest. All of this is similar to the cash, barter, and cash plus barter arrangements between syndicators and radio stations.

The decision process in radio network programming is difficult because the nature of the networks has not settled down since the advent of television. Some ideas, such as a revival of drama, have not worked well. Those in the decision-making positions are always trying to find new ideas and combinations of ideas that will prove profitable to all concerned.[5]

Public Radio

Public radio decision making is somewhat different from that of the commercial structure because most public radio stations are affiliated with one or two networks. There are independent public radio stations, most of them run by college students who make the same kind of decisions that seasoned professionals in commercial radio make. They decide on formats and features and, although they do not need to concern themselves with commercials, they must keep an eye on finances so that production costs do not exceed the budget.

Most of the larger, high-powered public radio stations affiliate with National Public Radio and/or American Public Radio. These two networks distribute mostly cultural and classical programming, and the public radio stations do not stray very far from that image in their local programming. Public radio station managers usually program classical or offbeat music, remotes of local concerts, and public affairs shows that seek out cultural and thought-provoking elements of the community.

The public radio networks differ somewhat in their programming philosophy. National Public Radio produces most of its own programs and emphasizes news and public affairs. American Public Radio redistributes programs produced by public radio stations and concentrates more on music than NPR. Both networks receive compensation from the local affiliated stations.

Public radio has only a small niche in the overall radio programming market, but it is one that usually reaches upscale influential people.[6]

Independent Television Stations

The development of TV programming can also be understood by looking at an independent commercial station that needs something to fill its airtime.

As with radio, the program manager of an independent TV station is only one member of a team that determines which programs the station will air. Here, too, others are needed to decide whether the costs can be covered by the income and still leave a reasonable profit margin for the investors. TV stations do not generally have a specific format as such, although they may emphasize certain types of programming, such as sports, movies, or foreign-language programming.

Because most TV stations are on the air more than a hundred hours a week, the program department has a large chore. It must not only develop and produce local programs that will attract an audience from within its signal range, but it must acquire rights to air products that have been produced for national syndicated distribution.

When a program is syndicated, it is shown on various stations throughout the country, but not in a network manner that assumes all stations will air it at the same time on the same day. Instead, it is sold on a station-by-station basis. An independent station in New York might air the program at 10:00 P.M. Wednesday, a station in Chicago at 8:00 P.M. Friday, a station in Detroit at 5:00 P.M. Tuesday, and all stations in Los Angeles might reject the program so that it would not air at all in that city. There are well over two hundred independent stations on the air serving approximately one hundred cities, so the potential market for syndicated programming is substantial.[7]

The syndicated programs that independent station managers select to fill airtime have originally been produced as movies, as network TV series, or as independent programs intended to be syndicated. A movie such as *Jaws* is an example of the first category; reruns of "I Love Lucy" are an example of a former network series; and "The Wheel of Fortune" is an example of a program produced strictly for the syndication market.

Syndicated shows are usually acquired from **distributors** who are, in essence, middlemen. They acquire distribution rights for a number of different programs and then try to sell as many programs to as many stations throughout the country as possible.

Sometimes production companies will act as their own distributor; however, if a company is small and has only a few programs to sell, it will realize higher economic gains by paying a distribution company to peddle the programs (along with programs from many other production companies) than it will if it hires its own sales staff. Some of the large movie companies like Paramount and 20th Century Fox handle their own distribution because they have adequate product to make it worthwhile.

distribution

Movies that come from motion picture companies either directly or by way of a distributor have traditionally been a significant source of programming for independents. However, by the time most movies get to independents, they have had abundant exposure in theaters, on videocassettes, on pay-cable, and on network TV. This is an advantage to independents because they know the track record of a movie and can promote it and sell ads for it based on its past. On the other hand, many people will have seen the film and may not be interested in viewing it again.

movies

When a station acquires these movies, it usually obtains the right to air them as often as it desires for a specific time, such as one year. The station tries to negotiate with the distributor so that the price it pays for the movies is low enough that it can make money by selling commercial time during the initial run and reruns.

Some film companies now offer independents **packages** of films that have not been shown on network TV. This is because networks have been less interested in films in recent years. These films have been shown on pay-TV and, therefore, do not draw well on network TV. When film companies offer these films to independent stations, they try to set up **"ad hoc" networks**—groups of independent stations that agree to air the movies at approximately the same time so that the films can profit from at least some degree of national promotion.[8]

The second type of syndicated material, old network programming, was for many years the greatest source of program material for independent stations. The programs had been successful and viewers were familiar with them. Because many people like to see their favorite programs over again, a steady audience could be obtained. The programs were also relatively inexpensive. Stations in different sized markets paid different amounts, but often series could be bought for several thousand dollars per episode. From a scheduling point of view, the independents would usually **strip** old network programs. Where the network aired a series on a once-a-week basis, the local station offered it on a strip schedule of Monday through Friday in a time period selected according to viewer interest and availability. In other words, a station might air reruns of "Gilligan's Island" five days a week at 4:00 P.M. to attract schoolchildren and might air reruns of "Star Trek" at 8:00 P.M. on Monday through Friday to attract adults who had been working during the day.

old network shows

Because of this type of scheduling, a large number of programs, usually about one hundred, must be available in order to make the airing schedule worthwhile. During the first several decades of TV this was no problem because network shows would run for several years. But as networks adopted a policy of canceling shows at a rapid rate, the number of programs produced shrank and, therefore, the number of shows available for syndication dwindled.

selling practices

As a result, a number of different practices arose regarding the selling of old network shows. First, the prices rose dramatically. Top-rated series could command six figures per episode and even series low in the ratings could commandeer $50,000 per episode. Independents complained about this, but there was little they could do because the market was responding to supply and demand.[9]

A practice known as **futures** was also developed. Some of the successful programs, such as "Happy Days," "Laverne and Shirley," and "Little House on the Prairie," not only set high prices, but also asked stations to commit to buying the programs long before they were actually available for airing on a rerun strip. In other words, an independent might buy a series in 1977, but not be able to begin showing it until 1980. Independents were not happy with this method of buying because they were not sure the programs would still be applicable several years hence. But if one station in a market would not buy this way, another would in order to try to ensure that it would have the highest ratings in the future.

To sweeten the pie a bit, some production companies began promising independents they would continue producing a particular series to provide approximately one hundred programs even if the series was canceled by the network before it reached that number. The price at which programs were being sold in syndication made this economically feasible.

Distributors also tried creative packaging. They would acquire several series of similar style or content that had had short network runs and offer them together under a unifying theme, such as family humor or high adventure. The price to independents was reasonable and the programs received extra, unexpected life.

For the long-running series, however, distributors became even more aggressive about setting prices. Sometimes they would offer a particular series in a city and set a floor price of over $100,000 per episode. They would then tell all interested stations that they could offer whatever they wanted above that price through a closed-bid process. The station submitting the highest bid over and above the floor would get the series.

Independents generally felt maligned by these practices, but there was little they could do because off-network programming continued to be their most popular product.[10]

independent productions

More and more, independents turned to a third source of programming—programs that were specifically made to be syndicated. These were produced with the idea that the main distribution vehicle would be independent stations.

The first of these was "Small Wonder," a comedy aired initially in 1984 and produced by a consortium of companies who owned independent stations—Gannett, Hearst, Metromedia, Taft, and Storer. It was successful enough that quite a few more comedies were produced specifically for syndication.[11]

As a result, some stations started checkerboarding these made-for-syndication comedies. **Checkerboarding** is a programming strategy that involves playing different programs at a specific time of day rather than stripping one show across the board. For example, a station might air "Small Wonder" at 6:00 Monday, "What's Happening Now" at 6:00 Tuesday, "Mama's Family" at 6:00 Wednesday, "Gidget" at 6:00 Thursday, and "What a Country" at 6:00 Friday.[12]

Various production companies also create material for the syndication market. Some of these are foreign companies making their programming available in the United States. Others are production companies that had an idea rejected by the networks but felt strongly enough about it to produce it and hope for syndication.

production companies

Some programs seen on independents are ones that production companies did not want to take to networks for one reason or another. Various talk shows come under this category because the talk show hosts might not have wanted to undergo the vicissitudes of network cancellation decisions. Magazine format programs that cover a variety of topics have also been quite successful in the syndication market. For some of these shows, the local station produces a small segment of the program in order to cover some local aspect of a topic covered in the syndicated material. Game shows are also prevalent in syndication fare.

There are other sources of outside programming besides the syndication market from which an independent program manager can choose. For example, several **independent networks** offer live sports events, such as golf, football, basketball and soccer, on an as-they-occur basis. Another network supplies news to independent stations and another offers special dramatic and variety programs and several beauty contests.

independent networks

Different methods of payment exist for syndicated programming. Sometimes a station pays a set cash amount for a program or series of programs and then sells commercials itself to fill all the spots. Other times stations engage in a barter deal with a syndicator. The syndicator sells and inserts two minutes of ads in a half-hour show, leaves the station four minutes to sell, and the station gets the program for free. Other deals involve cash plus barter. The syndicator sells one minute, leaving five minutes for the station, but the station pays the syndicator a small amount of cash. These types of payment arrangements called barter syndication are worked out on an individual basis between stations and syndicators.

payment practices

Religious programs are available to stations, sometimes free and sometimes for a fee. Wire services and independent news bureaus offer news programs to stations for a fee.

While outside sources can easily fill many hours of airtime, no TV station can be really profitable unless it produces some of its own programs. How does one start to develop programs? This can be answered by looking back to the program manager and following the steps he or she takes.

To assist in putting together local programs, the program manager normally hires an executive producer for the station. This person will have a successful track record in developing and producing programs. When the program manager has decided on the number of hours to be locally produced, he or she will, in concert with the general manager and sales manager, allocate funds for the productions needed.

The executive producer then must find several experienced people to act as producers for news, sports, entertainment, and public service programming. In smaller stations, one producer will wear several hats. The best sources of experienced people to become producers are those with whom the executive producer has worked in the past and those from local stations in the area. Through observation and recommendation, the executive producer will be able to choose people who are ready to move up to a more responsible post and those who have reached a dead-end at their present positions.

In the news area, the executive producer and producer must select on-air talent they feel will draw the greatest audience. They must also acquire field reporters, camera people, and sound technicians to handle local news.

The news producer must act as or hire an assignment editor to assure complete coverage of news events with an eye to economy. The news producer must also develop a format for presenting the news in the most interesting and informative manner possible. This involves working with the station's art department to design distinctive sets and graphics as a means of attracting larger audiences. When a team is operative, the news producer must constantly evaluate performance and recommend changes that will improve ratings.

An additional assignment of the news producer may be sports programming. He or she must survey the community for interest in professional and school team activities and develop shows around these interests. If no local team sport will generate a large enough audience, an option is to produce sports interview programs that feature local coaches and sports personalities, along with professionals who might be in town for other reasons. The biggest problem is competing with the top sports events in the country offered by the networks.

The news producer might also be head of public affairs programming, being constantly aware of community news happenings. Elected officials and candidates for office who can discuss community problems are good sources for this type of program. Citizens airing opinions about upcoming legislation as it will affect the community are another good source. Occasionally a local station will produce an expensive, hard-hitting documentary.

Into the news area also falls the station's editorial policy, which is usually set by the general manager in committee with sales, business, program, and news departments. Station editorials are discussed and approved by this group then turned over to the legal department, which evaluates the terminology from the standpoint of possible legal actions. As with radio, the station is obligated to allow those with opposing viewpoints time for rebuttal.

Another area of concern is entertainment programs that are produced by the station. The producer and executive producer must develop program ideas that will attract the local audience away from network programs. This is no easy chore. Some of the more successful local entertainment shows include variety shows, children's programs, and music video shows.

entertainment

Religious, instructional, and other public service programs are selected and produced to serve the community that constitutes the audience. Programming an independent station in an economical manner has become a truly challenging job.

Network-Owned or -Affiliated Television Stations

Unlike radio, there is a great deal of difference between commercial independent and affiliated or owned TV stations. Legally the difference is the same. A network has financial and programming control over its owned stations, simply supplies programs to its affiliates, and has no formal relationship with independents. However, because there are only about one thousand commercial TV stations, compared with about ten thousand commercial radio stations, the percentage of owned or affiliated TV stations is much greater than in radio.

differences from radio

Networks, like other companies, are allowed to own twelve TV stations. Networks also try to affiliate with stations in as many markets as possible. In markets that have only three stations, independents are usually nonexistent. If a market has only one or two stations, those stations will share networks. In other words, they will affiliate with more than one network. In a four-station market, the lone independent can sometimes be a strong station because it can select from all nonnetwork programming. By the time a market has five or six stations, the point of diminishing returns sometimes sets in for the independent stations.

Programming procedures differ greatly between commercial independents and affiliates. Whereas independents must obtain most of their programming from syndication sources, the affiliates receive the bulk of their programming from the network.

network programs

For the time that the network does not fill, affiliates can produce local programming or obtain programs from syndication. There have been instances when a program originally aired on CBS has been purchased by an ABC affiliate from a distribution company and run during nonnetwork time. When the prime-time access rule was instituted to limit the number of evening hours a network could program, a large new market opened up for syndicated programming. Affiliated station program managers bought this programming to fill the time that networks were no longer allowed to fill.[13]

local and syndicated programs

The programming that does come from the network is under a **network-affiliate contract,** which is a rather complicated document. This contract states that the network will pay the station for the network programs that the station airs. This may at first seem backward because the station is the one receiving the goods. However, the network sells most of the ads connected with the program, so it receives its revenue that way. The more stations that air the program, the larger the audience, and the higher the price the network can charge for the ads.

network-affiliate contract

The amount the network pays the station is decided by a complicated formula that considers the amount the station charges for ads, the time of day, the length of the program, and the number of commercials that the network plans to include in the program.

The network pays the station at a percentage of what the station could have earned had it filled the time with local ads. In prime time, this percentage is about 33 percent, and at other times, it is less. Of course, the network will pay the station a larger total for a one-hour than for a half-hour program. Usually the network will not fill one or two commercial minutes in a program so that stations can sell local ads. It then pays the station less than it would if all the commercial time had been filled. For example, a network pays a station more for airing a one-hour program with four minutes of network ads on Monday at 8:00 P.M. than it pays the station for airing a one-hour program with ten minutes of ads on Monday at 1:00 P.M.

The whole question of network payments to affiliates has been a sore point. Networks, with their shrinking audiences and shrinking dollars, feel that one way they can develop better bottom lines is to cut back on the amount of money they pay affiliates to run programs. Affiliates, of course, are unhappy with this concept.[14]

In addition to payment procedures, there are many other details in the network-affiliate contract. A station can refuse to clear a network program if it feels the program is unsuitable for its audience or if it wishes to air material of greater local importance. The station must indicate its acceptance or rejection of regularly scheduled network programs within two weeks of the time they are offered and must indicate the same for special network programs within seventy-two hours of offering time. If an affiliated station rejects a network program, the network is free to offer that program to another station within the market area.

The network pays the station less if the program is aired on a delay basis—later than the network suggests it be aired. The network holds liability responsibility for lawsuits that may arise from any network program. If a station changes ownership, the network has the right to decide if it wishes to offer continued affiliation to the new owners.[15]

advantages and disadvantages of affiliation

At first glance it appears that the commercial affiliated station program manager has an easier job than his or her counterpart at the independent station. The network provides most entertainment shows and some news and public affairs. This means that the program manager needs to oversee only the local programs, such as news, documentaries, public affairs, and instruction. Needing to find programs for a fraction of the program day permits the program manager of an affiliated station to be more selective. The added advantage of viewer habit in watching the network station provides more money to bid for and obtain the best available talent in the local market.

Less obvious are the program manager's problems of finding good airtime for the required programs. A network may program every day from 7:00 A.M. until 10:00 P.M., with the exception of several hours in the late afternoon and early evening and some time on the weekend. This means that the program manager's selections must air at times when the audience is minimal or when children constitute the major portion of viewers. Such limitations tend to stifle the program manager's creativity and make the station vulnerable to complaints by local community groups. The vulnerability is aggravated by the fact that the affiliated or owned station usually commands the larger share of the available audience than does an independent station and more local groups seek its airtime to tell their message to the public.

Another problem of the affiliated station program manager is the need to constantly review network programs as they relate to the morals and mores of the community the station serves. Programs that seem perfectly innocent in Los Angeles, New York, and other larger markets may offend many people in smaller towns. This can be a costly and time-consuming problem at best and can cost the station its license at worst. In other words, network affiliation is not a cure-all for the program department.

standards

Television Networks

Network prime-time TV programming is the most viewed, most controversial, and most maligned of all broadcast programming. Ideas for network shows come from personnel within the networks and stations, production companies, and on rare occasion, from the public at large.

The first material that is usually developed for a prime-time series is a written **story line**—the general idea for the series and ideas for some of the specific programs. Although these story lines can come from individuals, they are more likely to come from one of the many production companies, such as Universal Studios, Lorimar, Paramount, and MTM, that regularly submit ideas to networks.

story line

Any ideas that turn out well in the eyes of the network executives may be made into a **pilot**—one program taped or filmed as it might appear in the series. Pilots have fallen into some disfavor because they are so expensive to produce. If the series is not chosen and the pilot is not shown, the money spent on it is a total waste. Because both networks and production companies usually wind up investing in pilots, they are often aired as specials or summer one-shot programs in order to make them worthwhile. The concept of a pilot has also been corrupted because production companies, eager to have their ideas accepted, go all out on the pilot and produce it in a much more elaborate form than they can sustain for the entire series.

pilots

Producers and networks battle over the amount the network pays to the producing company in the form of **license fees** to air the series produced. Originally, the amount paid covered all the costs of the production company and left a small profit. However, as production costs have risen and network fortunes have fallen, the amount the network pays often leaves the producer with a deficit between $125,000 and $625,000 per show. If the series was successful, this money can be recouped through syndication or foreign sales, but often the producing company finds that the more it produces, the deeper in debt it falls.[16]

payment

FIGURE 11.2

The set used for the pilot of "Villa Allegre" was designed to represent Don Quixote's home.
(Courtesy of Michael Baugh, set designer)

scheduling

In the early spring of each year a committee of network executives begins making decisions regarding the next fall's program lineup. The committee considers all the programs presently in the lineup, as well as new ideas submitted to it from independent production companies.

These decisions are compounded by such factors as ratings of present programs, fads of the time, production costs of the series, the overall program mix the network hopes to attain, ideas that have worked well or poorly in the past, the kind of audience to which the program will appeal, and the type of programs that the other networks may be scheduling. Often networks attempt **block programming**—the programming of one type of program, such as situation comedies, for an entire evening. This is done in an attempt to ensure **audience flow**—the ability to hold an audience from one program to another.

The final schedule is determined after a great deal of juggling of time sequences and programs. Usually about 80 percent of the programs from the previous season will be killed and replaced by new shows. Many continued programs will have their time slots changed.

budget

Once a series is selected, the production company begins producing programs with an eye toward costs. These costs have risen every year in television's short history. An hour show can cost over a million dollars. Half-hour shows are cheaper, not only because of their shorter length, but because hour

FIGURE 11.3

The set used for the series "Villa Allegre." Many changes took place between pilot and airing. *(Courtesy of Michael Baugh, set designer)*

shows are generally dramas and half-hour shows are usually comedies. Comedies do not have expensive location shoots since they are usually shot onstage. Also, comedies are often shot on tape—a cheaper material than the film used for dramas.

Budget construction is a serious business that depends on the entire production team. The budget is broken down into two main categories: **above-the-line costs** and **below-the-line costs.** The first category covers the team hired to handle the show from inception to completion, including talent, producer, director, and secretaries. The second involves crew, facilities, and physical elements, such as scenery, videotape, film, and graphics.[17]

Budget costs begin almost as soon as an idea is approved. The producer must be hired and he or she, in turn, immediately starts to assemble a team. The writers, associate producers, director, and secretaries are needed first. Then come scenic and graphic artists, lighting director, makeup artists, hairdressers, and, of course, talent.

program production

The director, lighting director, and scenic artist must survey and select several facilities capable of handling the production either onstage or on location. This gives the producer some bargaining power in coming to terms on facilities. Settings must be approved by the creative teams and contracts let

Name of Show _____

Above the Line	Show Budget			Show Actual	
	No. of People	No. of Hours	$	No. of Hours	$
Talent					
Staff Announcers	___	___	___	___	___
Free-Lance Announcers	___	___	___	___	___
Other AFTRA Fees	___	___	___	___	___
Other Talent	___	___	___	___	___
AFTRA P & W	___	___	___	___	___
Miscellaneous	___	___	___	___	___
Producer/Director	___	___	___	___	___
Total Above the Line	___	___	___	___	___

Below the Line—Labor					
Technical—In Shift	___	___	___	___	___
Overtime	___	___	___	___	___
Stagecraft—In Shift	___	___	___	___	___
Overtime	___	___	___	___	___
Film Dept.—In Shift	___	___	___	___	___
Overtime	___	___	___	___	___
Miscellaneous	___	___	___	___	___
Contingency	___	___	___	___	___
Total Labor	___	___	___	___	___

Below the Line—Production Facilities					
Technical Facilities—Standard In-House				___	___
Extra				___	___
Stagecraft—Standard In-House				___	___
Extra				___	___
Stagecraft Sets and Props				___	___
Original Construction Cost				___	___
Weekly Continuing Construction Cost				___	___
Raw Stock Tape and Film				___	___
Film Processing				___	___
Misc. Equipment Rental or Purchase				___	___
Contingency				___	___
Total Facilities				___	___
Total Below the Line				___	___
Grand Total				___	___

FIGURE 11.4

Sample budget work sheet for TV production.

for their construction. Many preproduction meetings are necessary before the show is ready to go into final production. The company then moves into facilities for taping or filming.

Costs are broken down into three phases in final production. The first is setup and rehearsal. Full FACS (facilities) is the second and occurs when all systems are ready to complete the show. The third phase is strike, which is the breaking down and putting away of all sets and props. Now the show is ready for postproduction editing, the final phase before airing or release. Here the producer, director, and editor team up to polish the finished product to a luster that will ensure success, they hope. The program is then given to the network for airing, and the network pays for the right to air the shows for one broadcast season. The network is responsible for promotion which, now that there are many competing media, takes on an important role. After airing, the program is given back to the production company, which can place it in syndication.

Ideally, programs will be produced in blocks to fill a thirteen-week season. If the series is successful, another block of programs will be prepared. However, not all programs last even thirteen weeks. Those with the poorest ratings are usually pulled early in the season and replaced by a series that was an also-ran during the spring meetings or by a summer replacement that drew good ratings. Some years networks cancel so many series during the fall that they actually have the equivalent of several seasons during one year. Sometimes networks juggle programs so much during the year that they really have no season.

cancellation

The cancellation of programs is often controversial, regardless of when it occurs. In a few instances, write-in campaigns by viewers have saved series, but generally the network follows through on its cancellation plans. Usually shows are canceled because of poor ratings, but sometimes there are other reasons. Production costs may soar beyond the revenue that can be generated. Writers may run dry of plot ideas for a particular series, or a star may decide to leave the series. One network may decide to change its type of programming in some time slot in order to counterprogram another network.

This **counterprogramming** is another controversial area. Sometimes one network will select a program specifically to draw an audience away from another network. For example, if an NBC children's program is receiving significantly higher ratings than a CBS drama at 8:00 Sunday evening, CBS may change to a children's program just to attract some of the audience away from NBC. Likewise, if a 9:00 P.M. CBS variety show is outdrawing an ABC comedy in the same time slot, ABC may switch to a variety show to attract some of CBS's audience. This leaves the viewer faced with the dilemma of choosing between two similar programs. Of course, viewers who own VCRs can solve the problem by viewing one program and taping another.

counterprogramming

Preemptions annoy some TV viewers, who become upset when their favorite program is not shown because of a special news event or entertainment program that the network feels it should air. Often there is not unanimity among the TV executives who make the decisions regarding whether an event is important enough to preempt a regularly scheduled program.

preemptions

Another bone of contention concerning network programs centers around **reruns.** The recent trend has been for networks to fund fewer and fewer program productions each year and instead begin the "summer" reruns in the spring. Network executives claim this is necessary in order to have a profitable organization. Because production costs continue to climb, the only way to make the money stretch is to produce less. There are costs involved with reruns, such as the residuals that must be paid to some of the people involved in the original production, but the overall cost of a rerun is much lower than the cost of producing a new program.

Networks program a great deal of material in addition to their prime-time series, such as news programs, documentaries, sports, soap operas, children's programs, and specials. The ideas for and production of most of these programs come from the networks themselves, although soaps are sometimes supplied by production companies or sponsors. Sometimes these programs are canceled because of poor ratings, altered because they are too expensive, or keyed down because they are too controversial. This latter category receives the most publicity.

Producers of documentaries frequently find themselves in a squeeze between reporters who wish to expose injustices and managers who do not wish to alienate any of the hands that feed them. For example, during the era when cigarette companies advertised heavily on radio and TV, documentaries about the hazards of smoking were touchy to say the least.

Those making program decisions in both radio and TV live a fishbowl existence. If their decisions are popular, as well as financially successful, they will ride high; if not, there are always others waiting in the wings.

Public Television

There really is no such thing as an independent public TV station because all the stations are affiliated in some way or another with the Public Broadcasting Service. These PBS stations have both similarities and differences with commercial stations that are affiliated with NBC, CBS, or ABC. The public station program managers have the same problem of finding adequate airtime when non-PBS programs can be aired, but public stations, operating through the **Station Program Cooperative,** have a freer choice concerning which of the network shows they wish to air.

During the yearly decision process period of the Station Program Cooperative, stations have predetermined the nationally-distributed programs that they want and do not need to be concerned with the other PBS network offerings. Also, because commercials are not involved, the elaborate contracts of who pays whom how much and under what circumstances are not needed.

PBS program ideas are station generated to a much greater degree than are those of the commercial networks. The stations propose ideas for the Station Program Cooperative and the stations oversee the productions. Of course, there are programs that PBS acquires from sources other than local stations, particularly the programs from the BBC.[18]

Because the livelihood of a **PBS** show or series is not dependent on rating points translated as advertising dollars, the scheduling process is not as frantic as that found at the commercial networks.

The responsibilities of the local programming staff vary from station to station. For example, some stations like to emphasize formal education programs and others prefer general public affairs programming. Public TV station program heads, like their commercial counterparts are responsible for producing, acquiring, and scheduling whatever programming is needed.

For a typical public TV station, if such a thing exists, the daily programming might be as follows: from 7:00 to 10:00 A.M., the station airs children's programs obtained through PBS; from 10:00 until noon, it airs programs designed for use in schools as part of the curriculum. (Some of these programs might be produced locally and others are produced at other stations or by production companies.) From noon to 1:00 P.M., the station programs local news (locally produced); and 1:00 to 3:00 P.M. again would be taken up with classroom instruction. Children are generally out of school by 3:00, so from then until 6:00 P.M., there is a repeat of the PBS children's programming aired in the morning. The hour from 6:00 to 7:00 P.M. is devoted to college credit courses, and the evening hours until sign-off consist of PBS programming that includes a cooking show, a documentary, a drama, and a symphony.

Cable TV

Many of the decision processes of cable TV imitate those of the broadcast networks. Pay services have independent production companies produce made-for-TV movies and situation comedies.[19] Syndicators often find cable services interested in their products.[20] Cable programs are cancelled and/or rescheduled and as advertising becomes more important to cable, it, too, looks for significant ratings.

Although the words "network" and "affiliates" are used in the cable business, they have a slightly different meaning than they do in broadcasting. Some companies do own both satellite services and cable systems (for example, Time-Life owns Home Box Office and ATC, a cable company), but the financial relationship does not give the satellite service control over the cable system. Because cable systems have many channels, they affiliate with many networks. It is not at all uncommon for a TCI cable system to carry Home Box Office, ESPN, CNN, USA, MTV, and a myriad of other services.

Many of the local cable programming decisions are made by the members of the public who provide the access programming. Some cable systems even have advisory boards of community members who set programming policies.

Much of cable, both local and national, has adopted the concept of **narrowcasting** as opposed to broadcasting. With narrowcasting, the program producer assumes that only a limited number of people will be interested in the subject matter or program as opposed to the mass appeal sought by network

scheduling

local programming

typical programming

similarities to broadcast

affiliates

local input

narrowcasting

TV. In this way, cable takes on some of the characteristics of the specialized magazines—a channel or program that emphasizes health, movies, sports, culture, or music.[21]

Other Media

Patterns of program decision making are not set with some of the newer media, although many of them seem to be lining up at the same door where everyone else lines up—the motion picture production houses. This is true for MMDS, SMATV, LPTV, and videocassettes. None of these media have uncovered substantial programming forms, other than movies, that bring adequate revenue, although some self-help style videocassettes have done quite well.

movies

Hollywood only produces two hundred to three hundred movies a year, and many of these are far from blockbusters. As a result, consumers are often dissatisfied with what they think is inadequate programming on the part of many of the newer media.

data

The data services, such as teletext, obviously do not program movies because of their textual nature. However, they are groping to find just what type of material they can offer that will entice subscribers to pay.

Programming for any of the electronic media is both a challenging and creative activity. This year's hit can be next year's flop, and an idea that has languished for years can suddenly become a success.

Conclusion

Making decisions regarding the product of the telecommunications industry—its programs—is no easy task. Radio and TV programs are not necessities of life and are subject to the whims of the public. In addition, the decisions made by one telecommunications entity affect the decisions of others. The format and special features chosen by one radio station can affect decisions made by a station down the street so that each can carve its niche in the community. Programming decisions also affect the lives and careers of professionals in the field, as they both perform and organize program material.

Radio and television differ in their programming decision organization. In radio, all stations are fairly uniform despite the fact that some are independently owned and programmed—some are independently owned but receive occasional programming from networks, and some are owned by the networks and receive programming from them. The fact that radio networks have been relatively weak has made all radio station program decision making of the same ilk. The primary exception has been public radio stations affiliated with NPR and APR because public radio, by its nature, assumes some format features.

Television, however, is quite different. Independent stations must gather programming from a variety of sources, including independent networks, wire services, and syndication companies that distribute movies, old network shows, and independently produced programs. Their battles to obtain network reruns have been particularly treacherous and have led to production of programs specially made for syndication. Network-affiliated and network-owned TV stations, whether they be commercial or public, receive most of their programs from the networks under conditions spelled out in network-affiliate contracts. Both independents and affiliated stations must produce their own local programs, but while affiliated stations usually have the larger audiences for these programs, independent stations have the most flexibility in terms of time.

Because TV networks have been strong, their decision-making process has been geared to a high level of competition among both networks and the production companies vying to sell program ideas to networks. Also, because of the strength and high visibility of networks, they have been criticized for practices that foster allegedly unwarranted cancellations, needless counterprogramming, undesired preemptions, spiraling costs, and endless reruns.

Public television programming revolves, to a great extent, around the Station Program Cooperative. Stations program the PBS material and local programming, which they design and acquire themselves.

Cable TV has undertaken a narrowcasting approach to programming, but many of the decision-making processes are similar to those of commercial TV.

The newer media have a chance to experiment with program decision processes, but many have opted to make movies their mainstay.

Thought Questions

1. What type of local programming could radio and TV stations in your area offer that they are not offering now?
2. Should TV networks be allowed to program as much of an affiliate's day as they do? Why or why not?
3. Should there be a requirement that all network series run at least thirteen weeks before they can be canceled? Give reasons for your answer.
4. In what direction do you think program decision making among the newer media will go?

Regulation and Business

Telecommunications is a business. Like any other business, it must maintain an awareness of income and outgo, of profit and loss. Unlike many businesses, the product can be elusive, with programs disappearing from consumers the moment after they receive them. Some people say the product is not really programs at all, but rather people—people delivered before a TV set so they can watch commercials. Although some of the media forms do not involve advertising, in most cases the sponsor's dollar is the ultimate source of finance. Telecommunications is also a regulated business—victim or victor of the political vicissitudes of the nation. The interweaving of sponsors, audience, employees, and government make it a challenging and rewarding business.

PART

V

Regulatory Bodies

12

Overview

There are few countries of the world in which the government has such a hands-off policy toward telecommunications as in the United States. In some countries, the government totally controls both the finances and the programming of telecommunications entities. In other countries, the government oversees and directs overall philosophy and content.

In the United States, government bodies affect and regulate telecommunications but not to any great degree. In addition, there are groups and individuals who exert informal regulatory control through the democratic process. These groups and individuals can bring pressure to bear upon broadcasters that is sometimes more potent than formal regulation.

A discussion of the organization and functions of the various regulatory agencies, both formal and informal, includes these topics:

The method by which the three branches of government interrelate in the regulatory process

The structure of the FCC in terms of commissioners, offices, bureaus, staff, and functions

The technical functions of the FCC, including frequency allocations, call letters, and EBS

The FCC's role in the granting, renewal, transfer, and revocation of licenses

The Federal Trade Commission and its functions

The involvement in regulation of various agencies of the executive branch of government

The rationale for the original congressional intervention into broadcasting and the ongoing role of the House and Senate

The appeal function of the courts

Organizations that influence telecommunication decisions, particularly the NAB

Awards for excellence in programming given by various organizations

The broadcast standards function of stations and networks

The influence of citizen groups and individuals

Arguments for and against regulation

A broadcast license is a license to print money.

Lord Roy Thompson

Regulatory Interweaving

Many entities are involved in regulation of the telecommunications industry, both formally and informally. They range from Congress to individual citizens, from the Federal Communications Commission to broadcast producers. The regulatory groups intermingle, in part because of the manner in which the nation's forefathers established the basic government. The legislative branch writes the laws, the executive branch administers them, and the courts adjudicate them.

democratic process

This basic process carries over to telecommunications regulation. A broadcast station that feels it has been wronged by a decision of the Federal Communications Commission can appeal that decision to the courts and also lobby in Congress to have the offending law changed.

In addition, democracy accords many rights to individuals and to organizations. These, too, have their impact upon the regulatory process.

Various entities become involved in telecommunications regulation in varying degrees. The Federal Aviation Authority becomes involved only when improperly lit antenna towers may be a hazard to airplanes. On the other hand, the Federal Communications Commission's main function is the regulation of the airwaves.

varying involvement

The Federal Communications Commission

The Federal Communications Commission (FCC) is an independent regulatory body that was created by Congress because of the mass confusion and interference that had arisen when early radio stations broadcast on unregulated frequencies and at unregulated power. First Congress passed the **Radio Act of 1927,** which created the Federal Radio Commission (FRC) to deal with the chaos. Then in 1934 Congress wrote a new law, **the Communications Act,** which formally established the FCC with powers similar to its predecessor, the FRC.

origins

The FCC is composed of five commissioners appointed for five-year terms by the president with the advice and consent of the Senate. The president designates one commissioner to be chairperson, but generally no president has the opportunity to appoint many commissioners because their five-year terms are staggered. Each commissioner must be a U.S. citizen with no financial interest in any communications industry, and no more than four of the five commissioners are supposed to be from one political party. Usually commissioners have backgrounds in engineering or law.

commissioners

The commission maintains central offices in Washington and field offices in thirty districts. The commission staff is organized into seven administrative offices (managing director, general counsel, science and technology, plans and policy, Congressional and public affairs, administrative law judges, and the review board) and four bureaus (mass media, field operations, common carrier, and private radio).

organization

FIGURE 12.1

FCC organization
chart.

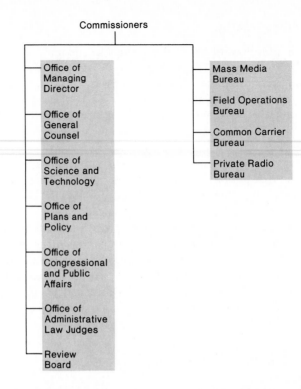

Commissioners

Office of Managing Director

Office of General Counsel

Office of Science and Technology

Office of Plans and Policy

Office of Congressional and Public Affairs

Office of Administrative Law Judges

Review Board

Mass Media Bureau

Field Operations Bureau

Common Carrier Bureau

Private Radio Bureau

Policy determinations are made by all the commissioners, with the chairperson then being responsible for the general administration of the commission's affairs. Most of the day-to-day work, such as handling interference complaints, public inquiries, and station applications, is undertaken by the staff.[1]

nonbroadcast functions

The FCC has myriad functions, many of which are not related to electronic mass media. For example, it has jurisdiction over airplane communications, ship-to-shore radio, police and fire communications, telephone and telegraph common carrier services, ham radio operations, military communication, and citizens' band radio. In wartime it coordinates the use of radio and TV with the national security program and may set up a service to monitor enemy propaganda. It is constantly encouraging new uses of radio waves, particularly those that will promote safety.

All of this makes for a heavy work load for the commission, particularly as new uses for radio waves surface. The sudden popularity of citizens' band radios in the mid-1970s swamped the commission with problems of interference and channel reallocation,[2] and the more recent interest in low-power TV stations and multichannel multipoint distribution service has done likewise.[3]

engineering functions

A great deal of what the FCC does in regulating radio and TV stations involves engineering. The FCC assigns **frequencies** to individual stations, determines the power each can use, and regulates the time of day each may operate. It then polices the broadcasters to make sure they stay within the frequency, power, and time regulations and to make sure unauthorized persons do not use the airwaves. In fact, about one-fourth of all FCC employees are employed in this fieldwork. The FCC also makes overall regulations to prevent interference among stations and regulates the location of station transmitters and the type of equipment used for transmission.

The FCC also controls the general allocation of frequencies, deciding which frequencies go to ship-to-shore communication, which to TV, which to FM radio, and so forth. Within the frequencies it allocates to radio and TV, the FCC creates a table designating the power and times that stations can exist in each section of the country and on each frequency. It also has the jurisdiction to set technical standards, such as those for color television or stereo TV. It can decide not to set a standard, however, as in the case of AM stereo. frequency allocations

The commission deals with **call letters** of all stations. Those stations west of the Mississippi begin with K and those east of the Mississippi begin with W (except for some of the early stations such as KDKA in Pittsburgh that had call letters before the ruling went into effect). A station can select or change the other letters of its call letters so long as the letters it chooses are not already in use by a competing station or do not in some way infringe on the rights of another station.[4] call letters

The **Emergency Broadcasting System** is also under the jurisdiction of the FCC. This is a national hookup that ties together all radio and TV stations so that information can be broadcast from the government to the citizenry during a national emergency. All stations are required to maintain the equipment necessary for receiving emergency notification and to test this equipment regularly. If a state of emergency is declared, some stations will remain on the air broadcasting common information and others will shut down and remain off the air until the emergency is over. EBS

Most of the technical responsibilities of the FCC are noncontroversial and broadcasters generally appreciate the FCC's role in this regard. The sensitive points arise in the area of licensing. The FCC has the power to grant, renew, transfer, and revoke licenses.

Any citizen, firm, or group interested in a radio or TV license must file a written statement of qualifications with the FCC. Most of the regular radio and TV frequencies have been allocated, but the granting of licenses is now brisk in such areas as LPTV and MMDS. licensing

One category of qualification for any license is character, which includes obvious matters such as felony convictions, participation in community organizations, and desire to be involved in the day-to-day operations of the stations. Aliens and foreign companies cannot own U.S. stations except in unusual circumstances. character

Applicants for licenses must also describe their financial and technical qualifications. Financially, they must have access to enough capital to build and begin operation of the station. finance

Applicants applying for a new AM station or a low-power TV station must arrange for an engineering investigation to establish that the stations will not interfere with other stations because of its frequency, power, or hours of operation. An applicant for an FM or regular TV station can consult the allocation tables already set up by the FCC to find a place for a station.[5]

programming

The applicant must also set forth a full statement for its proposed program service that the FCC can take under cautious consideration. The FCC has no power to censor program materials. In fact, the Communications Act specifically prohibits censoring by stating that:

> Nothing in this Act shall be understood or construed to give the Commission the power of censorship over the radio communications or signals transmitted by any radio station, and no regulation or condition shall be promulgated or fixed by the Commission which shall interfere with the right of free speech by means of radio communications.[6]

This indicates that the FCC cannot refuse a license to a financially, technically, and morally qualified candidate simply because it plans to broadcast some form of programming that the FCC finds objectionable. However, another section of the Communications Act states that:

> The Commission, if public convenience, interest, or necessity will be served thereby, subject to the limitations of this Act, shall grant to any applicant therefore a station license provided for by this Act.[7]

This **"public convenience, interest, or necessity"** statement has become a keystone of regulation.

CP

Once an applicant has filed a written statement of character and financial, technical, and programming qualifications, the FCC must decide whether or not it should grant permission to build the station. If there is only one applicant, this process is fairly simple. If the paperwork indicates that the applicant will be successful and conscientious, then a **construction permit** (CP) is issued and the applicant can begin to build the station.

lottery

If there are multiple applicants, which is usually the case, then the FCC must decide which applicant will receive the CP. This used to be a very laborious process whereby the FCC staff members sorted through all the applications trying to determine which group or person was most qualified. Now, for the most part, the FCC decides station allocations by **lottery.** In essence, the names of all applicants who meet the qualifications are "put in a hat" and the winner is chosen by chance.[8]

This applicant then receives the CP and can begin building the station. After construction is completed, the license and permission to begin conducting program tests are granted. Once a station receives a license, it usually experiences little supervision from the FCC until license renewal time.

license renewal

License renewal is the most controversial and publicized function of the FCC, mainly because the process has undergone radical changes over the years.[9] Early in its history, the FCC was very lenient with license renewal, refusing to renew a license only if there was blatant wrongdoing, such as continually broadcasting with excessive power.

Gradually over the years, the FCC began giving more attention to whether or not the station was serving the public convenience, interest, and necessity. In the 1960s, it changed the license renewal forms so that stations had to state how they planned to meet the needs of the community during the upcoming three years. This process was known as **ascertainment** and involved interviewing many community leaders to learn what they felt were the primary issues of the community. The broadcasters then set about designing programs to deal with these issues.

ascertainment

The results of these interviews and the proposed program ideas were then submitted to the FCC. Along with this material, copious information concerning station operation for the previous three years, including a **composite week's** list of programming, was also sent. These seven days from the previous three years were selected at random by the FCC. The FCC's main concern here was whether the programs listed adhered to what the station had proposed three years earlier.

license application

During this period, the FCC also seriously considered complaints written against the stations by members of the community. As a result of all this, before broadcasters were given license renewals, they were scrutinized much more carefully than in the past. In addition, community members became important in the station renewal procedure. Prior to the 1960s, license renewal had been a private affair between the stations and the FCC, but during the sixties and early seventies, community people became quite involved.

community members

In some instances, community groups asked the FCC to deny the renewal so that they could operate the station themselves. These were called **comparative license renewal** hearings because the group holding the station was compared to the group wishing to take over the station.

comparative license renewal

In other cases, citizen groups did not want to operate the channel, but rather wanted a new licensee that would be more sympathetic toward the ideas or causes of their group. These groups filed **petitions to deny** station licenses as they came up for renewal. These petitions formally asked the FCC to deny a particular station a license and listed the reasons the denial was requested. Such petitions came to be used more as bargaining points than actual attempts to have licenses changed. It was generally cheaper for the stations to give in to the citizens' demands than to undergo a lengthy hearing process in Washington.[10]

petitions to deny

Some licenses were changed and many were threatened during this period, but by the end of the 1970s, the pendulum began to swing the other way. The FCC and broadcasters began to fear that they had given too much to special interest groups and too little to station owners who had worked hard to establish and nurture their stations.

new legislation

New renewal license policies were initiated in 1981 that extended radio licenses to seven years and TV station licenses to five years. The license renewal forms were also changed to decrease the work undertaken by stations in order to renew licenses. In fact, radio station renewal forms were reduced to the size of a postcard. Ascertainment is no longer required, but TV stations are asked to make quarterly lists of five to ten community issues and tell how they are meeting those issues through programming.[11]

revocation

If the FCC finds that a station is not fulfilling its obligations in serving in the public convenience, interest, and necessity, then it can revoke the license. The FCC has done this for such causes as unauthorized transfer of control, technical violations, fraudulent contests, overcommercialization, and indecent programs.[12]

other penalties

The FCC can also issue lesser punishments. It can impose a cease-and-desist order, which really is no punishment at all but a letter telling a station to stop a certain action. It can fine stations up to $2,000 for every day that an offense occurs, up to a maximum of $20,000. It can issue a short-term renewal for six months or a year, which indicates to the licensee that it must mend its ways or the license will be revoked.[13]

license transfer

The FCC also becomes involved when a station is sold and the license is transferred from one party to another. It is the station management's prerogative to set the selling price and select the buyer, but the FCC does check on the character and financial, technical, and programming qualifications of the buyer. The FCC used to have a regulation that required new owners of a station to keep it for at least three years before selling it. This was to prevent people from buying stations just for the investment potential with no consideration for the community welfare. However, with deregulation, this regulation disappeared and now stations are traded much more frequently than they used to be.[14]

networks

In the area of licensing, the FCC has no direct control over the networks, but it can control them indirectly through their owned and operated and affiliated stations. For example, when the FCC limited the number of hours of network programming that a station could air during prime time, a half-hour was chopped off network programming. When CBS was found to have rigged some tennis matches, the FCC slapped CBS's hand by giving a short-term license renewal to the CBS-owned Los Angeles station, KNXT.[15] But the networks themselves are not licensed. Consequently, they go through the license-renewal process only vicariously by way of the stations carrying their programs.

The FCC, then, serves as the major government watchdog over broadcasting and is charged with implementing the provisions of the Communications Act and with developing its own policies and guidelines. The extent of its power and the degree to which it exercises its regulatory muscle changes as the electronic media, government, society, and the FCC commissioners themselves change.

The Federal Trade Commission

establishment

Another independent regulatory body that becomes involved in broadcast regulation from time to time is the Federal Trade Commission. This commission was established by the Federal Trade Commission Act and the Clayton Act. Both acts were passed in 1914, basically to prevent unfair competition resulting from the giant corporate trusts. In 1938 the **Wheeler-Lea Amendment**

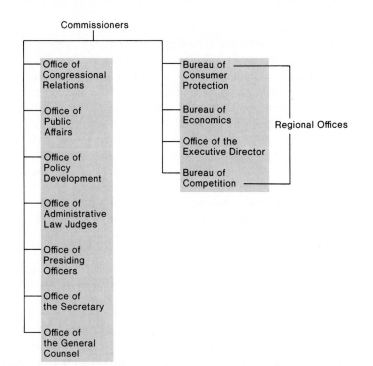

FIGURE 12.2

FTC organization chart.

to the FTC act was passed. It declared that "unfair or deceptive acts or practices in commerce are also illegal."[16] This opened the door for the FTC to look into matters where consumers were being deceived, and it served as the takeoff point for the FTC's involvement in broadcasting.

This involvement concerns mainly fraudulent advertising that is deceptive to consumers. If the FTC feels that a company's ads are untruthful, it has the power to order the company to stop broadcasting the ads. In some instances, it can make the company broadcast a message publicly disavowing the false claims in an untruthful ad. Needless to say, this is distasteful to advertisers.

function

Five FTC commissioners are appointed for a term of seven years by the president with the consent of the Senate. The president appoints the chairperson, who in turn handles most of the management and personnel aspects of the commission. Working directly under the chairperson is the executive director, who as the commission's chief operating officer oversees day-to-day operations of the FTC offices and bureaus. The commission organization includes people to advise the commissioners on such issues as legal matters, policy development, competitive practices, and consumer education. Throughout the country there are regional FTC offices to handle local complaints and problems and to refer them to Washington, if necessary.

organization

The FTC attempts as much as possible to be preventive rather than punitive in its relationships with broadcasting, as well as other forms of business. In other words, it tries to disseminate its philosophies to businesses in the form

guidelines

of written statements known as **Trade Regulation Rules, Industry Guides,** and **Advisory Opinions** so that businesses know how to avoid the wrath of the FTC. If, for example, the commission decides to issue guidelines on toy advertising, it will inform all toy companies that it is planning to create rules in their area. Executives of toy companies, as well as members of the general public, then have a chance to express opinions about what these rules should include. After the members of the FTC have heard the various opinions, they write the guidelines and publicize them as widely as possible among toy companies, related trade associations, and consumers. The Trade Regulation Rules, Industry Guides, and Advisory Opinions do not have the force of law; they are suggestions that the FTC hopes businesses will follow in order to avoid trouble.

If a company violates one of these guides or engages in other practices that the FTC feels hurt competition or deceive consumers, then the FTC can bring action against the company.

Obviously, the FTC staff cannot act as a constant watchdog over every company in the country. Although the staff does occasionally bring charges, it depends greatly on the public at large and businesspeople in general to bring complaints to it.

complaint procedure

Often investigations are initiated by letters from consumers sent to the Washington FTC office or to one of the regional offices. These letters are first reviewed to make sure that the complaints expressed actually come under the jurisdiction of the FTC. If so, the investigation begins with FTC personnel questioning officials of the company involved in the alleged misdeed. Usually the FTC wants proof from the company that its ad is accurate. For example, if the complaint is against a car manufacturer who says its car is 50 percent quieter than any competitor's car, the FTC will require scientific proof of this claim.

At the end of the investigation, the FTC may close the case for failure to find a violation or it may issue a formal complaint and a cease and desist order to the company involved. If the company does not feel it should have to stop its alleged violation, it can appeal to an FTC hearing examiner, then to the FTC commissioners, and from there to the U.S. District Court of Appeals, and eventually to the Supreme Court. Very few cases actually go to the Supreme Court; most are reconciled at the lower levels.

children's advertising

The FTC occasionally holds hearings on broad issues beyond one particular company's ads. In the late 1970s it planned a series of hearings on advertising on children's TV, a very controversial subject. When evidence suggested that at least one FTC commissioner had made up his or her mind against the ads before the hearings were to start, broadcasters complained. As a result, Congress refused to allocate the funds for the hearings and they did not materialize to any significant degree.

Although the FTC is not as involved with the electronic media as the FCC, it occasionally handles issues of great concern to media advertisers.[17]

The Executive Branch

Obviously, the President has influence over the media, both formally and informally. Formally, he can suspend broadcasting operations in time of war or threat of war and call into action the Emergency Broadcast System. The president also nominates commissioners and appoints both the FCC and FTC chairpersons. Most of the interaction between the president and the media, however, comes informally, with the president seeking positive coverage from radio and TV. This reached a peak during the Nixon administration when Nixon created an **Office of Telecommunication Policy** to advise him on the media and to make statements concerning media practices. This office's complaints about the type of coverage Nixon was receiving created an even wider gap between him and the press.

president

Partly as a result of this, President Jimmy Carter disbanded the OTP and formed the **National Telecommunications and Information Administration,** which he placed under the Department of Commerce. Although this organization advises the president on media issues, its placement in the Department of Commerce keeps it somewhat distant from the president.[18]

NTIA

The Department of State has a Bureau of International Communications and Information Policy that advises on media issues with possible international consequences. From time to time, other departments, such as defense and education, become involved with media-related issues.

Other agencies that are either part of the executive branch or are independent occasionally affect broadcasting. The Equal Employment Opportunity Commission watches over telecommunications companies to assure that they are fulfilling commitments to affirmative action. The Federal Aviation Authority becomes involved with antenna towers and lighting so that planes will not crash into antennas. The Food and Drug Administration occasionally becomes involved with the misbranding or mislabeling of advertised products. The surgeon general's office became a party to the 1972 ban of cigarette advertising on radio and TV.

other agencies

The Legislative Branch

Congress is heavily involved in telecommunications. It passed the Communications Act of 1934, which set up the basic broadcasting structure. Major changes in this structure must be approved by Congress as amendments to this act.

The whole basis for the government's intervention in radio frequencies was, and occasionally still is, debated in Congress. Ordinarily, the First Amendment would prohibit the government from infringing on the rights of citizens to communicate by whatever means they wish. However, Congress invaded this area because of the **scarcity theory.** Not everyone who wants to can broadcast through radio frequencies because this would cause uncontrollable interference. As a result, some body, namely Congress, needed to intervene and determine a mechanism for making decisions regarding who could and could not use the airwaves.[19]

scarcity theory

Once given this right, Congress needed to find a provision in the Constitution that covered the subject. The provision chosen was the one empowering Congress to regulate commerce with foreign nations and among the states. Although broadcasting was not commerce as such—goods did not change hands—the exchange of information by mail and wireless had already been accepted as a form of commerce, so broadcasting was added to this definition.[20]

In addition to overall regulatory decision making that is tied to the Communications Act, Congress passes other laws, such as the copyright law and the cable TV law, that have important impact upon those dealing with telecommunications. Congress approves FCC and FTC commissioners and budgets, and it monitors the FCC through both the House and Senate subcommittees on communications. These and other committees also often conduct special investigations into such aspects of telecommunications as the quiz scandals or management consolidations.

State legislative bodies can also affect telecommunications. Under the Constitution, federal laws such as the Communications Act take precedence over state laws. Occasionally state laws dealing with libel, advertising, or state taxes have an effect upon elements of the industry.

The legislative branch generally sets up broad policies that are handled on a day-to-day basis by other agencies. Needless to say, these broad policies are very influential in setting direction for the electronic media.

The Judicial Branch

The courts of the United States have had a significant impact on telecommunications, primarily through the appeal process. FCC decisions concerning licenses can be appealed through the U.S. Court of Appeals for the District of Columbia Circuit in Washington, D.C. This court can confirm or reverse the commission's decision or send it back to the FCC for further consideration. Appeals for FCC or FTC decisions that do not involve licenses can be taken to any of the eleven other U.S. Courts of Appeal, generally referred to as circuit courts. From any of these courts, final appeals can be taken to the Supreme Court, which can confirm or reject a lower court's decision. The Supreme Court can also refuse to hear a case, which means that the lower court's finding will stand.

Courts are also significant in dealing with decisions other than those arising from FCC or FTC decisions. Rulings on freedom of speech, obscenity, censorship, copyright, monopoly, and many other subjects often have an impact that determines what directions electronic media can or cannot go.

Broadcasting Organizations

Organizations to which media practitioners belong often serve as regulators in an informal manner. They publish policy statements, lobby with Congress, and hold meetings at which pertinent issues in telecommunications are discussed.

commerce

congressional laws

state laws

appeals procedure

court decisions

Regulation and Business

The largest, most influential organization is the **National Association of Broadcasters** (NAB). It was formed in 1923 to counter demands from the American Society of Composers, Artists, and Publishers (ASCAP) regarding increases in the amount ASCAP was planning to charge radio stations for airing music.[21] When that issue was resolved, the NAB acquired many other purposes, such as lobbying in Washington for actions favorable to broadcasters, conducting broadcast-related research, acting as a public relations arm for broadcasters, and holding conventions and workshops to facilitate communication and professional growth among industry members.

NAB

At one time, it also acted as a strong self-regulation force. It developed two codes—"The Television Code," and "The Radio Code." The codes were printed in two booklets that spelled out dos and don'ts for radio and television stations.[22] Stations were not required to follow these rules; they were merely advisory. But sometimes the inclusion of a provision in the NAB code prevented that provision from being legislated in a manner that would be compulsory.

codes

The codes contained two major types of statements—those dealing with programming and those dealing with advertising. The advertising portion was fairly specific about the amount of time that could be devoted to commercials during different times of the day and was the part of the code that could be utilized most tangibly.

This portion of the code came under attack, however, in 1979. The Justice Department, acting on its own, started an antitrust suit against the advertising provisions stating that these guidelines artificially limited the supply of advertising time, increased the cost of commercial time, and violated the **Sherman Antitrust Act.** In 1982 the U.S. District Court sided with the Justice Department. The NAB stated that the decision made absolutely no sense because it would mean that the public would be subjected to more commercials. Nevertheless, it announced that it would retire the time standards portions of the code. That, in effect, killed the self-regulatory function of the NAB.[23]

Justice Department action

The NAB still exists as an effective voice for broadcasters, but it is no longer the self-regulatory body it once was.

Self-regulation is promoted by other organizations, too. Advertisers have an organization called the **National Advertising Review Board,** which handles complaints about ads from individuals, groups, or an advertiser's competition. **The Radio-Television News Directors Association** has a Code of Broadcast News Ethics. Although these are both fairly formal methods of self-regulation, many other organizations have more informal guidelines established at meetings.

NARB

Most of the organizations represent subgroups of the radio and TV business, the proliferation of which is often a matter for criticism. For example, broadcast executives can choose to join the National Association of Television Program Executives, the Association of Maximum Service Telecasters, the Broadcast Financial Management Association, the Broadcasters Promotion Association, the International Radio and Television Society, the Association of Independent Television Stations, and/or the National Association of Farm Broadcasters.

other organizations

FIGURE 12.3

A panel discussion on student internships at a convention of the Broadcast Education Association. *(Courtesy of Livingston Hinckley)*

For news personnel there are such organizations as the Radio-Television News Directors Association and the Radio-TV Correspondents Association.

Engineers probably have more associations available to them than any other group. These include the Audio Engineering Society, the Institute of Electrical and Electronics Engineers, the Society of Broadcast Engineers, the Society of Motion Picture Television Engineers, and the Society of Cable Television Engineers.

Advertising organizations also abound, including the American Advertising Federation, the American Association of Advertising Agencies, the Advertising Research Foundation, the Association of National Advertisers, and the League of Advertising Agencies.

Cable TV's equivalent to the NAB is the National Cable Television Association. A cable organization for access producers is the National Federation of Local Cable Programmers. Many of the newer media have formed organizations specifically related to their businesses, such as The Society of Satellite Professionals and the MDS Industry Association.

For educators there are organizations such as the Broadcast Education Association and the University Film and Video Association.

The Academy of Television Arts and Sciences is for those involved in the creative activities of programming; Alpha Epsilon Rho is a national honorary broadcasting society; and Broadcast Pioneers is open to those who were involved in early radio and TV.

Other representative organizations include American Women in Radio and Television, Women in Cable, Television Critics Association, National Religious Broadcasters, Station Representatives Association, and National Black Media Coalition.[24]

No one can belong to all relevant organizations, but membership in selected ones can help a person keep up to date and in touch with others with similar interests and regulatory approaches.

FIGURE 12.4

The Emmy, awarded
for outstanding
television
programming. *(Courtesy
of the Academy of Television
Arts and Sciences)*

Awards

Like the organizations, broadcasting awards have an informal regulatory
function. Broadcasters covet awards and use them for personal and station
promotion. Therefore, they will frequently air programs that they feel might
qualify for awards.

Most famous of the TV awards are the **Emmys** that are bestowed by the
Academy of Television Arts and Sciences for many categories of national prime-
time programming. (Daytime awards are given by a different but similar organization, the National Academy of Televison Arts and Sciences.) Each year
the categories within which awards will be made are hotly disputed by committees of industry representatives. How many westerns should be on TV before
an award is given for best western? Is sound mixing on videotape different
enough from sound mixing on film to merit separate Emmys? Should writers
of short series be placed in the series category or the specials category? These
intramural disputes occur to some extent because of the changing nature of
TV programs, the huge outpouring of material, and the importance placed on
winning an Emmy.

The NCTA has established **ACE** (Awards for Cablecasting Excellence)
awards that are presented each year, and the National Federation of Local
Cable Programmers presents awards for access programs.

Emmys

cable

For commercials, the top statuette is the **Clio,** given annually at the American TV and Radio Commercials Festival Awards for best U.S. radio and TV commercials. The prestigious George Foster **Peabody** Awards for Distinguished Broadcasting are bestowed annually in news, entertainment, education, youth programs, documentaries, and public service.

Other awards include the Armstrong Award for excellence in radio broadcasting; the Freedoms Foundation Awards for programs to bring about a better understanding of America; the Golden Mike Awards for excellence in newswriting and presentation; the National Press Photographers Association Awards for best news stories; and the Ohio State Awards for educational, informational, and public affairs broadcasting.[25]

Network and Station Policies

Many networks and stations have their own written or unwritten codes. For example, some stations will not accept beer and wine ads and others publish a list of words that they consider obscene or indecent.

At the commercial broadcast networks, the business of making sure all programs and commercials adhere to good taste falls to the **broadcast standards** department, a group usually operating independently of programming or sales and reporting directly to top network management. This group reviews all program and commercial ideas when they are in outline or storyboard form and then screens them several times again as they progress through scripting and production. If any of the proposed ideas run counter to standards, the broadcast standards department requests changes before the idea can proceed to the next step. A broadcast standards employee almost always observes program rehearsals and tapings and may request costume or movement changes. The department is also responsible for making sure all copyrighted material used within programs has been cleared and that all advertising claims can be substantiated.

Naturally, all is not roses and moonlight between broadcast standards and program or commercial producers. The latter can appeal decisions to top management or try to convince the broadcast standards department to compromise on certain issues.

At local stations the broadcast standards function may not be as formal. Frequently the general manager performs the function on an "as needed" basis, and sometimes the function is delegated to whichever group or person handles station legal matters or public relations.

Within the cable TV industry, talk of self-regulation has arisen from time to time, particularly in regard to hard-R movies, but nothing has been implemented.

Network and station self-regulation occurs constantly and informally at all levels through the actions of individuals employed in broadcasting. A disc jockey who decides not to play a record he or she considers in poor taste is engaging in self-regulation, even though management provides no direction on the matter. Likewise, writers, producers, directors, actors and actresses, and editors are constantly basing decisions on their own ideas of propriety and appropriateness.[26]

Citizen Groups and Individuals

Community and national citizen groups can alter broadcasters' behavior and thereby execute informal regulation. Organizations such as the NAACP and NOW have been successful in changing both program fare and hiring practices as they relate to women and minorities. Action for Children's Television has had an important role in changing children's programming. Consumer groups have frequently organized boycotts against products advertised on certain programs, and broad consumer groups have established a power base to affect overall programming.

citizen groups

Professional critics occasionally serve as a catalyst for establishing self-regulation. Their individual influence is small since television programs live and die by ratings rather than by critical review, but their columns are sometimes used to substantiate points being made by other groups. Occasionally critics have been instrumental in saving a public affairs or documentary show that might have been dropped because of poor ratings.

critics

Broadcasters are often the object of unfavorable comments made by national opinion leaders. It is hard to tell whether Newton Minow's "vast wasteland" speech or former Vice-President Agnew's blast at broadcast journalism led to any self-regulation, but both certainly caused a stir within the industry.

opinion leaders

All of these forces—the branches of government, organizations, awards, media practitioners, and citizens—aid in shaping the regulation of telecommunications.

The Pros and Cons of Regulation

The relationship between the electronic media and the government is a forced marriage that has never enjoyed a honeymoon atmosphere. Broadcasters are ever fearful that their privileges will be regulated to oblivion, and government agencies are afraid that broadcasters will take advantage of their privileges to the detriment of the entire system. The upper hand is attached to a constantly swinging pendulum seeking a middle ground.

Sometimes government regulation protects the broadcaster. Obviously the technical regulations fall into this category, preventing the interference and chaos that reigned in the 1920s. Also, the system generally ensures that stations are licensed only to people of proper character so that one or two bad apple stations cannot ruin the entire broadcasting barrel.

protection of broadcasters

Licenses are also refused to station applicants who might promote unfair competition, and antitrust suits are filed against station owners who tend to monopolize. This protects the free marketplace and helps all broadcasters stay in business once they have obtained a license.

Government regulations generally attempt to protect station owners from being overpowered by the networks. The ruling that limits the networks to three prime-time programming hours per night is an example of an attempt in this direction. The ruling that disallowed one company from owning two networks, thus separating NBC Red and Blue into NBC and ABC, was also an attempt to protect station owners.

protection against rapid change

Regulation also protects against rapid, irreversible change. Both the FCC and FTC are criticized for laxity and slowness, but by setting down only vague guidelines and deciding most issues on a case-by-case basis, both organizations prevent cataclysmic changes that would jar the systems of individuals and companies. There is usually a great deal of advance warning and discussion before any sort of policy is set, and even what is set can easily be overturned if it does not work as intended. The net result is moderation and the kind of policy that survives a host of tests. This indirect method of regulation allows for compromises that can resolve conflicting points of view and for new interpretations of the Communications Act.

Criticism of government regulation abounds, particularly within broadcasting circles. Regulation is condemned because it drifts, stalls, and vacillates. Terms such as "interest, convenience, and necessity" on which major decisions are made are inadequately defined. Semantic problems plague the definition of "public service." Guidelines issued by the various agencies are often muddled to start with, altered frequently, and then reversed by the courts. The broadcaster trying to abide by the law becomes confused and befuddled and must wade through thick books that attempt to explain broadcast rules, regulations, and policies. FCC and FTC commissioners often have widely disparate views as to the extent to which their agencies should regulate, and many of their decisions are by a one-vote margin.

slowness

The snail's pace of decision making is a constant frustration to broadcasters. The length of time needed to obtain a TV construction permit or to secure permission to alter a technical device can try souls and pocketbooks. The long, drawn-out hearings, almost always held in Washington, are a great personal and financial inconvenience to the broadcasters involved. Some of the slowness is attributed to the ineptness of the bureaucracy involved and some to the overdeliberation of commissioners and judges. Many government agencies suffer from an overload of work that not only slows the process, but also prohibits thoroughness. There are so many problems inherent in broadcasting that the FCC can do little more than put out fires. Neither commissioners nor staff members have time or opportunity to develop policy or do long-range planning. Hence, many decisions are based on the facts of today without consideration for the plans of tomorrow.

ineffectiveness

Even in areas where the FCC and FTC claim to regulate in a systematic manner, they are often accused of being ineffective. For example, the FCC is supposed to see that stations broadcast in the public interest, but educational, religious, and public service groups do not really feel helped because the material they do get to air is low-budget and broadcast at undesirable times. The FCC does not regulate airing time or production cost for these groups, so, in effect, it is not aiding their cause significantly.

Federal Trade Commission rulings against false advertising can be particularly ineffective. By the time the FTC conducts an investigation and issues a cease and desist order for a particular commercial, the commercial has run its course anyhow and sold many products, so the company is quite willing to remove it from the airwaves. Sometimes legal proceedings concerning a type of advertising may go on for ten years, and all during that time the advertising continues. About the only real clout the FTC has regarding any particular ad is the bad publicity that might arise for the company involved. This effect can be unfortunate, too; if the company is innocent, it still suffers from the bad publicity.

Many people feel that the regulating agencies overstep the thin line that separates watchdog and dog watched. The primary criticisms are against the FCC's incursions into programming decisions. The Communications Act specifically prohibits censorship, but some of the actions taken by the FCC in the name of "public interest" are controversial enough to raise cries of censorship. overstepping bounds

Regulation suffers from people problems that often seem insolvable. Commissioners who come to the FCC have little prior experience with broadcasting, and yet some of them use their FCC position as a stepping stone to high-paying executive positions within broadcasting. Senators and representatives depend on the support of broadcasters to become elected and, hence, are often heavily influenced by broadcasting lobbies. Commissioners, because they know their decisions can be appealed to the courts, often make conservative decisions that are unlikely to be overturned. people problems

In the area of self-regulation, there are those who mourn the death of the NAB codes. One of the primary complaints about broadcasting is the proliferation of commercials. If broadcasters are prevented from keeping their own house in order on this issue, then how can public indignation be handled? self-regulation

Self-regulation in the area of the newer media presents particularly interesting problems. Should X-rated movies be allowed on cable or, for that matter, on cassettes? Perhaps there should be self-regulation standards for those services that are on the basic tier of cable and that all subscribers must receive if they want to receive any cable TV, but no self-regulation for the upper-tiered pay services that subscribers are free to select or reject. Interactive text services may need regulation to protect consumers from fraudulent money schemes or rigged surveys that could cause scandals akin to the quiz show scandals of the 1950s. Unfortunately, the potential for service and variety inherent in the new media also brings the potential for abuse.

The extent to which citizens should be involved in station operation is also debated. On one hand, stations exist for the public, so the public should be allowed to have its say regarding policies and operations. On the other hand, many citizens do not understand what is involved in operating a station and, as a result, make unreasonable demands. citizen groups

The informal vehicles of self-regulation—organizations and awards— can be criticized for over-proliferation, wheel-spinning, and inequality, and yet they can point with pride to their accomplishments.

Despite all the pros and cons of both formal and informal regulation, broadcasters have managed to exist with it through periods of growth and rapid change, and it is likely to remain with them in some form during the years to come.[27]

Conclusion

Broadcasters interrelate with all three branches of government—legislative, executive, and judicial—and with independent regulatory bodies. Congress gives the broad stroke to regulation by enacting and amending the Communications Act. It also approves the appointment of commissioners and holds hearings on communications-related issues.

Within the executive branch, the president appoints commissioners and can call into effect the Emergency Broadcast System. Most of the President's influence is informal, however.

The courts are the place of appeal for those who feel they have been dealt with unjustly by other branches of the government or by elements of society. The courts have heard cases on such subjects as freedom of press, obscenity, and copyright.

The main government agency interrelating with the electronic media, however, is the FCC, which issues specific guidelines and handles the main regulatory chores. Some of these, such as prevention of interference, are technical in nature and have not changed much over time. Others, such as licensing, are more philosophical and have been subject to changes as society's outlook alters. The areas of granting, renewing, transferring, and revoking licenses have caused broadcasters varying degrees of consternation over the years.

While the government can regulate in a formal manner, there are other institutions that can regulate informally. Organizations to which media practitioners belong are among these. Some of them have codes that serve as guidelines for behavior. The NAB is the largest organization for broadcasters, but its codes have been struck down by the Justice Department.

Some stations and networks have their own internal regulatory forces, often called broadcast standards departments. Citizen groups, critics, and awards can all strongly encourage telecommunications forces to produce programming to meet specific needs.

Although regulation is often criticized in our democratically run country, its values are evident and its influence is much less than in other countries.

Thought Questions

1. Should citizen groups be able to challenge TV station licenses? Why or why not?
2. Should FCC commissioners be required to have broadcasting experience? Explain.
3. Was the Justice Department correct in bringing suit against the NAB Code? Why or why not?

Laws and Regulations

13

Overview

The regulatory bodies discussed in Chapter 12 create laws and regulations governing the telecommunications industry. These range from informal suggestions to constitutional amendments. They are executed, enforced, and interpreted.

Anyone involved in broadcasting and its related areas must keep up on the latest changes in regulatory procedures in order to operate effectively. Innumerable laws and regulations are relevant to the industry. This chapter tries to cover the most important by discussing:

The role the FCC has assumed over the years regarding license renewal

Changes that have occurred regarding the acceptability of cross-ownership

Issues that fall under the guidelines of the First Amendment

Censorship versus clear and present danger

Electronic media obscenity cases, including "seven dirty words" and "topless radio"

The history of libel cases, including the Westmoreland case

The changes that have allowed TV cameras in courtrooms

Editorial regulations from the Mayflower Decision to the current public broadcasting ruling

The equal opportunity provisions for political candidates under Section 315 of the Communications Act

Some cases that have arisen under Section 315, including those related to splinter party candidates, libel situations, major candidate debates, and entertainers as candidates

How the fairness doctrine arose from editorial and Section 315 decisions and its evolution through such cases as Red Lion, Banzhaf, and Friends of the Earth

The present status of the fairness doctrine

The most recent trends in laws and regulations

A function of free speech under our system of government is to invite dispute. It may indeed best serve its highest purpose when it induces a condition of unrest, creates dissatisfaction with conditions as they are, or even stirs people to anger.

William O. Douglas
former Supreme Court Justice

Regulation Interweaving

various bodies

The government regulatory bodies that oversee telecommunications create and refine many laws and regulations affecting the conduct of the industry. Sometimes they are laws drafted by Congress; sometimes they are regulations or advisory statements issued by such agencies as the FCC or FTC; and sometimes they are policies or precedents set as the result of specific court cases.

Often all of these interrelate. Congress passes a law that the FCC executes, but a station may be unhappy with this execution and appeal to the courts. Sometimes the courts decide the FCC had improperly executed the law and request a new implementation. Other times the courts side with the FCC and the station's recourse is to go to Congress to try to have the basic law amended.[1]

This type of interrelationship can be seen in the actions that have arisen regarding license renewal.

Licensing

The power to renew licenses was given first to the Federal Radio Commission in 1927 and then to the Federal Communications Commission in 1934. Both of these agencies renewed licenses almost automatically in the early years. Two cases did arise, however, in the late 1920s and early 1930s where a license was not automatically renewed.

Brinkley case

One concerned Dr. J. R. Brinkley who broadcast medical "advice" over his Milford, Kansas, station. He specialized in goat gland operations to improve sexual powers and in instant diagnosis over the air for medical problems sent in by listeners. These problems could always be cured by prescriptions obtainable from druggists who belonged to an association Dr. Brinkley operated.

Shuler case

The other culprit was the Reverend Robert Shuler, who used his Los Angeles station to berate Catholics, Jews, judges, pimps, and others in his personal gallery of sinners. He professed to have derogatory information regarding unnamed persons who could pay penance by sending him money for his church.

The Federal Radio Commission decided not to renew either of these licenses and, although both defendants cried censorship, the Court of Appeals and U.S. Supreme Court sided with the FRC on the grounds that with only a limited number of frequencies available, the commission should consider the quality of service rendered. In ruling on the Shuler case, the court wrote that "if stations were possessed by people to obstruct the administration of justice, offend the religious sensibilities of thousands, inspire political distrust and civic discord, or offend youth and innocence by the free use of words suggestive of sexual immorality, and be answerable for slander only at the instance of the one offended, then this great science, instead of a boon, will become a scourge, and the nation a theater for the display of individual passions and the collusion of personal interests. This is neither censorship nor previous restraint."[2]

Aside from these two cases, the early FCC was quite lenient in license renewal, being more concerned with granting original licenses. But in 1945 the FCC decided to philosophize a bit about license renewal in order to clarify public convenience, interest, and necessity, while at the same time issuing temporary renewals to six stations so that it could have time to decide whether they were, indeed, serving the public interest.

In 1946 the commission issued an eighty-page document detailing its ideas on license renewal: "Public Service Responsibility of Broadcast Licensees." This document was almost instantly dubbed "the **Blue Book**" by broadcasters, in part because of its blue cover but more sarcastically because blue penciling denotes censorship. The document stated what has come to be known as the **promise versus performance doctrine.** It stated that promises made when stations were licensed should be kept, and the performance on those promises should be a basis for license renewal.

Blue Book

Broadcasters probably would not have objected to that philosophy, but the document went on to detail proper broadcasting behavior. It was particularly adamant about avoiding the evils of overcommercialization, broadcasting public affairs programs and local programs, and maintaining well-balanced programming. The document was so lengthy that broadcasters looked at it as an affront to freedom of speech and a violation of the section of the Communications Act that stated the commission was not to have the power of censorship. On these grounds, broadcasters fought the Blue Book, and its provisions were never truly implemented.[3]

Licenses continued to be renewed almost perfunctorily. Stations did receive rebukes for such violations as failure to make proper entries in logs, failure to broadcast station identifications frequently enough, failure to have engineering instruments calibrated properly, failure to authenticate sponsorship of programs, failure to present controversial issues properly, and failure to give equal time to political candidates. For these violations, stations were fined, issued cease and desist orders, and issued short-term probationary licenses of less than three years, but these types of activities were few and far between.[4]

In 1960 the FCC softened its tone and issued a much briefer statement of policy that listed fourteen elements usually necessary to meet the public interest:

1960 statement

1. Opportunity for local self-expression
2. Development and use of local talent
3. Programs for children
4. Religious programs
5. Educational programs
6. Public affairs programs
7. Editorializing by licensees
8. Political broadcasts
9. Agricultural programs
10. News programs
11. Weather and market reports
12. Sports programs
13. Service to minority groups
14. Entertainment programs

It also warned broadcasters to avoid abuses to the total amount of time devoted to advertising as well as the frequency with which programs are interrupted by commercials, but it did not specifically define what constituted abuses.[5] In addition, the FCC changed the license renewal forms so that broadcasters had to undertake **ascertainment**—a process by which they had to interview community leaders to obtain their opinions on the crucial issues facing the community so that they could design programming to deal with those issues.

For several years after the 1960 changes, licenses continued to be renewed rather perfunctorily. Over a thirty-five-year period dating from the 1920s, only forty-three licenses were not renewed out of approximately fifty thousand renewal applications.[6]

In 1964 a black group from Jackson, Mississippi, led by the United Church of Christ, asked the FCC if it could participate in the license renewal hearing of station **WLBT-TV.** The group felt the station was not serving its viewership properly and was presenting racial issues unfairly. The FCC refused this hearing because prior to this time, only people who would suffer technical or economic hardship from the granting of a license were permitted to testify at hearings, but it did issue WLBT a short-term license.

The UCC appealed to the courts stating that ordinary citizens should be able to be heard concerning license renewals. The courts agreed and ordered that a hearing be held. After this 1966 hearing, which some felt was a sham, the FCC decided WLBT should receive a full three-year license. The UCC once again appealed and the courts decided WLBT should have its license withdrawn and also scolded the FCC for its administrative procedures during the hearing. WLBT was turned over to a nonprofit group, but the significance of this case was that it established the precedent for citizen participation in license renewal.[7]

In 1969 a group of Boston businesspeople successfully challenged the ownership of TV station **WHDH,** which had been operated by the Boston Herald Traveler for twelve years. This was a very complicated case and involved rumors of improprieties on the part of all three companies that had originally applied for the station shortly after the freeze was lifted. The station had been awarded to the Herald Traveler in 1957, but various allegations had kept the ownership longevity in turmoil. In 1962 the Herald Traveler was given a four-month temporary license in the hope that the situation would be clarified within that time.

Four years later, an FCC hearing examiner recommended that the license be kept by the Herald Traveler, mainly because it had done a good job with programming during the four years that were supposed to have been only four months. Three years later (or seven years after giving WHDH a four-month license), the commission itself revoked the Herald Traveler license and awarded it to Boston Broadcasters, Inc., an organization that included some of the original unsuccessful applicants for the station back in the 1950s.[8]

WLBT-TV case

WHDH case

This January 1969 decision sent shivers through the entire broadcasting community. Never before had a station license been transferred involuntarily unless the station licensee had first been found guilty of excessive violations. From all that was written about the case, most broadcasters concluded that the FCC did not base its decision on the fact that the Herald Traveler was an undesirable licensee, but rather on the opinion that the businesspeople were better. Broadcasters immediately began pounding on the doors of Congress to urge passage of legislation that would prohibit license challenges. Meanwhile, the WHDH decision was taken through the appeals courts and upheld.

The net result of the WHDH and WLBT cases was that a raft of renewal challenges were filed. In some instances, groups requested a **comparative license renewal** hearing because they wanted to operate the stations themselves.

In other cases, citizen groups filed **petitions to deny** license renewal mainly so that they could bargain for what they wanted. A 1969 case that became a model for bargaining between citizens and stations was that of **KTAL** in Texarkana, Arkansas. The citizens complained because KTAL, although licensed to Texarkana, was neglecting that city by moving its studios to Shreveport, Louisiana, a much more lucrative market. The groups met and worked out a thirteen-point plan that would assure that KTAL served Texarkana. The citizen group then withdrew its petition to deny the renewal. The FCC applauded the efforts of the citizens and station and said it would examine the record of KTAL when it came up for renewal again to make sure it was, indeed, serving Texarkana.[9]

KTAL

License renewal was also no longer perfunctory as evidenced by the fact that even public broadcasting stations had renewal troubles. In 1970 the **Alabama Educational Television Commission** applied for what it assumed would be routine renewals for its public television stations. However, the FCC refused renewal because of citizen petitions stating that the stations had systematically deleted all Public Broadcasting Service programs dealing with blacks or Vietnam. The Alabama commission was allowed to reapply for its stations with the understanding that it would mend its ways. Nevertheless, the case marked the first time such harsh action was taken against any public broadcasting stations.[10]

Alabama public TV

One license renewal case that has dragged on for years involves **RKO** and the sixteen stations that it owned. Several of the RKO stations had been under attack by citizen groups during the early 1970s and petitions to deny renewal had been issued against them. Then, in 1980, the FCC refused to renew the licenses of three of the RKO television stations—WNAC in Boston, WOR in New York, and KHJ in Los Angeles. Its reason for doing this was that RKO's parent corporation, General Tire and Rubber, had admitted to the Securities Exchange Commission that it had bribed foreign officials. The FCC felt that RKO, therefore, did not meet the character requirements necessary to operate radio and TV stations.

RKO

The license refusals were appealed and the courts agreed that WNAC should lose its license. The license was turned over to another company, New England Television Corporation, which changed its call letters to WNEV-TV.

The WOR license went through an interesting evolution. The state of New Jersey did not have any VHF stations and had long objected to that fact. Senator Bill Bradley from New Jersey introduced a bill into Congress that directed the FCC to grant a five-year license to the owner of any VHF station that would agree to relocate the station in a state having none. This bill, which became law in 1982, was obviously aimed at RKO, giving it the opportunity to salvage its license by moving to New Jersey. WOR did just that and received a five-year license renewal by the FCC. Before the station came up for license renewal in 1987, RKO sold it to MCA.

The courts sent the KHJ case back to the FCC for further deliberation. This was frustrating to both those for and against RKO because the KHJ case had actually begun in 1965 when a company called Fidelity challenged the license of the station. As part of the deliberation process, the FCC decided applications could be made for the other RKO licenses, just in case RKO was found an undesirable licensee in the KHJ hearings. The hearings also uncovered that RKO Radio had overbilled advertisers $7.9 million between 1980 and 1984—a fact that did not help RKO's case.

During the course of the hearings, RKO tried to sell its remaining thirteen stations, something the FCC encouraged. However, most of the deals broke down because of complications. For example, in 1985 Group W wanted to buy KHJ and proposed that, because of RKO's questionable status, the license be given to Fidelity for a brief period of time and then sold to Group W. The FCC did not approve this plan, however, and a year later Group W got tired of waiting and dropped out. Disney then offered to buy KHJ but, because of FCC deliberations, did not meet with instant success.

Meanwhile, in 1987, an FCC administrative law judge ruled that RKO was unfit to hold any broadcast licenses, sending the whole situation into a deeper mire.[11]

1970 and 1977 statements

Overall, however, the threat of losing a license abated during the late 1970s and early 1980s. In 1970 the FCC issued an opinion stating that if a licensee could demonstrate that its program service had been substantially attuned to the needs and interests of the community, that licensee would be preferred over a newcomer when applying for renewal. In 1977 the FCC issued another statement to the effect that stations did not need to provide superior service in order to have a license renewed, but that the service needed to be above the level of mediocrity.[12] Both of these statements led broadcasters to breathe a sigh of relief, for they indicated that the FCC would act favorably toward incumbents.

renewal legislation

The renewal legislation that extended radio station licenses to seven years and TV station licenses to five years also bolstered stations, as did the simplifying of the renewal process. However, there are still movements afoot to obtain legislation to eliminate comparative renewal.

License renewal obviously has had a checkered history. Future paths are as yet uncharted, but a philosophy of deregulation appears to be predominant as the 1990s begin.

Another area of licensing that has been subject to fluctuations is what has come to be known as **cross-ownership.** Many of the early owners of radio and TV stations were companies that also owned newspapers. This was considered against the public interest, especially if a town had only one TV station and it was owned by the newspaper. The fear was that the public would learn only one version of the news. On the other hand, the companies with news interests can operate stations more efficiently because their news functions can serve a double purpose. The FCC wrestled with this problem for many years, never finding a politically acceptable solution. Finally, the Justice Department became interested in the subject and urged the FCC to take action.

cross-ownership

In 1975 the FCC issued guidelines on cross-ownership that prohibited companies from owning both a newspaper and a broadcast station in the same town. Later, it also prohibited such companies from owning cable TV systems. However, it grandfathered any cross-ownership that already existed, so there still are companies that own combinations of newspapers, radio, TV, and cable TV in particular areas. Eventually this will disappear as changes in ownership occur, unless the FCC issues new more lenient guidelines—something it is considering doing.[13]

A new wrinkle in cross-ownership is rearing its head with the newer media. The telephone companies are pressing to own cable TV in areas they serve. At present this is not permissible, but hearings and court judgments may change that.[14]

The philosophical shifts that have occurred in licensing procedures have all taken place under one law—the Communications Act of 1934. This act, like much of the legislation of this country, allows for change and broad interpretation.

The First Amendment

The First Amendment to the U.S. Constitution is basic to much of what occurs in the telecommunications industry. This amendment states:

> "Congress shall make no law respecting an establishment of religion, or prohibiting the free exercise thereof; or abridging the freedom of speech, or of the press; or the right of the people peaceably to assemble, and to petition the Government for a redress of grievances."

freedom of speech and press

The **freedom of press** and **freedom of speech** aspects of this amendment have a great effect upon broadcasting and have been subject to many court cases and formal and informal regulations.[15]

Issues are raised under many banners, one of them being **censorship.** If newspeople feel the government is trying to censor or withhold information, they consider this a breach of freedom of the press. The government often replies that such withholding of information is a **clear and present danger** to the country. If the press were told everything, the government would not be able to compete with other governments of the world.

censorship

Clear and present danger extends to individuals, also. In one of its decisions, the Supreme Court noted that freedom of speech does not give someone the right to falsely yell "Fire" in a crowded theater because this would create a clear and present danger to those in the theater. On the other hand, governmental agencies are reluctant to label words as dangerous if freedom of speech is involved. During the 1960s, the NAACP wanted the FCC to censor speeches being made over Georgia TV by a candidate for the Senate on the grounds that the speeches contained racially inflammatory remarks that were a danger to the TV stations and to the people of Georgia. The FCC refused to issue prior restraint in that case, saying that the speeches did not literally endanger the nation.[16]

source disclosures

News reporters often feel freedom of the press is being violated if they must disclose sources from which they have obtained news or if they must surrender outtakes from their stories or testify as to their "state of mind" while they were researching a story.[17]

The First Amendment has been used for a myriad of other subject areas. Producers, writers, directors, and actors of the Hollywood community sued

family hour

under the First Amendment when **family hour** was instituted. They claimed that the FCC, NAB, and networks had conspired to censor prime-time programming by requiring that the content be suitable for children. They succeeded in having the family hour concept repealed.[18]

cable franchising

Cable TV companies successfully used the First Amendment to win the right to build a cable system in a city, even though they had not been granted a franchise. The rationale was that cable TV operators are recognized First Amendment speakers and should not be denied the right to cablecast.[19]

polling interviews

Journalists won the right to continue to interview people within three hundred feet of polling places by invoking First Amendment rights.[20]

The issues related to the First Amendment that have plagued telecommunications practitioners the most deal with obscenity, libel, access to the courts, editorials, political candidates, and fairness.

Obscenity

changing definition

Obscenity is one of the major areas where the First Amendment comes into play. It is outlawed by the U.S. Criminal Code, but there have been many instances where this criminal code and the First Amendment clash. Often it is difficult to determine when someone's freedom of speech should be abridged because what they are saying is obscene. Part of the problem surrounding this is the changing definition of obscenity. What was considered obscene in one decade may be perfectly acceptable in the next decade. Also, what is obscene in Nebraska may be fairly commonplace in Hollywood.

"seven dirty words"

A number of obscenity cases have arisen in radio and television. One of the most famous arose when a Pacifica Foundation public radio station in New York, **WBAI**, aired a program on attitudes toward language. It was aired at 2:00 in the afternoon and included a comical monologue segment that spoofed seven dirty words that could not be said on the public airwaves. A father driving in the car with his son happened to hear this monologue and complained to the FCC. The FCC placed a note in WBAI's license renewal file, which led the station to appeal through the courts all the way to the Supreme Court. The high court determined that the words were not actually obscene, but they should not have been broadcast during the daytime when children might hear.

In another instance, the FCC fined a station $2,000 for airing what became known as **"topless radio."** A number of stations throughout the country aired talk shows to which listeners called and recounted explicit sexual experiences. After one station was fined, and this fine was upheld by the courts, this form of talk show disappeared from the airwaves.[21]

topless radio

In 1987 the FCC sent letters to three radio stations asking them to respond to allegations of airing indecent and obscene material. In doing this, the FCC strengthened its "seven dirty words" Pacifica stand by stating that language or material that depicts sexual or excretory activities in terms offensive to contemporary community standards would be in violation if broadcast at a time of day when there is reasonable risk that children are in the audience.[22]

"seven dirty words" strengthened

The newer media are more likely to see action regarding obscenity than commercial broadcast radio and TV because some of them are geared toward smaller audiences or include provisions that make it more difficult for the mass audience to be exposed to them. For example, the Playboy Channel must be specifically subscribed to before it comes into the home. Likewise, someone must make an effort to rent or buy a videocassette to show on a home recorder.

newer media

In Utah a group tried to use a state obscenity statute to keep sexual acts and nudity off cable TV in the state. However, the courts ruled that this statute could not be used, clearing the way for Utah airing of such services as the Playboy Channel.[23] Video stores have also been taken to court for stocking sexually-oriented cassettes, but have been successful in winning the cases by using the First Amendment.[24]

Utah case

The area of obscenity in the telecommunications area will undoubtedly receive more attention as time progresses.

Libel

Most libel laws and principles that apply to electronic media have their roots in the written press.[25] In early days of radio and television, there was debate as to whether libel applied at all. Libel is defined as defamation of character by published word, whereas **slander** is defamation by spoken word. Slander carries less penalty than libel, ostensibly because it is not in a permanent form to be widely disseminated. Some people felt radio and television should be under slander laws because their words were spoken rather than printed. But because these spoken words are spread far and wide, broadcast defamation has come under the libel category.

slander

Libel was not a big issue with the broadcast media until the late 1970s. Part of this was due to the fact that radio and television engaged in very little investigative reporting, so did not open themselves for libel suits. Radio and television newscasts and documentaries are also limited in terms of time, so generally deal only with well-known issues and people. People who are **"public figures"** have great difficulty winning a libel suit because the rules and precedents applied to them are much stricter than for ordinary citizens. In order to win a libel suit, a public figure must prove that a journalist acted with **actual malice.** As a result of all this, very few public figures bothered to bring libel suits against the broadcast media.

public figures

By the late 1970s, however, TV was a major source of information and a dominant force in society that some felt had become overly arrogant. What was said about people on TV definitely affected their reputations and livelihoods. As a result, libel cases were on the upswing. In the early 1980s, almost 90 percent of the people who filed libel suits against broadcasters won. This was soon reduced to 54 percent as broadcasters began taking these cases more seriously.[26]

cases

Cases brought to the courts included the following: a Chicago businessman who said ABC's "20/20" made him look like an arsonist because he had an interest in some buildings that burned; a Philadelphia mayor who claimed the CBS station there had broadcast, incorrectly, that he was the target of a federal investigation; a former army officer who felt he was depicted as a liar on a "60 Minutes" segment dealing with charges of army cover-ups of atrocities during the Vietnam War; a right-wing presidential candidate who sued NBC but wound up being told to pay NBC; and singer Wayne Newton, who claimed that an NBC newscast made it appear that he had strong ties with the Mafia. Some of these cases have been settled either in or out of court and others are still winding their way through the appeals process.[27]

Westmoreland

In 1985 a very important case that had been called "the libel case of the century" came to a quiet close. It had been brought against CBS by General William **Westmoreland.** The suit centered around a documentary aired in January of 1982 called "The Uncounted Enemy: A Vietnam Deception." In this documentary CBS accused Westmoreland of purposely deceiving his military superiors and President Lyndon B. Johnson with estimates of Vietnam enemy troop strength that were much lower than was really the case. Westmoreland filed a libel suit asking for $120 million in damages, but because he was a public figure, he had to prove actual malice on the part of CBS. The trial began in 1984 with both Westmoreland's reputation and the credibility of network reporting at stake. The trial progressed for several months, then was settled out of court. CBS issued a statement saying Westmoreland had fulfilled his patriotic duties as he saw them and Westmoreland withdrew from the suit. Because the trial did not go to jury and because it ended in a rather bewildering manner, the effect on the body of libel law was negligible. Most observers saw the case as a victory for CBS and predicted it would have a chilling effect on other public figures tempted to sue the media.[28]

Such seemed to be the case because succeeding libel cases, most of them against newspapers, were generally decided in favor of the media.[29]

Access to the Courts

A clash between the First and Sixth Amendments has led to difficulties for broadcast journalists in the courts. Although the First Amendment guarantees freedom of press, the Sixth Amendment guarantees **fair trial.** For many years lawyers and judges felt a fair trial was impossible if cameras were present in the courtroom.[30]

In 1937 the American Bar Association adopted a policy, known as **Canon 35,** that barred still cameras and radio from courtrooms. Later TV cameras were included under the same policy. This was not a law, simply an ABA policy. Most judges abided by this policy, so it effectively kept cameras out of courtrooms for many years. The rationale was that cameras would detract from the proceedings, cause disruption, and lead defendants, lawyers, and jurors to act in ways that they would not act if no cameras were present.

Canon 35

In the early days of television there was reason for this rationale because of the bulky equipment and lighting needed for TV coverage. However, in the 1970s, when unobtrusive ENG equipment became available and when TV was no longer such a novelty, broadcasters began pressuring for access to the courtrooms.

In 1972 the American Bar Association liberalized Canon 35, redesignating it **Canon 3A.** It recommended that individual states and individual judges be given discretion as to whether or not to allow cameras into their courtrooms. One by one states began allowing cameras in courtrooms on an experimental basis. Nothing went awry, so by the mid–1980s all states were allowing cameras into the courtrooms under specific conditions. Each state and, in some instances, each judge set specific rules for media coverage of trials. For example, cameras had to be located in fixed positions; juries could not be photographed; and lights could not be used.

Canon 3A

The right of broadcast media to be present in the courts was further affirmed in 1981 when two Florida defendants claimed that televised coverage of their trial had denied them due process. The Supreme Court ruled against them, stating that the Constitution does not prevent states from allowing TV cameras in the courts.[31]

Florida case

The federal courts and the Supreme Court itself have been more and more receptive to the idea of cameras in their courtrooms and have allowed them in limited ways.[32]

Supreme Court

Controversies still arise regarding court trial coverage. Certain subjects, such as rape and child molestation, are sensitive and usually decided on a case-by-case basis.[33] Overall, the broadcast media has covered trials in a tasteful manner that has not proven disruptive.

Editorializing

Editorializing was first ruled on in 1941 in a case involving **WAAB,** a Boston radio station that for several years had been expressing its views on political candidates and controversial issues. This case became known as the **Mayflower Decision** because the Mayflower Broadcasting Corporation filed a challenge to the license renewal of WAAB, stating that the station had not served

Mayflower decision

in the public interest and that its license should be transferred to the Mayflower Broadcasting Corporation. Part of Mayflower's case stated that WAAB had been editorializing and that these one-sided presentations were not in the public interest. The FCC rejected Mayflower's bid, but on grounds having nothing to do with editorializing. The FCC stated in its report on the case that it disapproved of editorials and felt broadcasters should not be advocates.

Cornell petition

This decision, although resented, was not challenged by broadcasters because few of them were interested in editorializing. However, during the 1940s several top members of the National Association of Broadcasters became interested in the issue and the **Cornell University** radio station petitioned the FCC to reconsider the Mayflower Decision. As a result, after hearings held from 1947 to 1949, the FCC finally reversed itself, stating that stations should be encouraged to editorialize.

opposing viewpoints

However, in order to limit one-sided points of view, the FCC ruled that stations had to make a positive effort to see that **opposing viewpoints** were also broadcast. Stations were also to make sure that the people chosen to refute editorials were qualified, so that the rebuttal had the same chance of convincing the public as the original editorial.

This decision did not have any immediate impact on commercial broadcasting. Stations did not immediately rush into editorializing, since they traditionally had not engaged in it. The stations were also unsure about what exactly the FCC meant when it said they had to seek opposing viewpoints. Over the years, editorializing guidelines have been defined and refined by court cases and FCC rulings. It is now generally accepted that all editorials do not need to be refuted, but if the character or integrity of a group or person is going to be attacked, that person or group must be notified and given reasonable opportunity to reply. Also, if the station plans to endorse a political candidate editorially, it must notify his or her opponents within twenty-four hours and provide opportunity for reply.[34]

public broadcasting

For many years public broadcasting stations were forbidden to editorialize because they received federal funds. However, a public radio chain, Pacifica Foundation, took this restriction to court and in 1984 the Supreme Court ruled that prohibiting public stations from editorializing was a violation of the First Amendment.[35]

Equal Time

Section 315

Section 315 of the Communications Act has come to be known as the equal time provision. In reality, it deals with **equal opportunity** rather than equal time. Section 315, as written in the 1934 Communications Act, reads as follows:

> If any licensee shall permit any person who is a legally qualified candidate for any public office to use a broadcasting station, he shall afford equal opportunities to all other such candidates for that office in the use of such broadcasting station, and the Commission shall make rules and regulations to carry this provision into effect, provided that such licensee shall have no power

of censorship over the material broadcast under the provisions of this section. No obligation is hereby imposed upon any licensee to allow the use of its station by any such candidate.[36]

This provision is in effect only during periods of election campaigns and deals only with candidates for political office. The people it covers must be legally qualified candidates who have publicly declared that they are running for political office. It does not apply to any particular person until he or she has officially declared for office. In other words, although it may be common knowledge that a senator is planning to run for reelection, opponents cannot demand equal time for the senator's appearances until he or she makes an official public declaration. Because there are deadlines for declaring one's candidacy for office, a person must become an official candidate well before the actual election.

official candidates

Because Section 315 actually mentions equal opportunity, it guarantees much more than that all candidates for a particular office will be given the same number of minutes of airtime on a particular station. The airtime given must approximate the same time period so that one candidate is not seen during prime time and another in the wee hours of the morning. The cameras and other facilities that the first candidate is able to use must be available to opponents. All candidates must be able to purchase time at the same rate and if a station decides to give free time to one candidate, it must give equal free time to all opponents.

equal opportunity

However, if one candidate purchases time and other candidates do not have the money to purchase equal time, the station does not need to give free time to the poorer candidates. Because some candidates usually have larger campaign chests than others and because some candidates prefer to concentrate their campaigning on radio and TV, while others choose to emphasize other publicity methods such as direct mail or billboards, candidates rarely have equal time on any particular radio or TV station. The station is not required in any way to equal the actual time as long as equal opportunity for the time was afforded each candidate and the candidates were the ones who chose not to avail themselves of the time.

purchasing time

The provisions apply equally to all candidates, whether they be from the major Republican and Democratic parties or from smaller ones such as the Libertarian and Progressive parties. As long as candidates have met the requirements for the office for which they are running, they are covered under the provisions of Section 315.

all parties

Through the years, interesting cases have arisen in regard to Section 315. Shortly before the 1956 election campaign between Dwight Eisenhower and Adlai Stevenson, Eisenhower, who was the incumbent president, was given free time on all networks to talk to the American people about an urgent crisis in the Middle East. Stevenson's forces immediately demanded equal free time. However, the FCC decided that Eisenhower's speech was exempt from Section 315 because it did not deal with normal affairs, but with a crisis situation. This has become a precedent that has been followed in other election years with other presidential, crisis-oriented speeches.

Eisenhower's crisis speech

In 1959 **Lar Daly,** a candidate for mayor of Chicago, protested that the incumbent mayor, who was running for reelection, had been seen on a local news show and Daly demanded equal time on that station. Had he been an ordinary candidate, he might have been granted time and the issue dropped. However, he was an eccentric who dressed in an Uncle Sam outfit—complete with red, white, and blue top hat. Daly ran, unsuccessfully, for some office in every election. Daly was granted his equal time, but Congress amended Section 315 so that the equal time restrictions would not apply to candidates appearing on newscasts, news interviews, news documentaries, and on-the-spot coverage of news events.

The actual wording of this amendment was as follows:

Appearance by a legally qualified candidate on any—

1. bona fide newscast
2. bona fide news interview
3. bona fide news documentary (if the appearance of the candidate is incidental to the presentation of the subject or subjects covered by the news documentary), or
4. on-the-spot coverage of bona fide news events (including but not limited to political conventions and activities incidental thereto),

shall not be deemed to be use of a broadcasting station within the meaning of this subsection. Nothing in the foregoing sentence shall be construed as relieving broadcasters, in connection with the presentation of newscasts, news interviews, news documentaries, and on-the-spot coverage of news events, from the obligation imposed upon them under this Act to operate in the public interest and to afford reasonable opportunity for the discussion of conflicting views on issues of public importance.[37]

This amendment enabled broadcasters to continue to perform their usual news-oriented functions without being inundated with requests from minority candidates. However, the minority candidate issue still remains a thorny one. When large numbers of candidates are running for a particular office, radio and TV stations often find themselves devoting an inordinate amount of airtime to political commercials and programs. Stations that would like to give free airtime to major candidates so that the public could hear their views often refrain from doing so because of the huge amount of broadcast time that would then be taken up by splinter party candidates.

In 1956 WDAY-TV in Fargo, North Dakota, granted time to a splinter candidate running for the Senate. In keeping with the regulation of Section 315, the station did not censor the candidate's speech, which later became the subject of a successful libel suit against both the candidate and the station. The state supreme court declared that the station should be immune from damages because it could not, by law, censor the speech. The U.S. Supreme Court agreed, stating that Section 315 grants a station immunity from liability in the case of such libelous material.

The splinter candidate issue was raised again in 1960 when Democrat John F. Kennedy and Republican Richard M. Nixon were running for president. The networks wanted the two to debate on television and the two candidates agreed, but Section 315 would have necessitated equal debating opportunity to all the other candidates on the presidential ballot. Because Congress felt that this was a special situation, it suspended Section 315 temporarily for the 1960 presidential and vice-presidential offices only.

major candidate debates

Broadcasters hoped that similar suspensions would be forthcoming, but the issue took a different turn. The FCC, under the 1959 Communications Act amendment that allowed on-the-spot coverage of bona fide news events, decided in 1975 that networks and stations could cover candidate debates if these were on-the-spot news events. The broadcasters themselves could not arrange the debates, nor could the candidates, but if some other group, such as the League of Women Voters, set up the debates, the media could cover the event in much the same way as they cover awards ceremonies or state fairs.

In 1976 and 1980, presidential debates were shown in this way—sponsored by the League of Women Voters and telecast as news events. Then in 1983, the FCC ruled that debates could be sponsored by broadcasters. The League of Women Voters appealed, but the court sided with the FCC, stating that broadcasters should have the same right to hold debates and exclude candidates as a civic group.[38]

In 1971 Congress again amended Section 315 so that stations had to sell time to candidates at the lowest rate that they would sell that time to advertisers who regularly purchased large quantities of time. This significantly lowered the amount of money that candidates paid for announcement time.

lowest rate

In recent years, entertainers and other TV personalities have been entering the political arena at an increasing rate. This has created a dilemma for broadcasters in terms of Section 315. If, for example, a candidate has a legitimate profession as a station weathercaster, must the station give equal time to other candidates each time the weathercaster is seen or heard on radio or TV? The answer is "yes," but the practical reality is "no." In 1972 NBC offered free time to candidates because it inadvertently showed a movie in which actor Pat Paulsen, who was then a declared candidate for president, had appeared for thirty seconds. Overall, networks and stations stopped showing Pat Paulsen movies so that they would not be confronted with the Section 315 provision.

entertainers as candidates

The issue, of course, became more acute when Ronald Reagan ran for governor of California and then president of the United States. Although Section 315 did apply to the movies in which he appeared, problems did not arise because the films, for the most part, were not aired during the election period and because his opponents made light of the situation, joking that showing the films might actually lose him votes.

Waivers from opponents can also handle this particular situation. If a weathercaster is thinking of running for city clerk, he or she may obtain waivers from other candidates by promising not to mention the election on the air or use the weathercasting as a political platform. In such waivers, the opponents promise not to seek equal time or opportunity from the station or stations involved.

In general, people who appear regularly on radio or TV will not run for office if they cannot obtain waivers. However, if they feel strongly about entering the political arena, they can quit their jobs or ask for another job of an off-air nature. They can also try to convince their employer that the station should give the equal time to the opponents. A Sacramento TV reporter is taking Section 315 through the courts because of difficulties he encountered when trying to run for local office.

length of spot

During the 1976 presidential campaign, Gerald Ford's campaign committee tried to buy spots on WGN in Chicago. The management's position was that any candidate who wanted to appear on either the WGN radio or WGN-TV station had to buy time in units of at least five minutes. The philosophy behind this was that no candidate for something as complicated as a major political office should make announcements of a thirty-second or sixty-second nature, but should explore the issues for at least five minutes. The FCC, however, ruled that candidates should be allowed to buy whatever length of time they wanted and so ordered WGN to change its policy.

campaign beginning

During the 1980 campaign, Jimmy Carter's aides wanted to buy a half hour of network prime time early in December of 1979 in order to launch Carter's campaign for reelection. The networks refused, saying that December was too early to start an election campaign. Carter forces appealed the decision and the Supreme Court sided with them, indicating that it is the candidates, not the broadcasters, who can decide when a campaign has begun.[39]

Each election year the courts are inundated with new cases regarding Section 315. Many of the ramifications from Section 315 are still largely uncharted, despite the thick tomes on political broadcasting that are issued by the FCC and the National Association of Broadcasters (NAB).

The Fairness Doctrine

The fairness doctrine has a foggier history than either editorializing or equal time. It did not appear as a concrete doctrine, but rather as a series of actions and rulings dealing with presentation of controversial issues. In many respects its roots are located in both the editorializing decisions and Section 315.

editorial statement

At the end of its 1947 to 1949 editorial hearings, the FCC issued a statement, part of which read as follows:

> . . . the licensee must operate on a basis of overall fairness, making his facilities available for the expression of the contrasting views of all responsible elements in the community on the various issues which arise.[40]

This statement did not raise any ripples at the time, but in later years it was interpreted to mean that fairness applied to forms of programming other than editorials.

Likewise, when Congress amended Section 315 in 1959, it stated that Section 315 amendment nothing in the news exemptions should be construed as relieving broadcasters from the obligations imposed upon them under this chapter to operate in the public interest and to afford reasonable opportunity for the discussion of conflicting views on issues of public importance.[41] This phrase, too, seemed harmless at the time but was later used in conjunction with fairness issues.

The 1960s was the decade in which fairness came to the fore. The political climate of unrest and distrust bred controversies that might have remained dormant in more settled times. The first cases to raise the issue revolved around station and network documentaries of the early 1960s. Several cities complained to the FCC that they had been depicted unfairly in the documentaries, but the FCC disagreed, saying that the producers had given opportunity for all points of view.

documentaries

Since the issue of fairness was frequently being raised, the FCC issued a 1963 advisory statement to stations on the subject, followed by a 1964 fairness primer. Although both of these were weak in terms of actual guidelines, they did make reference to the 1949 statement that had appeared in the editorializing document. They also established that fair presentation of material extended beyond editorializing and into other areas of programming.

fairness primer

Following this, the FCC seemed more inclined to favor entities that claimed fairness had been violated. It ruled that a Colorado station needed to give time to debt adjusters who claimed that they had been maligned in a documentary. It also ruled that two Alabama radio stations needed to present counterarguments to those presented in a syndicated program that dealt with a nuclear test ban treaty.

The case that has become the hallmark of the fairness doctrine is the **Red Lion** case. In 1964, station WGBC in Pennsylvania, operated by the Red Lion Broadcasting Company, broadcast a talk given by the Reverend Billy Hargis that charged Fred J. Cook, author of books critical of Barry Goldwater and J. Edgar Hoover, with Communist affiliations. Cook demanded that WGBC give him the opportunity to reply. The station said it would if Cook would pay for the time, but Cook objected and contacted the FCC, which ordered the station to grant Cook the time whether or not he was willing to pay for it. The decision was appealed through the courts and upheld by the Supreme Court on the grounds that free speech of a broadcaster does not embrace the right to snuff out the free speech of others. The significance of this decision went beyond the dispute between Cook and the Red Lion Broadcasting Company because it upheld the constitutionality of the fairness concept and of the FCC's right to implement it.

Red Lion case

One creative application of the fairness doctrine was a case in 1967 brought up by **John F. Banzhaf III.** In 1964 the surgeon general's office had determined that there was a link between cigarette smoking and lung cancer, a point that was considered controversial. Banzhaf, several years later, petitioned WCBS-TV in New York saying that cigarette commercials showed cigarette smoking in a positive light, so free time should be given to antismoking groups to present the other side. WCBS replied that within its news and public

John F. Banzhaf III case

affairs programs it had presented the negative side of smoking and did not feel that it needed to give equivalent commercial time to the issue. Banzhaf took the matter to the FCC, which sided with him—as did the appeals courts. Therefore, until 1972, any station that aired cigarette commercials had to provide time for anticigarette commercials. In 1972 Congress passed a law that banned cigarette ads on radio and TV and, therefore, the anticigarette commercials also stopped.

All of these fairness decisions of the 1960s led broadcasters and the public to believe that future decisions by the FCC and the courts would favor those who felt fairness should be strictly interpreted, and complaints about unfair presentation of issues poured into the FCC. However, once again the pendulum began to swing. No doubt, part of the reason for this was that broadcasters were now being more careful about fairness issues and were voluntarily complying with some of the demands. At any rate, in general, the decisions of the 1970s began to make a turnaround.

Friends of the Earth

For example, in 1970 an ecology group, **Friends of the Earth,** tried to emulate Banzhaf's case by requesting that ads for gasoline be countered by antipollution announcements. However, this attempt was not successful. The FCC stated that the cigarette ruling was not to be used as a precedent because that case was unique.

Vietnam views

In another case, two different groups wanted to buy time to present their views against the war in Vietnam. One group said that it would be acting to counter an armed forces recruitment spot and the other wanted to respond to the president's views. Both were turned down, one by a Washington, D.C., station and the other by CBS, so both groups went to the FCC. The FCC also turned them down, stating that broadcasters, not public interest groups, should decide when an issue required that opposing viewpoints be expressed, and that broadcasters should select the person or persons to deliver those viewpoints.

"The Selling of the Pentagon"

CBS presented a documentary, **"The Selling of the Pentagon,"** that showed how the military used public relations tactics to convince the public that the Vietnam War and other military projects were worthwhile. The military establishment, enraged by the program, went to the FCC to complain that the presentation had been one-sided and edited in a distorting manner. After much study, the FCC dismissed the charges by indicating that CBS had set aside an hour for opposing viewpoints and that there was no evidence of deliberate distortion intended to misinform.

Fairness Report

In 1974 the FCC issued a Fairness Report that came closer to a "doctrine" than anything that had been associated with the various rulings that came to be known as the fairness doctrine. The Fairness Report was an attempt to clarify for broadcasters what their duties and obligations were in this area. For example, it stated that a licensee had an obligation to devote a reasonable amount of program time to the discussion on controversial issues and that overall programming should be balanced concerning specific subjects. It said that commercials would be subject to fairness only if they directly addressed controversial subjects. The report included under fairness all aspects of political campaigns not included in Section 315, such as statements on behalf of candidates by their supporters.

During the 1980s, the number of complaints brought forth under the fairness doctrine waned and the doctrine itself became the focus of attention. In 1987 Congress passed a bill that would have codified the fairness doctrine and made it a law rather than an FCC policy. However, President Reagan vetoed the bill. Shortly thereafter the FCC said it was going to stop enforcing fairness and, in effect, abolish the whole fairness concept. This meant stations could air programming about controversial issues without having to worry about complaints or contrasting points of view. Although this appeared to kill the fairness doctrine, the whole issue could rear its head again in the 1990s.

abolishment

If the fairness doctrine appears somewhat confusing, it's because it is. For better or worse, the future of this concept is likely to be as confusing as its past.[42]

Other Regulations

Of course, there are many other laws and regulations that affect telecommunications practitioners. For example, the U.S. Criminal Code outlaws **lotteries** sponsored by radio or TV. Stations are allowed to have contests, though, and sometimes cases arise as to whether a station is conducting a contest or a lottery. Generally something is considered a lottery if a person pays to enter, if chance is involved, and if a prize is offered.

lotteries

The **copyright law** and court cases involving copyright violations have the potential to greatly affect the industry, as do regulations that the FCC may or may not make regarding **financial interest and domestic syndication.**

others

Most of the laws and regulations facing those in the telecommunications business are in a constant state of flux, but they do give guidelines for operation while, at the same time, allowing room for change.

Trends and Issues in Regulation

Regulation policies are in a constant state of flux—a situation having both advantages and disadvantages. A fluid state allows for changes to keep up with the times, but it also creates internal inconsistencies among the various regulatory bodies. It inhibits rigidity on the part of both elected and appointed officials, but it also causes confusion and uncertainty for those who must deal with regulation.

The present trend involves **deregulation,** as can be seen in the increasing hands-off attitude toward licenses and license renewals. And yet, it was not many years ago when the government was very heavy-handed in its role in license renewal. The constantly swinging pendulum, which seems to follow the government's philosophical approach, could easily change back to heavy regulation, especially if broadcasters begin to destroy the public's confidence in their ability to serve the public interest, convenience, and necessity.

degree of regulation

Although deregulation seems to be the order of the day, government appears to be increasingly intent on eliminating cross-ownership. In this way, communities are provided with varying points of view on issues, although newspaper owners make the most efficient operators of broadcasting stations because news functions can double up.

cross-ownership

First Amendment

The First Amendment sometimes appears to be a dumping ground for all the problems plaguing society. And yet it serves as the protectorate of individual and collective freedom. Just how tightly it is enforced varies with the times. When the relationship between the press and the president is a cordial one, censorship seems less of a threat than when the relationship is antagonistic. The public's need to be informed often correlates more with current events than with philosophical principles. If there is doubt about the effectiveness with which the government is being operated, information needs seem more intense.

obscenity

Laws and informal guidelines are intended to protect the public, especially the young, from obscenity. Yet sexual taboos are disappearing at a rapid rate in both information and entertainment programming. This area of regulation is in a particularly rapid state of flux.

reputations

Reckless reporting must be avoided so that individual reputations are not damaged unjustly. And yet it is the duty of the press to investigate the behavior of those in power. Power corrupts and absolute power corrupts absolutely, and in a representative government this must be prevented. However, when the press becomes arrogant and overplays minor errors in order to obtain a scoop, it must be curtailed. This is another constantly swinging pendulum that manages to maintain a fair degree of equilibrium.

courts

Present trends indicate that the First Amendment is being given precedent over the Sixth Amendment in terms of freedom of the press versus fair trial. Again, in the recent past this was reversed and cameras were not allowed in the courtroom. The electronic media often has difficulty resisting sensationalism, and an overly free hand in the courtroom could prove detrimental to those accused.

editorials

Broadcasters have never been known for their hard-hitting editorials, and the current trend appears to be toward less controversy rather than more. Public broadcasting has not done anything radical with its power to editorialize.

elections

Because television is so important in a modern election campaign, care must be taken to ensure equal opportunity for all candidates. Yet the thicket of problems that this creates can become a nightmare for all involved. Here again flexible rules become important because of the changing nature of media and the changing nature of election campaigns.

fairness

The future of the fairness doctrine will be interesting to follow. Broadcasters oppose it, saying that it has a chilling effect on them. They claim they will not bother to address controversial issues if they must go to the trouble and expense of contending with innumerable opposing points of view. Yet, with the fairness doctrine absolved, the risk of indoctrination increases because of the power of the electronic media. This is an issue where the degree of responsibility shown by telecommunications practitioners will, in the long term, probably determine the degree of regulation needed.

Future directions of regulation are impossible to predict, especially in light of the variations that have occurred in the past. Future generations will formulate those laws and regulations that best work for the society they create.

Conclusion

All the branches of government are intrinsically involved in the laws and regulations that govern telecommunications.

The FCC has been the main overseer of station licensing. As such, it has not only granted licenses but has also revoked them and issued warnings to stations for malfractions. These have included fines against obscenity as in the case of "topless radio," disapproval and then approval of editorials as in the Mayflower and Cornell cases, and decisions of when equal time provisions were violated as in cases regarding length of political commercials and presidential crisis speeches. On occasions the FCC has issued guidelines, such as the "the Blue Book," cross-ownership rules, or fairness statements, that broadcasters have felt exceeded its bounds.

The executive branch is the one most likely to call clear and present danger if the president feels reporters are acting irresponsibly. The president can veto communication bills as in the case of the fairness bill.

The legislative branch occasionally amends the Communications Act. This has had particular effect upon equal time. The amendment to Section 315 that removed equal time restrictions from bona fide newscasts had a great effect upon election campaigns, as did the amendment that required stations to sell time to candidates at their lowest rates. The fairness doctrine also grew out of a Section 315 amendment and a bill was sent forth codifying fairness. Other criminal and civil laws passed by Congress, such as those dealing with lotteries and copyrights, affect broadcasting.

The courts and telecommunications have a double-edged interrelationship. On one hand, a battle has been waged regarding the right of radio and TV to have access to the courts for news coverage. On the other hand, the courts have made many decisions that directly affect broadcasters.

Most of the license renewal battles of the 1960s and 1970s wound up in the courts, including WLBT, WHDH, and RKO. Obscenity rulings, such as WBAI's "seven dirty words" and Utah's cable TV channels, were eventually decided in the courts. Likewise, libel cases, including the Westmoreland case, often begin or end in the courts. Some Section 315 decisions have been appealed to the courts, such as determination of when a campaign begins and who can sponsor presidential debates. The fairness doctrine has kept the courts quite busy with cases that include Red Lion, cigarette advertisements, and Friends of the Earth.

Although the various branches of government disagree with each other from time to time, equilibrium is maintained by the checks and balances system.

Thought Questions

1. Should one company be allowed to own both a newspaper and a TV station in the same town? Why or why not?
2. Give some examples of what you feel would constitute censorship and what you feel would be clear and present danger.
3. If you were going to make a list of rules for TV journalists to abide by regarding access to the courts, what would they be?
4. Should the fairness doctrine be reinstated? Why or why not?

Advertising and Business Practices

14

Overview

Telecommunications is a business, and various elements of telecommunications—such as broadcast networks, cable TV systems, advertising agencies, and radio stations—are all individual businesses. If they lose money over a significant period of time, they cease to exist. Nonprofit entities such as public TV stations must also operate with an eye to the dollar. Profit is essential to telecommunications entities, for although they exist to serve the public, they must also serve their shareholders or their public service will be in vain.

The following subjects point out ways in which these profit-oriented businesses function:

Advertising as the primary, but not exclusive, means of income for telecommunications entities

Rate card variables, such as number of commercials, length of ads, type of facility, local or national status, time of day, and extent of use

Various selling practices, such as barter, per-inquiry, run-of-schedule, co-op, rate protection, rate cutting, program participation, and spots

The role of local and national sales staffs

The role of station representatives

The role of advertising agencies

The process of commercial production

Finding grants and writing grant applications

Consumer sales techniques including tiering, pay-per-view, and cassettes

Categories of expenses of telecommunications entities

Balance sheets and profit and loss statements

The type of thinking that goes into profit determination

Common criticisms of advertising

Sources of Income

All telecommunication entities are dependent upon money in order to operate. Most of these entities depend on **advertising** as their primary source of income. This has been true since the very early days of radio and, although advertising practices have changed over the years, they have not diminished in importance. Simply stated, radio and television stations and networks obtain money from advertisers and use it to produce and transmit programs. The intent is for programs to be watched by as large an audience as possible so that the ads, in turn, reach as large an audience as possible.

advertising

Advertising is not the only source of income in the telecommunications field, however. Sometimes stations rent their facilities to bring in limited income. Public broadcasting engages in **underwriting** that does not involve commercials, but does involve the use of money from corporations. This money is used to produce programs and the corporations are mentioned on the air. A large share of public broadcasting's money comes from the government and hence from taxes. Much of public broadcasting's income, whether it be from the government or from corporations, is obtained by means of grant applications for particular equipment, projects, or programs. The art of writing grant proposals and applications is well respected within the public broadcasting ranks.

other sources

Cable TV is utilizing advertising at an increasing rate but the great bulk of its money comes from subscriber fees for both the basic and pay services.[1] Videocassettes, MMDS, and other newer media receive money directly from the public also.

Without a doubt, however, advertising is still the king income producer for radio and television. The process through which advertising generates income is a fairly complicated one that involves rate cards, varying selling procedures, station representatives, and advertising agencies.[2]

Rate Cards

Basic to most broadcast sales is the rate card, a chartlike listing of the prices that the station charges for different types of ads. One of the many variables that affects the prices listed on the rate card is the number of commercials the advertiser wishes to air. Stations rarely accept only one ad at a time—it would be too expensive in relation to the time taken by the salesperson to sell the ad. What is more, a buyer would not realize satisfactory results if his or her product was mentioned only once. Usually a station requires that an ad be aired at least twelve times. It will try to induce the advertiser to buy even more airtime by offering frequency discounts, which are lower prices per ad as the number of ads increases. For example, a company that places an ad twelve times on a radio station might be required to pay $25 per ad, or a total of $300. If it placed the ad twenty-four times, it would pay $20 per ad, or a total of $480. And if its ad ran thirty-six times, each ad would be reduced to $18, for a total of $648.

number of commercials

Another variable is the length of the ad—usually fifteen, thirty, or sixty seconds. Obviously, a one-minute commercial will cost more than a thirty-second commercial, but usually not twice as much. For example, twelve one-minute ads might cost $25 each, and twelve thirty-second ads might cost $15 each. This is because station costs, such as handling, are more expensive for two separately produced thirty-second commercials than for one sixty-second commercial.

type of facility
Another major cost variable is the type of facility for which the ad is purchased. A thirty-second ad that costs $15 on a cable TV local origination channel might cost $25 on a radio station; $1,000 on a radio network; $750 on an independent TV station; $1,000 on a network-owned TV station; $15,000 on a cable TV network; and $60,000 on a broadcast TV network. The underlying basis of all rate cards is the number of station listeners or viewers. A 3,000-watt FM station would not charge as high a rate as a 50,000-watt AM station, and a small-town TV station would have a much lower overall rate card than a similar station in a metropolitan area.

national or local
Costs to national advertisers may be higher than costs to local advertisers. The rationale here is that the national advertiser can profit from reaching all the people in a station's coverage area while a local advertiser might not. For example, a national automobile manufacturer has the potential for selling cars to people in an entire city that a station covers, but the people in the southern part of the city are not likely to respond to an ad for a car dealer in the northern part of the city, especially if there is a similar dealer in the southern part of the city. For this reason, part of the station's coverage area is virtually useless to the local car dealer advertiser; therefore, the station charges the local advertiser a lower rate than it charges the national advertiser.

time of day
Audiences are of varying sizes at different times of the day, so this factor also becomes a variable on rate cards. A radio station or network would probably have the greatest number of listeners from 7:00 to 9:00 A.M. and from 4:00 to 6:00 P.M., when people are in their cars. This is referred to as Class AA time and would cost the most money. The hours between 9:00 A.M. and 4:00 P.M. and 6:00 to 10:00 P.M. would have a lower fee—these hours are referred to as Class A time. The lowest rate of all is Class B time, which would be 10:00 P.M. to 7:00 A.M., when many people are sleeping. Television stations are a different story, with most viewers congregating between 7:00 and 11:00 P.M., the Class AA time. Classes A and B vary from station to station, depending on programming. Some stations fine-tune their times to AAA, AA, A, B, and C degrees. Television stations often sell ads on particular programs rather than for particular times. An ad bought for 9:00 P.M. Tuesday during "The Tuesday Night Movie" might cost more than an ad bought at 9:00 P.M. Wednesday during "This Week's Report." In this case, the rate card lists programs rather than times. Television networks usually do not employ a rate card at all but sell ads on the basis of what the market will bear.

other factors
Other variables taken into account when rate cards are established include the number of months or years an advertiser utilizes a particular station; the average age of the audience listening or viewing; and whether an advertiser

CLASS "AAA"				CLASS "AA"		
3 PM — 12 MIDNIGHT, Monday - Friday				6 AM — 10 AM, Monday - Sunday		
10 AM — 12 MIDNIGHT, Saturday - Sunday						
	60's	30's			60's	30's
1x	$90	$76		1x	$70	$60
6x	$80	$68		6x	$65	$55
12x	$70	$60		12x	$60	$50

CLASS "A"				CLASS "B"	
10 AM — 3 PM, Monday - Friday				12 Mid — 2 AM, Monday - Sunday	
				Flat Rate:	$25
	60's	30's			
1x	$60	$50		CLASS "C"	
6x	$55	$45		2 AM — 6 AM, Monday - Sunday	
12x	$50	$40		Flat Rate:	$15

FIGURE 14.1

Rate card of a small radio station.

is willing to be flexible concerning the advertising times or whether a fixed position is wanted to assure that an ad will be run at a particular time. Sample rate cards are reproduced in figures 14.1 and 14.2—one for a small station, the other for a large station. These show some of the variances, such as number of commercials, length of commercials, and time of day.[3]

Selling Practices

Theoretically, a station time sales representative approaches a potential advertiser, convinces him or her to buy a group of ads at the price established on the rate card, and then sees that a bill is sent to the advertiser for the correct amount. However, in the real world of economics this theoretical case is not always the practiced one.

For example, many radio stations, particularly smaller ones, engage in what is called **barter** or **trade-out**—they trade an advertiser airtime for some service the station needs. Perhaps the station owns a car that needs occasional service; in order to receive this service free from Joe's Garage, the station will broadcast twelve ads a week for Joe's Garage with no money changing hands. Similar barter arrangements are often negotiated with restaurants, stationery suppliers, gas stations, audio equipment stores, and the like.

barter

Some stations sell ads on a **per-inquiry** basis, which means that the station gets paid only if consumers respond to the ad. A typical sales pitch for such an arrangement might be, "Hurry right down to your nearest Foody Market, buy at least $10 worth of groceries, mention WXXX to the checkout clerk and receive one toothbrush of your choice absolutely free." For each toothbrush given away, the station would receive a set amount of money.

per-inquiry

One of the most common methods of selling commercial time is called **run-of-schedule** (ROS). The salesperson and advertiser decide on how many ads should be run and sometimes make up a package that consists of ads in different times—AA, B, and C. The station is the one that decides on the

run-of-schedule

Advertising and Business Practices   349

FIGURE 14.2

Rate card of a large radio station.

FREQUENCY ANNOUNCEMENTS

Announcement rates are based on the number used during an established twelve-month period, and become effective from the beginning of service on firm contracts or as earned.

60-Second Announcements

(150 Words, Live)

Times	AAA	*AAA Combo	AA	A	B	**C
1	$320	$250	$205	$131	$100	$44
15	309	247	194	124	95	42
50	296	237	179	116	89	39
150	275	220	166	107	84	36
300	261	209	159	100	79	34
500	253	202	154	97	77	33
750	246	197	151	95	75	32
1000	240	192	149	93	74	31

30-Second Announcements

(75 Words, Live)

	AAA	*AAA Combo	AA	A	B	**C
1	$256	$205	$164	$105	$80	$35
15	248	198	155	99	76	34
50	238	190	143	92	71	31
150	220	176	132	86	67	29
300	209	167	127	80	63	28
500	203	162	124	78	61	27
750	195	156	121	76	60	26
1000	193	154	120	75	59	25

10-Second Announcements

(25 Words, Live)

	AAA	*AAA Combo	AA	A	B	**C
1	$160	$130	$103	$66	$50	$22
15	155	124	98	62	47	21
50	148	118	89	58	45	20
150	138	110	83	54	42	19
300	130	104	80	50	40	18
500	126	101	78	48	39	17
750	124	99	76	47	38	16
1000	121	97	75	46	37	15

*Combo: Rates apply only to the number of AAA announcements that are ordered and broadcast in combination with an equal or greater number of announcements of the same or greater length within other time classifications except Class C.

Frequency Announcements Rate based on total number of announcements used during an established twelve-month period. Rate applies as earned.

Weekly Package Plan Rate based on a consecutive seven-day period.

WEEKLY PACKAGE PLAN

Weekly Package Plan rates apply only to the number of 60-second, 30-second and 10-second announcements broadcast for one product within a consecutive seven-day period on run-of-station schedules. Such announcements are subject to immediate preemption for frequency announcements without notice. Further, without liability to the Station, Weekly Package Plan announcements are subject to omission by Station without charge to Advertiser, or to rescheduling by the Station in a time period considered equivalent by the Station.

Frequency announcements may be combined with Weekly Package Plan announcements to earn lower plan rates, but Weekly Package Plan announcements may not be combined with frequency announcements to earn lower frequency rates.

60-Second Announcements

(150 Words, Live)

Weekly Plan	AAA	*AAA Combo	AA	A	B	**C
6	$291	$233	$176	$117	$94	$40
12	284	227	171	112	89	38
18	275	220	166	107	84	36
24	268	214	161	102	79	34

30-Second Announcements

(75 Words, Live)

	AAA	*AAA Combo	AA	A	B	**C
6	$233	$186	$142	$94	$75	$32
12	226	181	137	90	71	30
18	220	176	132	86	67	29
24	214	171	128	82	63	27

10-Second Announcements

(25 Words, Live)

	AAA	*AAA Combo	AA	A	B	**C
6	$146	$117	$89	$60	$48	$20
12	143	114	86	57	45	19
18	138	110	83	54	42	18
24	135	108	81	51	40	17

TIME SIGNALS

(Flat and not combinable with other announcements)
(12 Words of Commercial Copy or Six Seconds Transcribed)

AAA	*AAA COMBO	AA	A	B	**C
$104	$83	$62	$41	$30	$13

**C May be counted toward Frequency Announcements or Weekly Package Plan, but may not count toward Combo.

specific times the ads should run based on the availability of commercial time. For this privilege, it gives the advertiser a discount so that he/she does not pay what the rate card designates for the times selected. In some instances an advertiser who buys ROS will get an even better deal because the station will run an ad that was supposed to be in B time in A time if an A time spot is available. The station does not charge the advertiser extra for this.

fixed buy

Somewhat the opposite of ROS is a **fixed buy,** wherein the advertiser very specifically states the exact time that each ad should run. For this, the advertiser pays the premium price. Of course, this often involves some negotiating. If two advertisers want 9:00 A.M. Monday, something has to give.

co-op

Co-op advertising involves shared costs, usually by a national and a local advertiser. A commercial for an "Instant Pleasure" camera may end with, "To purchase this sensational camera, make a short trip to Lou's All-You-Need Photo Stop at 160 Main Street." The cost of this ad would be divided on an

Regulation and Business

agreed-upon formula between Instant Pleasure and Lou's. This type of co-operative advertising has sired an illegal practice called **double billing.** Because local advertisers pay a lower rate than national advertisers in order to compensate for the fact that the entire coverage area of the station is not of value to them, a station makes out two bills for cooperative advertising. One is at the higher national rate and the other is at the lower local rate. Because the national advertiser is far away and doesn't usually know the exact commitments that were made, the local advertiser and the station can conspire to have the national advertiser pay more than it actually should because it pays as though the entire commercial were charged at the higher national rate. The station and the local advertiser then divide this extra pay. Double billing is illegal and, when uncovered, it has been used against stations at license renewal time. As a result, it has lessened in recent years.

Some stations offer **rate protection** to their regular customers so that even if the station's rate card costs increase, these good customers can still buy ads at the old prices.

rate protection

Although it is poor business practice, some salespeople, in order to obtain business, simply cut the prices on the rate card and offer ads at lower prices than those indicated.

rate cutting

Regardless of the financial arrangements, there are two primary ways to buy advertising—**program buying** and **spot buying.** In the early days of radio and television most advertisers bought programs, paying all the costs to produce and air such programs as "Lux Radio Theater," "Colgate Comedy Hour," "Kraft Television Theater," and so forth. The advertiser had the advantage of constant identification with the program and its stars. Program buying is rarely done anymore for several reasons. One reason is that costs of television production have soared to the extent that not even the largest companies can afford total underwriting of a program week after week. Also, after the quiz scandals of the late 1950s, the networks were leery of such overriding program control by advertisers and began to take greater control of content. Only occasional specials are now totally paid for by one advertiser. Instead, multiple advertiser participation is common. When one hears, "The following portion of 'Sock It to 'Em' is brought to you by the Widget Manufacturing Company," several advertisers have joined together to pay the costs of the program. This kind of program buying is common with network TV but rare in local radio, except for a few local public service or sports programs.

program buying

What is common in the local market and on network radio is spot buying. The advertiser bears none of the cost for or identification with a particular program, but rather buys airtime for certain times of the day. Hence, a disc jockey whose show runs from 6:00 to 9:00 A.M. may present ads from several dozen advertisers, all falling under the umbrella of spot advertising. In radio, spots can be aired at almost any time because essentially all ads are spots. In television spots are aired at more definite times, especially on affiliated stations where network shows come already brimming with ads. In this case, the TV station sells most of its spots for station break time or for insertion in locally produced shows. Of course, some network shows do leave a few ad spaces

spot buying

available for local spots, which usually sell well in the local market. Independent TV stations have more spots available than affiliated stations but, again, many of the programs that they air come complete with ads. Radio network news is generally sold on a spot basis, but some of the features are purchased through the program buying process. Regardless of the selling practices used, stations must be sure that their programs fill both a station need and an advertiser need.

Sales Staffs

Radio and TV time selling—like insurance, shoes, and stock—needs a salesperson to interact between the product and the customer. Exactly how this function is executed varies greatly from station to station.

local salespeople

How it works in a small radio station is not difficult to follow because most of the advertising comes from the local area. Usually the disc jockeys are also salespeople, spending four hours a day announcing and playing records and four hours going to the local hardware stores, grocery stores, and car dealers selling ads. Often the general manager of a small station doubles as a salesperson too.

Within larger radio and TV stations the selling process is more indirect. TV talent and big name disc jockeys usually do not sell time. Instead, members of a sales force contact local merchants and are generally paid on a salary plus commission or commision-against-draw basis. These salespeople are expected to service ads as well as sell them so that advertisers will repeat their business. Servicing ads includes making sure the ads are run at the appropriate times and facilitating any copy changes that may be needed.

TV stations and larger radio stations are interested in securing ads from companies that distribute goods nationally—companies such as Procter & Gamble, General Mills, and General Motors. It would be very expensive for each station to send salespeople to the headquarters of these companies, so

station reps

stations generally hire **station representatives** to obtain these ads. Station representatives are national sales organizations that operate as an extension of the stations with which they deal. They are middlemen between the company that wants to buy ads and the radio or TV stations that want to sell ads. Usually these reps handle national sales for dozens of stations around the country, some representing just radio stations, some just TV, and some a combination of the two. For obvious competitive reasons, a station rep will not service two stations in the same listening area. The network-owned and -operated stations have their own corporate reps, as do some large independent chains such as Westinghouse and Taft, but most of the stations of the country utilize station representative companies such as Petry, Katz, or Blair.

A station rep earns income from commission on the sales obtained for the station. The percentage of commission varies because it is negotiated by each station and rep, but it generally falls between 5 and 15 percent.[4] TV stations generally pay a lower commission because they sell millions of dollars worth of ads each year, while radio stations pay a higher percentage because the price per ad is lower. Needless to say, good communication must be maintained between station and rep so that the rep can work in the station's best interest. In addition to selling ads, some station representatives help their stations with programming decisions, audience research, and station promotion.

Radio networks maintain a sales force to obtain national ads for their programs. Some radio network programs have empty commercial holes so that station affiliates can insert their own spots.

TV networks maintain large sales forces that are devoted to selling and servicing major national advertisers. The competition among these sales forces is keen. Each network wants to ensure that all its programs sell—and sell at the highest rate possible. Network commercial time is a limited commodity. Only a certain number of spots are available and these cannot be expanded the way a newspaper can add pages when ad demand is heavy. A network has the same amount of time available at Christmas, a heavy season, as it does the rest of the year, so the sales force must be careful to sell in such a way that the overall needs are met.

network sales forces

Because the buyers and sellers of TV network time are generally a small, close-knit fraternity, it behooves a network salesperson to emphasize service. One way to do this is to make sure the client buys enough variety in the network schedule that if some of the programs fail, the client still has time in successful portions of the schedule. Generally network management is also deeply involved with sales and makes sure that a knowledgeable team presents programs and program ideas to potential sponsors.

For many years, TV network advertising sold out rather easily at high prices. However, as the newer media have eroded the audience, network sales have gone soft and networks have found they cannot raise prices and still attract advertisers.[5]

Advertising Agencies

When salespeople or station representatives are vying for ads from major advertisers, they generally do not deal only with the company, but rather work through the company's **advertising agency.** An ad agency handles overall advertising strategies for a number of companies. It usually advises them about newspapers, magazines, direct mail, and other forms of advertising, as well as radio, TV, and cable. Ad agencies came into existence over a hundred years ago to act as middlemen between companies who needed to advertise their product in markets where they were distributed and advertising outlets that wanted to obtain as many ads as possible.

Most major advertising agencies are generally termed **full-service** agencies. They establish advertising objectives for their clients and determine to whom the product should be sold and the best way to reach these consumers. They conduct research to analyze the audience or test the advertising concept and then attempt to obtain the best buy for the money available. They design the advertisement or oversee the commercial and determine when a company should initiate a campaign and how long it should continue it. After ads have

functions

run, advertising agencies handle postcampaign evaluations. In all these processes they work closely with marketing and management executives within the companies they represent.

payment

For its efforts, the advertising agency generally receives 15 percent of the billings. In other words, a billing of $1,000 to its client for commercial time on a TV station means that $850 goes to the station and $150 is kept by the ad agency to cover its expenses. Because dollar amounts are fairly constant from agency to agency, the main "product" an ad agency has to sell to its customers is service. Advertisers frequently change agencies simply because one agency has run out of creative ideas for plugging its product and a new agency can initiate fresh ideas.

Because agencies generally deal with all media, the broadcast salesperson who approaches an ad agency must convince the people handling accounts that broadcasting is the best deal for their clients and that the particular network or station he or she represents offers an exceptionally good deal. Cable TV representatives must, of course, tout their product over that of radio, TV, or other media.

organization

Some of the largest of the ad agencies are Young and Rubicam; J. Walter Thompson; Leo Burnett; Batten, Barton, Durstine, and Osborne; Grey Advertising; and McCann-Erickson.[6] Ad agencies, like many other companies, have been purchased and consolidated. Most of these agencies are divided into departments that include account executives, media specialists, TV program buyers, copywriters, art directors, and marketing research specialists. Therefore, a team of people works on the ad campaign of any one company. Not all agencies are large, however. In fact, some are one-person operations with only a few clients who like the overall attention they receive from the person directly responsible.

modular agencies

Not all major advertisers employ an ad agency. Some maintain their own **in-house services,** which amount to an ad agency within the company. Other companies will not hire a full service agency but will hire what are sometimes referred to as **modular agencies.** These agencies handle only the specific things a company asks for—perhaps only the designing of an ad or the postadvertising campaign research. Usually this type of work is paid for by a negotiated fee rather than a 15 percent commission. There are also media buying services that concentrate only on buying radio and television spots.

Although there are many variations, the most common procedure is for a company to hire an advertising agency, which then makes the basic advertising decisions and sees that they are carried out. As far as broadcasting is concerned, these ad agencies deal with networks, station representatives, and occasionally individual stations to ensure that the ads of their clients receive the best treatment. Throughout the process, both company officials and broadcast executives are informed of the successes and failures of the advertising campaign.

Commercial Production

A company buys time on a station or network in order to air a commercial extolling the virtues of itself as a company or of one or more of its products. Production of commercials, then, is a very important facet of the whole advertising scene.

Companies rarely produce their own commercials because they simply are not endowed with the equipment or know-how to do so. They are in the bread-making, car-selling, or widget-manufacturing business, not the advertising business.

Small businesses that place ads on small radio stations generally have the stations produce the ads. The people who sell the time to the merchant will write and produce the commercial at no extra cost. If they are good salespeople, they will spend time talking to the merchant to determine what he or she would like to emphasize in the ad and will check the ad with the merchant before it is aired. Production is usually quite simple, frequently just written copy that the disc jockey reads over the air. Often the commercial is prerecorded with music or sound effects added and occasionally the merchant will talk on the commercial. At any rate, the whole process is kept at a simple, inexpensive level.

radio productions

Large companies that advertise on radio usually have much more elaborately produced ads that include jingles and top-rated talent. The production costs are generally in addition to the cost of buying time.

Television commercials are still more elaborate. Usually they are slick, costly productions that the advertiser pays for in addition to the cost of commercial airtime. Local stations sometimes produce commercials for local advertisers and charge them production costs—the cost of equipment, supplies, and personnel needed to make the commercial. A great deal of local TV advertising and most network commercial production is handled through advertising agencies.

TV productions

As part of its service to the client, the advertising agency will decide the basic content of the commercial, perhaps trying to come up with a catchy slogan or jingle. Usually a **storyboard** is made—a series of drawings indicating each step of the commercial.

The advertising agency then puts the commercial out to bid. Various independent production companies state the price at which they are willing to produce the commercial and give ideas as to how they plan to undertake production. The ad agency then selects one of these companies and turns the commercial production over to it.

Some production companies maintain their own studios and equipment while others rent studio facilities from companies that maintain them or from stations or networks. It is possible that a station may lease its facilities for a commercial that winds up appearing on a different station.

OPEN ON 7-YEAR-OLD GIRL WEARING
A POTATO RING ON EACH FINGER.
DEMURELY, SHE TAKES ONE OF THE
RINGS OFF HER FINGER ...

MUSIC THROUGHOUT.

SINGERS: "Give a ring..."

PUTS IT IN HER MOUTH, AND
CRUNCHES.

SFX: LOUD CRUNCH.

SINGERS: "Give a crunch,
          Give Crunchi-O's."

CUT TO GORILLA FACE MASK.

SINGERS: "Give a ring..."

MASK IS REMOVED REVEALING
BEAUTIFUL WOMAN. SHE WINKS AT
CAMERA AS SHE BITES INTO
CRUNCHI-O.

SFX: LOUD CRUNCH.

SINGERS: "Give a crunch.
          Give Crunchi-O's."

CUT TO JEWELER LOOKING THROUGH
EYEPIECE AT POTATO RING.

SINGERS: "Give a ring..."

HE DECIDES RING IS GOOD ENOUGH
TO EAT. AND EATS IT.

SFX: LOUD CRUNCH.

SINGERS: "Give a crunch.
          Give Crunchi-O's."

DISSOLVE TO CRUNCHI-O CANNISTER.

MUSIC UNDER.

ANNCR. VO: "Introducing Crunchi-O's.
            The new potato snack from
            Nalley's...."

DISSOLVE TO CU OF PRODUCT.

ANNCR VO: "made in the shape of
           golden bite-size rings ...
           for a crunchy, crisp
           potato taste that's
           really different."

CUT TO TEENAGER REACHING
INTO CANNISTER FOR CRUNCHI-O.

SINGERS: "give a ring..."

HE TOSSES CRUNCHI-O OVERHEAD AND
WAITS TO CATCH IT ON THE WAY DOWN.

SINGERS: "Give a crunch..."

SUDDENLY, HUNDREDS OF CRUNCHI-O'S
POUR DOWN, FORCING TEENAGER TO
COVER UP.

SINGERS: "Give Crunchi-O's."

CUT TO MEDIUM CU OF CRUNCHI-O
CANNISTER.

MUSIC UNDER.

SUPER: AMERICA'S ONLY
       POTATO RING.

ANNCR VO: "Crunchi-O's.
          America's only
          potato ring.
          New from Nalley's."

FIGURE 14.3

A commercial storyboard. *(Courtesy of Della Femina, Travisano)*

(a)

FIGURE 14.4

(*a*) The Bing Crosby family doing a spot for Minute Maid orange juice. (*b*) Johny Philip Morris, who was part of the "Call for Philip Morris" cigarette ads on radio. (*c*) Betty Furness advertising for Westinghouse during the 1952 nominating conventions. *(a courtesy of the Coca-Cola Company; b courtesy of KFI, Los Angeles; c courtesy of Betty Furness)*

(b)

(c)

The time, energy, pain, and money that go into producing a commercial often rival what goes into TV programs themselves. One commercial often takes days to complete and involves fifteen or twenty people and costs well over $50,000. Take after take is filmed or taped so that everything will turn out perfectly. Once the commercial is "in the can," it may air hundreds of times over hundreds of stations. Hence, the obsession with perfection.

approaches

Commercials can take many different approaches. Some attempt to be minidramas that show how the advertised product can solve a problem. Others place the accent on humor. In an attempt to establish credibility, the ad agency might decide to have a well-known celebrity pitch the product with a testimonial of its virtues. Some ads demonstrate products in use. Others involve interviews with satisfied customers. Some ads use no people whatsoever but employ animation, special effects, or unusual design. Most ads try to appeal to some basic human instinct such as security, sex, love, curiosity, or ego inflation.[7]

past commercials

Commercials have enabled television and radio to create their programs, but the "program content" of commercials has also had an effect on the vernacular and life-style of American society. Remember"Where's the beef?"; "How are you fixed for blades?"; "Leave the driving to us"; "Look ma, no cavities"; "Everything's better with Blue Bonnet on it"; "Greasy kid stuff"; "Double your pleasure, double your fun"; "Ring around the collar"; "Try it, you'll like it"; "I can't believe I ate the whole thing"; "The asked-for motor oil"; "Reach out, reach out and touch someone"; and "You deserve a break today"?[8]

Grantsmanship

Occasionally programs underwritten by **grants** are aired on commercial stations. For the most part, however, grants are a source of income for public broadcasting.

Many grants come from foundations that have been set up either by the government or by private corporations for the sole purpose of issuing grants, including the National Endowment for the Humanities and the Ford Foundation. Others come from within public broadcasting itself. For example, the Corporation for Public Broadcasting awards grants to member stations for particular programs or projects.

finding grants

Anyone who works in public broadcasting would be well advised to understand grantsmanship even though many public stations have individuals or sometimes departments charged with the responsibility of keeping abreast of grant possibilities and writing grant applications.

writing grant applications

Books that categorize grants are available and most institutions that award grants issue information about their specific grants.[9] Anyone who is trying to obtain grant money should keep up to date on all this literature and match specific program or project ideas with specific grant objectives. The grant application must be written in careful accordance with the instructions and must be submitted by the deadline stipulated. The budget is usually a very important part of the application. If the budget is too extravagant, it will lose credibility; however, it must be ample enough to ensure that the project can be successful. Sometimes an application can be changed slightly and submitted to several agencies, giving it a better chance of being funded. Often projects that have partial funding from one source stand a better chance of obtaining additional funding from another source because the partial funding makes them appear viable.

follow through

If the grant is awarded, careful bookkeeping must take place so that the funds are spent in the manner specified. When the project is complete, a final report is usually required by the agency that awarded the grant.

Selling to the Consumer

direct dealings

Although broadcast programming is intended for the consumer, many broadcast entities have little direct contact with this consumer. Radio and TV signals travel through the airwaves and those in the community who wish to tune them in, do so. In this regard, the structure of cable TV, videocassettes, and some of the other new media is quite different. In these instances the media have first-hand contact with the consumer because the consumer must buy the product.

While the sales force of a radio or TV station is pounding the pavement contacting advertisers, the sales force of a cable TV system is pounding the pavement trying to convince viewers to sign up for the service.

cable tiers

Sometimes cable TV salespeople have a more complicated selling job because of the number of channels involved. Early cable systems charged an installation fee and a monthly fee for bringing in the local channels. With the advent of HBO, cable systems added a special additional fee for the commercial-free movie services. As cable programming proliferated, cable systems adopted what became known as **tiering.**[10] Subscribers can select the tiers of programming they wish to receive. Different cable systems tier in different ways, but a typical system of thirty or more channels might offer the following:

Tier 1—local stations and cable access channels: $6.00

Tier 2—superstations and distant imports: $3.00

Tier 3—basic cable channels (ESPN, USA, Nickelodeon, CBN, CNN, MTV): $5.00

Tier 4—one pay service (HBO, Showtime, Disney Channel, Movie Channel): $12.00

Tier 5—two pay services: $20.00

Tier 6—three pay services: $28.00

Subscribers have a choice of how many tiers they wish to buy. For some cable systems they need to subscribe to tiers 1, 2, and 3 to receive any of the pay services, and for other systems they can select any of the tiers they wish. In this example, the total bill for someone desiring all the channels would be $42.00 per month. Cable selling is further complicated on systems that offer **pay-per-view.** For this, the customer must be sold on specific programs, usually with short notice.

Videocassette recorder companies sell their products to customers in much the same way as other consumer product companies do—through a distribution network of wholesalers and retailers. One group of manufacturing companies sells the VCRs through video and audio stores. Film companies and others distribute the cassettes either for sale or for rent through various video stores. In order to induce people to buy, the companies engage in advertising campaigns and special sales.

Overall, salespeople are very important to the economic well-being of the electronic media. Without them, money would not exist to produce the programming.

Income and Expenses

Advertising revenue is by far the largest source of income for telecommunications companies, although funds also come from grants and subscribers.

Other minor forms of income include renting facilities to independent producers who wish to use them for taping programs or commercials and the money recouped from selling programs produced. Radio stations and networks do not rent facilities or sell programs as often as TV operations do, but there are occasions when outside organizations need audio recording capabilities and a station with a studio free is willing to rent. Because most local radio programs are live disc jockey shows or news, they do not lend themselves well to syndication. However, occasionally a special public service series that one station produces can be sold to other stations.

TV facilities are frequently rented for use, both within the facility and on remote location. Program matter can also be sold for rerun purposes or for airing in foreign countries—probably the most lucrative market of all for program syndication. In fact, most network-run programs do not meet expenses when they are initially aired on the network; they must wait until they are rerun in this country or aired abroad before they turn a profit.

The sources of income for telecommunications entities are fairly clear-cut: advertising plus grants, consumer fees, facility rental, and program sales.

However, expenses are quite varied. Over half the expenses of radio stations are for salaries of disc jockeys, newscasters, salespeople, management, and staff personnel. The rest of the outgo is for such things as rental of studio space, equipment maintenance, new equipment, office supplies, taxes, music license fees to ASCAP and BMI, travel and entertainment of potential advertisers, station promotion, fees paid to obtain rating reports on the size of the station's listenership, commissions to advertising agencies and station representatives, utilities and power, interest on loans, insurance, automobile expenses, and audio tape.

The general types of expenses of TV stations are similar to those of radio stations. About one-third of a TV station's expenses are for personnel, the dollar amount of which is larger than for radio stations. The TV station expenses for equipment, utilities, and power are also much higher than those of radio.

Balance Sheet

Assets		Liabilities and Shareholders' Equity	
Current assets:		**Current liabilities:**	
Cash and cash equivalents	$335,086	Current maturities of long-term debt	$ 2,984
Notes and accounts receivable	387,727	Accounts payable and accrued liabilities	348,023
Inventories	151,880	Income taxes	63,165
Program rights	94,214	Total current liabilities	$414,172
Prepaid expenses	32,860		
Total current assets	$1,001,767		
		Amounts due after one year:	
Investments	20,990	Long-term debt	96,666
		Liability for program and talent rights	12,556
Property, plant, and equipment:		Other	52,063
Land	25,226	Total	$161,285
Buildings	148,139		
Machinery and equipment	235,969	Deferred income taxes	30,280
Leasehold improvements	25,508		
	432,842	**Shareholders equity:**	
		Preference stock	544
		Common stock	72,472
Less accumulated depreciation	206,794	Additional paid-in capital	219,651
Net property, plant, and equipment	$228,048	Retained earnings	480,872
			773,539
Excess of cost over net assets of businesses acquired	53,376	Less common stock in treasury	27,666
		Total share	745,873
		Shareholders equity	$1,351,610
Other assets	47,429		
	$1,351,610		

FIGURE 14.5

Balance sheet for a TV network.

Local cable TV systems experience much higher start-up costs than do broadcast TV stations because of the huge investment in stringing cable. Once this is completed, however, cable TV expenses are largely for personnel, utilities, and programming. Expense patterns have not been set for most of the newer media.

cable system expenses

Network expenses (both broadcast and cable) involve large amounts of money that must be paid for transmitting program material via leased telephone wires, microwave facilities, and satellites. Programs purchased from independent producers constitute another major expense. Broadcasters pay large amounts to affiliates for airing programs. Within the cable realm, affiliates pay most networks for programs, so the expense is in the cable system's ledger.[11]

network expenses

Stations and networks keep track of their income and expenses as do most businesses, through **balance sheets** and **profit and loss statements.** The representative financial accountings shown in figures 14.5 and 14.6 illustrate sources of income and outgo.

balance sheets and profit and loss statements

Monthly Profit and Loss Sheet

Income

Local time sales	$8,451.87
Agency sales	829.90
Political sales	30.00
Total income	$9,311.77

Expenses

Payroll and payroll taxes	$4,010.55
Commissions to employees	588.04
Licenses and taxes	57.80
Office supplies and expenses	174.19
Travel and entertainment	97.48
Advertising	230.85
Sales promotion	147.61
Audience measurement	170.20
Agency commissions and talent fees	206.36
Utilities, office	37.06
Power, transmitter	58.55
Telephone	314.62
Satellite time	150.73
Rent	725.00
Interest	328.83
Royalties	465.85
Music programs	50.00
Dues and subscriptions	38.90
Operating supplies	100.36
Insurance	102.06
Professional services	145.00
Automobile expense	108.21
Repairs and maintenance	28.25
Miscellaneous	75.00
Total expenses	$8,411.50
Excess of income over expenses	$900.27

FIGURE 14.6

Profit and loss sheet for a small radio station.

radio example

Determination of Profit

Although profit and loss statements show whether a company is making a profit, they do so after the fact, when it may be too late. It would be sound policy to know before embarking on a project whether it will make money. An accurate projection often cannot be made because of the fickle tastes of the public, not to mention the difficult arithmetic of assessing profitability.

The most direct connection between expenses and income can be seen in a small radio station. Because disc jockeys are usually also salespeople, they must, in effect, pay for their own salaries and part of the overhead. Therefore, if a disc jockey is earning $500 a week and the average station ad is $50, he or she will be helping the station maintain a profitable position by selling over $600 in ads a week, or approximately twelve ads. Likewise, the station manager must earn salary plus overhead in order to operate a profitable station.

The connection is not so direct in television. The nature of the programming it airs and the complexity of the expenses make it difficult for a TV station to know when it is making money. As an example, assume that an independent station in a large city has bought seven ninety-minute Elvis Presley films, each costing $20,000, for a total of $140,000. The contract stipulates that for the $140,000 the station can air the programs as often as it wants for one year. The station decides to have an Elvis Presley week and air each program three times—9:00 A.M., 3:00 P.M., and 7:00 P.M. In effect, each program now costs $20,000 divided by three, or approximately $7,000. According to station policy, each ninety-minute show can have twenty minutes of commercials; however, the ad rates are going to be different for each showing.

A typical one-minute ad rate for 9:00 A.M., when very few people are watching TV, would be $250. In the afternoon more people are watching, so the rate might be about $400. In the evening the whole family is available to watch, and the rate might jump to $800. Therefore, twenty minutes of ads at 9:00 A.M. would be $5,000—not yet break-even for the $7,000 cost. At 3:00 P.M., twenty minutes of ads would yield $8,000; at 7:00 P.M., twenty minutes of ads would yield $16,000.

Because the total revenue is $5,000 + $8,000 + $16,000 (or $29,000), the station has seemingly made $9,000 on each $20,000 program. However, 15 percent of the collected $29,000 (or $4,350) goes to the advertising agency and another 10 percent (or $2,900) goes to the station representative. Overhead will run at least 15 percent of the $20,000 program cost (or $3,000). So $4,350 + $2,900 + $3,000 = $10,250, which must be added to the $20,000 cost, bringing it to $30,250. Because only $29,000 was collected, the station actually lost money.

The station is now faced with the problem of how to recoup its money and make some profit. It could have another Elvis Presley week later in the year without paying anything for the programs, but this would probably not sell very well because it would be such a recent rerun. The station could change its mind and air the programs four times a day instead of three, but this might not be too popular with viewers. The station also might air six or seven of the best programs later in the year and try to collect a little additional revenue this way.

Obviously, other situations arise for network-affiliated stations, cable TV systems, and the networks themselves, but this type of financial analysis must become second nature to managers if they are to make sound decisions that prevent the flow of red ink.

Advertising under Fire

Despite the fact that advertising achieves its purpose—that is, it sells products and it provides the money for program production—its very existence is highly controversial.

The overall structure of the broadcasting-advertising relationship is frequently questioned. Many countries depend on government tax money to operate their broadcasting systems; commercials, if they exist at all, provide only supplementary income. Of course, this leads to greater government control because he who pays the piper picks the tune. There are those who claim the piper analogy holds for American TV, too, but the heads of large companies and advertising agencies control program content instead of the government. They point to specific instances of advertiser censorship and to the overall censorship that occurs simply because advertisers refuse to sponsor certain types of programs.

As the new media develop, the question of the role of advertising becomes acute. Many people originally subscribed to cable TV or subscription TV pay services so that they would not have to watch commercials. As cable increases its advertising orientation, the consumer is winding up paying for the programming and also seeing commercials.

On the other side of the ledger advertisers are critical of one new media, the videocassette, because it allows people to zip through commercials on fast forward when they are playing back shows. Advertisers feel they should not be charged such high rates when, in fact, people are not exposed to the commercials.

There are many criticisms from the public about the way commercials are aired. The most frequent is that there are simply too many commercials, often referred to as **"clutter."** As stations and networks see their dollars shrinking, they are tempted to expand their inventory of ad possibilities, adding to the clutter.[12]

The frequency with which commercials interrupt programs is another gripe often voiced. In some countries commercials are grouped only between programs or all at one specific time of day. Of course, rates cannot be as high for clustered commercials because they lose impact under that sort of airing situation. For this reason, advertisers are generally opposed to the clustering that now occurs. Ads appear to be more numerous than they actually are because most commercials are much shorter than they used to be. Where one minute used to contain one advertisement, it now contains as many as four. This is because of the **split-thirty** tactic by which a company advertises two of its products (e.g., shampoo and conditioner) within thirty seconds.[13]

The frequency with which certain commercials are aired also grates on some people's nerves, and they become irritated with the repetitious sales pitches. The loudness of some commercials is also criticized. The commercial seems to blare out in comparison to other programming.

In some instances, the products advertised on TV are seriously questioned. There are those who claim that advertising nonprescription medicines encourages people to use them to excess. Likewise, beer, wine, and personal sanitary products are considered by some groups to be inappropriate TV fare.[14] A debate raged over whether or not condoms should be advertised, but they were eventually accepted, primarily because of the threat of AIDS[15] Cigarettes, of course, have been banned from radio and TV, but when this happened their consumption did not decrease markedly. The cigarette companies merely shifted advertising dollars to other media, raising the question of the fairness of banning advertising in only two media.

There are those who see advertising as having an overall negative effect on the structure of our society. It makes our society dominated by style, fashion, and "keeping up with the Joneses" while at the same time it retards savings and thrift. It fosters materialistic attitudes that stress inconsequential values and leads to waste of resources and pollution of the environment. It leads people way beyond what they need and want into purchases of total frivolity and waste and creates a society in which soft drinks and sugary snacks are considered prerequisites for health and vigor. Advertising also fosters monopoly because the big companies that can afford to advertise can convince people that only their brands have merit.

effect on society's structure

A great deal of broadcast advertising criticism is aimed at the commercials themselves. They are accused of being misleading, insulting, abrasive, and uninformative. Because fifteen seconds, thirty seconds, or even a minute is not enough time to explain the assets of a product in an intelligent manner, commercials frequently try to gain attention so that the listener or viewer will remember the name without really knowing much about the product. Of course, just as a writing school will discuss its successful graduates but never its unsuccessful ones, the liabilities of a product or service are never discussed.

time constraints

Words such as "greatest," "best," and "most sensational" may make for good copy, but they have very little concrete meaning. Worse yet are terms such as "scientifically tested" and "medically proven," which are not followed by any description of what constitutes the test or proof. Sometimes advertisers claim that "our product is guaranteed to give 50 percent more satisfaction" or "ours is whiter and brighter"—50 percent more than what? whiter and brighter than what? Statistics can lie too: "three out of four doctors" might represent a total sample of four doctors. Secret ingredients such as "DXK" and "KQ108" are thrown about without indication of what these ingredients contain.

wording

Sometimes demonstrations are not what they appear to be. One classic case involved a Rapid Shave commercial in which the shaving cream appeared to shave sandpaper. Actually, what was being shaved was a sheet of plastic covered with sand. Likewise, a Campbell's soup commercial once had marbles at the bottom of the bowl so that the vegetables would rise to the top and make for a richer looking soup. This practice was quickly outlawed—no one puts marbles in soup. Even when demonstrations are valid, they may not be applicable. For example, a watch put through a hot and cold temperature test may operate perfectly at 110 degrees and −20 degrees, but not be satisfactory at normal temperatures.

demonstrations

Testimonials come in for their share of criticism because frequently the well-known stars employed for the commercials have no way of knowing whether what they are reading from the cue cards is correct or not. Sometimes the implications given by or about the stars are misleading—"John Q. Superstar runs ten miles a day and eats Crunchies for breakfast." The implication, of course, is that the Crunchies enable him to run ten miles—in all probability a false premise.

Commercials have also been criticized for fostering **stereotypes** of women and minorities—the dumb housewife who can't get her wash clean without help from a detergent genie, or the lazy Mexican who doesn't want to do his work. Many times commercials are accused of exhibiting poor taste by being loud, repetitive, or ugly. However, because they attract attention, albeit negative, they often succeed in selling products.[16]

Commercials on children's programs have been particularly controversial in recent years. Children are much more vulnerable than adults to sales pitches. They usually cannot distinguish truth from exaggeration or fact from fiction. If a favorite cartoon character or program host says Crackle Bars should be eaten every day for breakfast, young tots will become believers and badger parents until Crackle Bars are served for breakfast. A child seeing a close-up of a small truck on TV will think the truck is as large as the TV screen, but will be disappointed to discover its actual size.

Other aspects of children's advertising that have been decried are the quantity of ads on children's programs and the deceptive special effects (quick cuts, sound effects, exaggerated camera angles) that have been employed to make products, toys in particular, appear exciting and irresistible.

For a period of time advertising to children was both reduced and improved, due largely to the efforts of Action for Children's Television. This group was largely responsible for the FCC's 1974 guidelines that stated, among other things, that stations should take special measures to provide auditory and/or visual separation devices between program material and commercials; that program hosts should not sell products; and that the display of brand names and products should be confined to commercial segments. However, in the current deregulatory frame of mind, the FCC seems to have washed its hands of regulations regarding advertising directed at children. ACT is up in arms once again in relation to quality and quantity of children's ads, particularly programs "starring" toys, which it considers to be program-length commercials.[17]

Although advertising is severely criticized from many quarters, it has its positive elements. It spurs the economy, it supports program costs so that the public can receive broadcast radio and television without paying for it directly, and it informs people about products available to them.

Conclusion

Business practices of each telecommunications entity are different, but generalizations about what might happen at a typical small commercial radio station, large commercial TV station, commercial network, public television station, and cable TV system fairly well summarize the various facets of advertising and business.

Most small radio stations are totally dependent on advertising for income. Their rates are much lower than those of TV stations, with commuter hours being the highest rated times. Small radio stations give quantity discounts and have particularly good deals for local advertisers who are likely to buy repeated spots. Sometimes they offer time well below the published rate card price, and often they barter by accepting goods or services instead of money. Members of the sales force usually have other jobs, such as disc jockey or general manager, and most of the selling is to the local community. The radio station staff builds lasting relationships with the advertisers and usually writes and prepares the commercials. Small radio stations' largest expense is salaries. The books that are kept are accurate but informal as the financial status of the station is fairly easy to determine. Viewers are likely to complain if the station's ads are too frequent or too loud.

A large TV station receives most of its income from advertising, but may also rent out its facilities or sell some of its programs. The most expensive time on its rate card is evening prime time, and it is interested in selling mostly thirty-second ads. The station may try to push run-of-schedule buying as a means of filling its commercial time. Most of the buying is on a spot basis, although occasionally an advertiser will pick up the tab for an entire program. A sales staff covers the local territory, but a large TV station also hires a station representative to obtain national ads. Occasionally the station sales manager deals directly with an advertising agency. Most of the commercials are delivered preproduced, but if the station does tape a commercial for an advertiser, it charges extra for that service. The main category of expense is salaries, but equipment, utilities, and power are also high. Determination of profit is difficult, so managers must plan carefully. Viewers are likely to become irritated by commercial interruptions and may complain to stations about misleading or vague commercials or commercials that foster stereotypes.

Commercial networks, too, are dependent upon advertising for income. They do not publish formal rate cards, but work closely with advertising agencies to supply programming (for the ad agencies' customers) that will sell the most ads at the highest rates possible. The commercials that appear on network TV are meticulously and expensively produced, usually by independent production companies, not the network itself. Transmitting and purchasing programs are primary expenses for the networks. The heat of advertising criticism is directed toward networks because they are so visible. Criticism includes the questioning of the very existence of an advertising-based entertainment structure. Other criticisms are directed at the advertising of personal products, the negative effects of advertising-induced materialism on society, the validity of testimonials and demonstrations, and the inclusion of commercials within children's television programs.

Public television receives most of its money from sources other than advertising—primarily federal funds. It does not have salespeople armed with rate cards who talk to potential advertisers, although it does attempt to obtain underwriting of program costs from some of the same companies that the commercial broadcasters approach. What public television does have is people who write and submit grant applications in order to obtain funding. Operational expenses of public television are closely akin to those of commercial TV stations.

Cable TV systems are looking to advertising as a source of income, but at present their main income is from the money subscribers pay, usually on a tiered basis, to receive the service. Therefore, their sales staffs primarily include people selling to consumers rather than people selling to companies or ad agencies. Cable system expenses have changed over the years. At first the start-up costs for laying cable were significant, but now costs are more traditional. Cable systems pay the networks for programming. The question of cable's involvement with advertising when it also charges the customer is being discussed.

All well-run companies keep a careful eye on their financial health by means of reports such as balance sheets and profit and loss statements. However, they must also keep a watchful eye on their service to the citizenry.

Thought Questions

1. Explain why you think broadcasters should or should not be made to hold strictly to their rate cards in selling all advertising time.
2. Should ads be allowed on radio and TV regardless of product? Support your answer.
3. Should ads be banned on children's TV programs? Why or why not?

Audience Measurement

15

Overview

Few aspects of telecommunications are criticized as vehemently as ratings, and yet they are all but worshipped within the executive realms of broadcasting. Ratings have arisen because of the desire of various telecommunications groups to know how many people are watching or listening to programs. Advertisers invest a great deal of money in commercials and commercial time and want to know that their messages are reaching an audience. Stations generally charge advertisers at a rate based on the number of viewers or listeners. Both stations and networks want to know public reaction to their programs so that they can anticipate changes in trends or know when a program has passed its prime.

When companies advertise in print media, such as newspapers and magazines, they are also interested in knowing the audience size. With the print media it is possible to count the number of copies sold; however, it is much harder to count the number of people watching TV or listening to the radio in the privacy of their homes or cars. As a result, audience measurement systems have been developed.

The following major topics concerning audience measurement are discussed in this chapter:

Hoopers, Crossleys, and other early rating systems

The various meters that Nielsen has used

Nielsen's sampling techniques and changes

The reports Nielsen publishes

AGB and the peoplemeter

The methodology and reports of Arbitron and its use of ADIs, TSAs, and MSAs

Other measurement services including heat sensors, pretesting, attitude research, and specialized service

How ratings, shares, homes using TV, average quarter-hour persons, average quarter-hour ratings, gross impressions, gross rating points, cume ratings, coverage, circulation, costs per thousand and costs per rating point are calculated

Criticisms of the methodology of ratings including sample size, uncooperative respondents, mechanical breakdown, human error, human bias, "hypoing," secrecy, computer error, and company discrepancies

Criticisms of the interpretation of ratings, mainly in terms of their emphasis

There are only two rules in broadcasting: keep the rating as high as possible and don't get in any trouble.

anonymous television executive

Studies and investigations of ratings by Congress, the Electronic Media Rating Council, the Committee on Nationwide Television Audience Measurement, the All-Radio Methodology Study, the Cumulative Radio Audience Method, and others

Arguments in defense of measurement methodology, secrecy, financial base, and interpretation

Early Rating Systems

fan mail

Systems for determining audience size began very early in radio history, the first "rating method" being **fan mail.** Research showed that one person in seventeen who enjoyed a program wrote to make his or her feelings known. This "system" was effective enough while the novelty of radio lasted, but it was never representative of the entire audience.

free prizes

When radio became more commonplace, stations would offer a free inducement or prize to those sending in letters or postcards. This "system" was not considered an accurate measurement of the total number in the audience, but it could give comparative percentages of listeners in different localities. For example, if two stations in two different cities made the same offer, the number of replies to each station would tell which station had the larger audience.

Crossleys

Early in the 1930s, advertising interests joined together to support ratings known as the **Crossleys** (Cooperative Analysis of Broadcasting). This was a **recall** type of rating—people in about thirty cities were telephoned and asked what programs they had been listening to at various times. Crossley ratings were discontinued in 1946 when commercial companies offered similar services.

Hoopers

Hooper ratings were also started in the early 1930s and were similar to Crossley ratings except that respondents were asked what programs they were listening to at the time the call was made—a methodology known as **coincidental telephone technique.** Part of Hooper was purchased by the A. C. Nielsen Company in 1950 and part still exists, providing customized services for radio.

The Pulse

Another radio rating service, **The Pulse,** Inc., that began in 1941 utilized **face-to-face interviewing.** Interviewees selected by random sampling were asked to name the radio stations they had listened to over the past twenty-four hours, the past week, and the past five midweek days. If they could not remember the stations they had heard, they were shown a roster containing station call letters, frequencies, and identifying slogans. This was generally referred to as the roster-recall method. The Pulse was a dominant radio rating service for many years but went out of business in 1978.

BMB

In 1946 the **Broadcast Measurement Bureau** (BMB) was started and supported by the stations themselves. It used **postcard questionnaires** to obtain county-by-county information about the audiences of each station. It published two reports before going out of business for lack of station support.[1]

FIGURE 15.1

A 1936 audimeter utilizing punch tape.
(Courtesy of A. C. Nielsen Company)

Nielsen

The name most readily associated with ratings today is Nielsen. This company was established in 1923 by A. C. Nielsen, Sr., primarily to conduct market research for industrial and consumer goods companies. Its primary research during the 1930s involved drugstores. Nielsen had a sample of drugstores save their invoices; the company then analyzed these and sold the information to drug manufacturers so they could predict national sales.

In 1936 Nielsen acquired a device called an **audimeter** from two M.I.T. professors. This device provided a link between a radio and a moving roll of punched paper tape in such a way that a record could be made of the station that was tuned in on the radio. Nielsen perfected this device and in 1942 launched a National Radio Index—a report that indicated how many people listened to various programs. For this report, Nielsen connected the audimeter to radios in one thousand homes. Specially trained technicians had to visit these homes at least once a month to take off the old punched tape and put on new—a very involved process. Nielsen analyzed the information on the tapes and sold it to networks, stations, advertisers, and others interested in knowing how many people were listening to radio programs.

By 1948 the audimeter had been perfected to the point that ordinary people could remove the tape and mail it in to Nielsen. The tape was eventually changed to film, which made the operation even easier.

In 1950 Nielsen began attaching its audimeter to television sets and preparing reports about the television audience as well as the radio audience. Then in 1964, due to economic considerations and the changing nature of radio, Nielsen dropped its radio research and concentrated specifically on television.

The sophistication of the audimeter increased and by 1960 most of them were connected directly to phone lines that led to a central computer in Florida. These were called **Storage Instantaneous Audimeters** (SIAs). With them, no

radio audimeters

television audimeters

SIAs

FIGURE 15.2

(*a*) A Nielsen storage
instantaneous
audimeter. (*b*) A
Nielsen recordimeter
and diary. (*c*) A
Nielsen peoplemeter.
*(Courtesy of A. C. Nielsen
Company)*

(a)

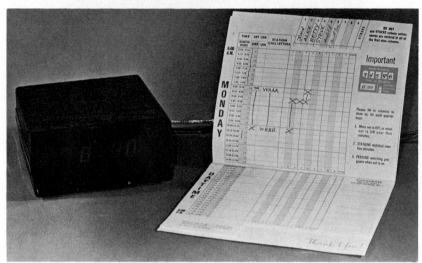

(b)

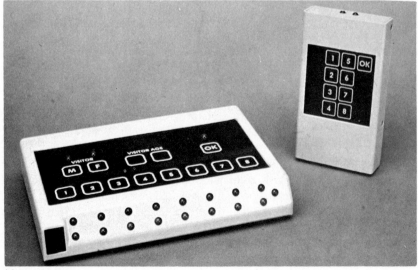

(c)

FIGURE 15.3

A portion of the computer room at Nielsen's Florida Facility. *(Courtesy of A. C. Nielsen Company)*

one had to mail anything to Nielsen. The computer dialed all SIAs several times an hour and gathered information from them.

The SIAs could only indicate whether or not a TV set was on and what channel it was tuned to. It could not tell whether or not anyone was actually watching the set. To try to solve this problem, Nielsen had a different sample of people keep **diaries.** In these diaries, people listed what programs were watched by which members of the family. These people had a different kind of device, called a **recordimeter,** attached to their sets. This was used as a check to make sure the entries in the diary were not overly different from when the TV set was actually on. It operated somewhat like a mileage counter on a speedometer in that digits turned over every six minutes that the set was in use. People keeping the diary had to indicate the "speedometer" reading at the beginning of each day. They also indicated demographic factors about members of the family—age, sex, income, education, and so on. In this way demographic characteristics could be matched with the programs people watched.

In 1987 Nielsen introduced an entirely new type of machine called the **peoplemeter.** This machine includes a hand keypad that looks something like   peoplemeters
the remote control for a TV. Each member of the family is assigned a partic- ular button on the keypad. They are to turn this button on when they begin watching TV, push it periodically while they are watching, and turn it off when they are not watching. The peoplemeter gathers both the information that previously had to be obtained from audimeters (is the set on and to which channel is it tuned) and diaries (exactly who is watching what programs). Therefore, Nielsen stopped using both the audimeters and the diaries as of fall 1987.

Obviously, there is more to the ratings process than mechanical devices attached to sets. Throughout the years, the techniques by which Nielsen gathers and analyzes data have changed. At first everything was analyzed by hand, perhaps with the aid of an old-fashioned adding machine. It took months for the reports to be issued after the data had been collected. With the advent of the computer, this changed. Data sent over phone lines from the SIAs or peo-plemeters could be analyzed almost instantly. This led to **overnight reports** on the previous night's programming that could be delivered to executives by noon the next day. As more companies obtained their own computer systems, they could interconnect their computers with Nielsen's and obtain the information on screens in their own offices.

The methods of determining the sample homes has changed somewhat during the years. Throughout its history, Nielsen has tried to reflect in its sample the **demographics** of the entire United States. This is no easy task and involves a great deal of census data, which, of course, has been changed be-cause of the availability of computers. Nielsen uses the census data to identify a core of counties throughout the country that will provide the main demo-graphics of the United States. Through a random sampling technique Nielsen identifies specific districts in the county, specific blocks in the districts, and specific households in the blocks. Representatives then try to solicit the co-operation of the households selected with about a 70 percent success rate. Where they are not successful they choose alternate homes, which supposedly have similar demographic characteristics. Once the households are selected, a machine is attached to the TV set and the people in the home are paid, in cash or gifts, to cooperate and do what is needed to obtain the ratings data.

The demographics of the country change fairly rapidly, for example, the divorce rate goes up or down, the number of Spanish-speaking people changes, the average income takes a dip. For this reason, Nielsen must be constantly changing its sample in relationship to how fast the demographics change. Most people do not remain in the sample for more than five years.

Another change has been in the number of machines needed per house-hold. When Nielsen began gathering data, most families had one radio set in the parlor and the whole family gathered around to listen. Over the years this pattern changed greatly and now many homes have multiple TV sets viewed by different members of the family at differing times. Nielsen has had to con-nect meters to all of the household sets and gather data accordingly.

Nielsen originally used only people with listed phone numbers, but as the number of unlisted phones increased significantly, the company began so-liciting samples from people with both listed and unlisted numbers.

The sample size has changed over the years as the number of households have increased. The original sample of 1000 was increased to 1200 in the 1950s and 1700 in the 1980s. The peoplemeter now samples 2000 homes, which is about .002 percent of the 87 million American homes.

Changing samples and increasing them raises the cost of doing the rat-ings. This cost is paid for by the customers who subscribe to the ratings ser-vices—networks, advertising agencies, stations, cable TV systems, and other companies concerned with advertising in the electronic media. Nielsen has raised its prices from time to time but it has to be sensitive to what the market will bear. It also charges differing amounts to different entities. While net-works may pay millions of dollars per year for the reports they desire, stations in small markets may only pay thousands per year.

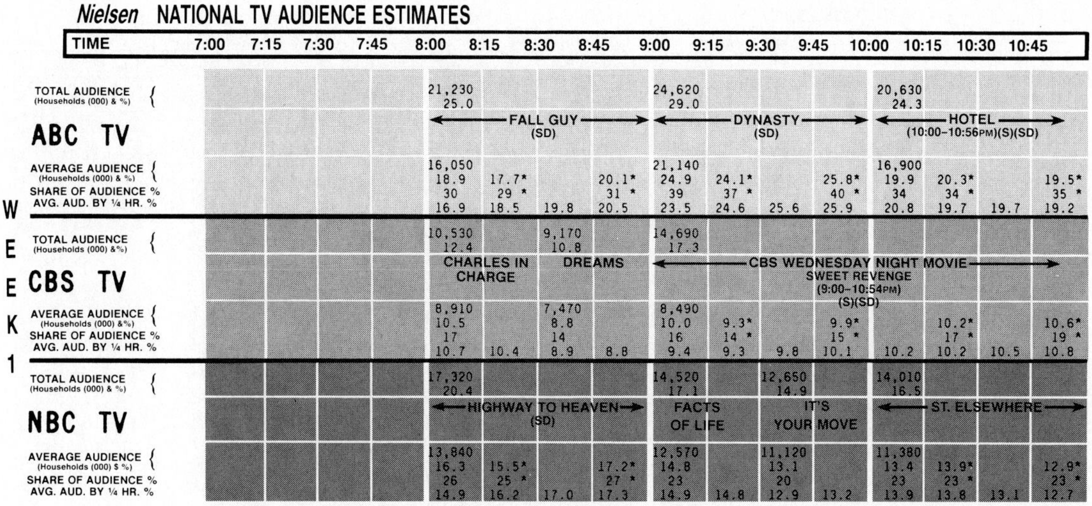

Nielsen NATIONAL TV AUDIENCE ESTIMATES

TIME	7:00	7:15	7:30	7:45	8:00	8:15	8:30	8:45	9:00	9:15	9:30	9:45	10:00	10:15	10:30	10:45

ABC TV

- TOTAL AUDIENCE (Households 000 & %): 21,230 / 25.0 (8:00); 24,620 / 29.0 (9:00); 20,630 / 24.3 (10:00)
- Programs: ←—FALL GUY—→ (SD); ←—DYNASTY—→ (SD); ←—HOTEL—→ (10:00–10:56PM)(S)(SD)
- AVERAGE AUDIENCE (Households 000 & %): 16,050 (8:00); 21,140 (9:00); 16,900 (10:00)
  - 18.9 17.7*   20.1*   24.9 24.1*   25.8*   19.9 20.3*   19.5*
- SHARE OF AUDIENCE %: 30 29* 31*   39 37* 40*   34 34* 35*
- AVG. AUD. BY ¼ HR. %: 16.9 18.5 19.8 20.5 23.5 24.6 25.6 25.9 20.8 19.7 19.7 19.2

W E E K 1

CBS TV

- TOTAL AUDIENCE (Households 000 & %): 10,530 / 12.4 (8:00); 9,170 / 10.8 (8:30); 14,690 / 17.3 (9:00)
- Programs: CHARLES IN CHARGE; DREAMS; ←—CBS WEDNESDAY NIGHT MOVIE—→ SWEET REVENGE (9:00–10:54PM)(S)(SD)
- AVERAGE AUDIENCE (Households 000 & %): 8,910 (8:00); 7,470 (8:30); 8,490 (9:00)
  - 10.5 8.8 10.0 9.3* 9.9*   10.2*   10.6*
- SHARE OF AUDIENCE %: 17 14 16 14* 15*   17*   19*
- AVG. AUD. BY ¼ HR. %: 10.7 10.4 8.9 8.8 9.4 9.3 9.8 10.1 10.2 10.2 10.5 10.8

NBC TV

- TOTAL AUDIENCE (Households 000 & %): 17,320 / 20.4 (8:00); 14,520 / 17.1 (9:00); 12,650 / 14.9 (9:30); 14,010 / 16.5 (10:00)
- Programs: ←—HIGHWAY TO HEAVEN—→ (SD); FACTS OF LIFE; IT'S YOUR MOVE; ←—ST. ELSEWHERE—→
- AVERAGE AUDIENCE (Households 000 $ %): 13,840 (8:00); 12,570 (9:00); 11,120 (9:30); 11,380 (10:00)
  - 16.3 15.5*   17.2*   14.8 13.1 13.4 13.9*   12.9*
- SHARE OF AUDIENCE %: 26 25*   27*   23 20 23 23*   23*
- AVG. AUD. BY ¼ HR. %: 14.9 16.2 17.0 17.3 14.9 14.8 12.9 13.2 13.9 13.8 13.1 12.7

FIGURE 15.4

A sample Nielsen Television Index. *(Courtesy of A. C. Nielsen Company)*

The rise of all the newer media has also changed the nature of ratings. Originally, ratings were used only for commercial and public TV stations and networks because any other uses of the TV set were insignificant. But as the newer media, especially cable TV and videocassettes, began reaching significant penetration levels, Nielsen had to begin taking them into consideration. Its meters had to accommodate many more channels and had to be able to determine when programs were being recorded off the air and whether or not they were every played back.

The end result of all the data gathering is the reports that are compiled by Nielsen—these, too, have changed over the years. The first reports handled national radio and then local radio was added. Television followed the same pattern because at first all that was large enough to measure was the television programs that were being broadcast nationally. Local station measurement began in the 1960s.

Nielsen's report of programs shown nationally is called **Nielsen Television Index** (NTI). When it had both audimeters and diaries, Nielsen actually issued two reports, the NTI and the NAC (National Audience Composition). The NTI gave the information from the audimeters in terms of total number of people viewing. The NAC came out later and broke the viewing down into the various demographics collected from the diaries. Peoplemeter data can give all of this at once.

Reports on local stations are called the **Nielsen Station Index,** sometimes referred to as the **sweeps.** For these, Nielsen divides the country into over two hundred designated market areas (**DMAs**). These are geographic areas that do not overlap but view the same general stations. All areas of the United States are surveyed at least four times a year; larger markets are covered up to seven times a year. The total number of households surveyed for all the DMAs is over 200,000. Although some of the larger markets are surveyed by

Margin notes: newer media · reports · NTI · sweeps

Audience Measurement   375

LOS ANGELES, CA WK1 11/01–11/07 WK2 11/08–11/14 WK3 11/15–11/21 WK4 11/22–11/28

WEDNESDAY
8.30PM–10.30PM

		DMA HH								DMA RATINGS																						
		RATINGS				MULTI-WEEK AVG	SHARE TREND			PERSONS		WOMEN						FEM PER		MEN						TNS	CHILD

STATION / PROGRAM (WEEKS 1 2 3 4 / RTG SHR / MAY '84 FEB '84 NOV '83)

R.S.E. THRESHOLDS 25·% (1 S.E.) 4 WK AVG 50·%

8.30PM
KABC FALL GUY
KBSC SUBSCRIPTION
KCBS E–R
KCET THE BRAIN
KCOP 8 O'CLOCK MOV
KHJ AVG. ALL WKS
 TIC TAC DOUGH
 LAKERS BKBL
 LAKERS WRAP UP
KMEX MUY ESPECIAL
KNBC HIGHWAY–HEAVEN
KTLA CH 5 MOV THEA
KTTV AVG. ALL WKS
 RITUALS
 ENT TONIGHT 30
KWHY SUBSCRIPTION
HUT/PUT/TOTALS *

9.00PM
KABC DYNASTY
KBSC SUBSCRIPTION
KCBS AVG. ALL WKS
 CBS WED–MOV
 ELLIS ISLAND 3
KCET AVG. ALL WKS
 NON–FICTION TV
 EL SALVADOR
 GIDEONS TRUMPT
 WAR ON AGING
KCOP 8 O'CLOCK MOV
KHJ 9 OCLOCK NWS
KMEX EL MALEFICIO
KNBC FACTS OF LIFE
KTLA CH 5 MOV THEA
KTTV MERV SHOW
KWHY SUBSCRIPTION
HUT/PUT/TOTALS *

9.30PM
KABC DYNASTY
KBSC SUBSCRIPTION
KCBS AVG. ALL WKS
 CBS WED–MOV
 ELLIS ISLAND 3
KCET AVG. ALL WKS
 NON–FICTION TV
 EL SALVADOR
 GIDEONS TRUMPT
 WAR ON AGING
KCOP 8 O'CLOCK MOV
KHJ 9 OCLOCK NWS
KMEX TRAMPA PARA
KNBC AVG. ALL WKS
 FACTS OF LIFE
 IT'S YOUR MOVE
KTLA AVG. ALL WKS
 CH 5 MOV THEA
 MOV PREVIEWS
KTTV MERV SHOW
KWHY SUBSCRIPTION
HUT/PUT/TOTALS *

10.00PM
KABC HOTEL
KBSC SUBSCRIPTION
KCBS AVG. ALL WKS
 CBS WED–MOV
 ELLIS ISLAND 3
KCET AVG. ALL WKS
 EL SALVADOR
 GIDEONS TRUMPT
 NOW TELL–WAR
 BATTLE–WARSAW
KCOP NWS 13
KHJ AVG. ALL WKS
 I SPY
 JOKERS WILD
KMEX TRAMPA PARA
KNBC ST. ELSEWHERE
KTLA NWS AT TEN
KTTV CHANNEL 11 NWS
KWHY SUBSCRIPTION
HUT/PUT/TOTALS *

WEDNESDAY
8.30PM–10.30PM

For explanation of symbols, see page 3.
For RSE explanations, see page 2.

FIGURE 15.5

A sample Nielsen Station Index. *(Courtesy of A. C. Nielsen Company)*

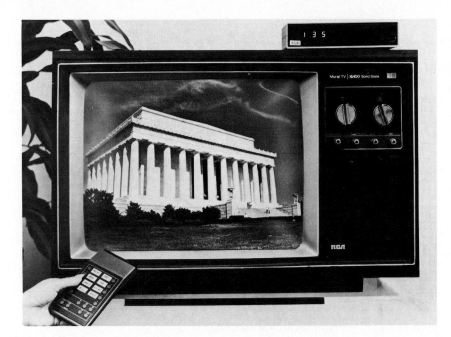

FIGURE 15.6

AGB's peoplemeter.
(Courtesy of AGB Television Research, Inc.)

peoplemeters, many of the DMAs still use diaries. The people in local areas are selected so they will be random, but still represent the demographics of the particular local area being surveyed. These people are then asked to participate and shown how to fill out the diaries. At the end of the survey week they mail the diaries to Nielsen, where the data is keypunched into a computer and analyzed. Approximately half the people actually complete and return usable diaries. The reports that are generated give detailed breakdowns according to demographics and specific times of day.

Another report that Nielsen has instituted is the **Nielsen Homevideo Index,** which deals specifically with cable, videocassettes, and other new video forms.

In addition, Nielsen will do specialized reports for just about any entity that wishes to have the data programmed in a particular way to uncover particular trends or traits. This, of course, costs above and beyond the payment for the regular reports.[2]

Homevideo Index

AGB Research

AGB is a relative newcomer in ratings in the United States. It is a British firm that developed a peoplemeter in the late 1970s, which it introduced in Europe. It is now the primary European ratings service. In 1985 it came to the United States and tested its peoplemeter in the Boston area. AGB then decided to stay and compete with Nielsen as a national ratings service.

peoplemeters

AGB operates in a manner very similar to Nielsen—it places meters in representative homes, gathers data by use of phone lines, analyzes the data with computers, and makes reports available to customers quickly.[3]

data

Arbitron

Another rating service is Arbitron Ratings Company, formerly called American Research Bureau, or ARB, which measures both local television and radio audiences. It was originally formed in 1949 and in 1967 became a subsidiary of the computer firm Control Data Corporation.

Like Nielsen, it publishes a television station index at least four times a year—February, May, July, and November. Larger markets are surveyed more frequently.

ADIs   Arbitron calls the geographic areas it surveys areas of dominant influence, or **ADIs.** To determine ADIs, Arbitron has divided the nation into 210 nonoverlapping viewing areas. Counties are placed in an ADI based on the stations most listened to in that county. Population size does not determine ADIs. In other words, a large area like Chicago is considered one ADI while a small area like Lafayette, Indiana, is another. Sometimes very small areas are combined, such as Albany-Schenectady-Troy, New York. The largest ADI is New York City, with over six million households, and the smallest is Glendive, Montana, with under six thousand. ADIs of sparsely populated areas may cover several states. For example, the Salt Lake City ADI covers not only all of Utah but parts of several other states.

diaries   Arbitron uses diaries for most of its television station ratings. Like Nielsen, homes are selected to conform to demographics, then letters and phone calls ask participants to cooperate for a small monetary incentive. Diaries are filled out to keep track of the viewing patterns of each family member and demographics are given for all people involved.

People mail the diaries to Arbitron's Beltsville, Maryland, headquarters, where data is keypunched and analyzed. Arbitron, like Nielsen, has a 50 percent return rate of diaries. Shortly after the survey is completed, reports are sent to the clients who have ordered them. These reports are broken down into audience composition for each quarter hour and are organized by age, sex, and other demographic factors. The Arbitron reports come out at approximately the same time as the Nielsen reports. Most stations will subscribe to both so that they can select the data that shows them in the most favorable light and present this to advertisers.

meters   Some of Arbitron's larger markets are metered with a meter similar to the Nielsen SIAs. Arbitron is also experimenting with a peoplemeter called **ScanAmerica.** In addition to pushing buttons to indicate whether or not they are watching TV, people would run scanners past all the groceries they purchased. In this way, Arbitron would be able to correlate program watching with types and brands of products purchased.

Unlike Nielsen, Arbitron does not have a service emphasizing network programming, although network overnight results can be gleaned from looking at the data compiled for the large metered markets.

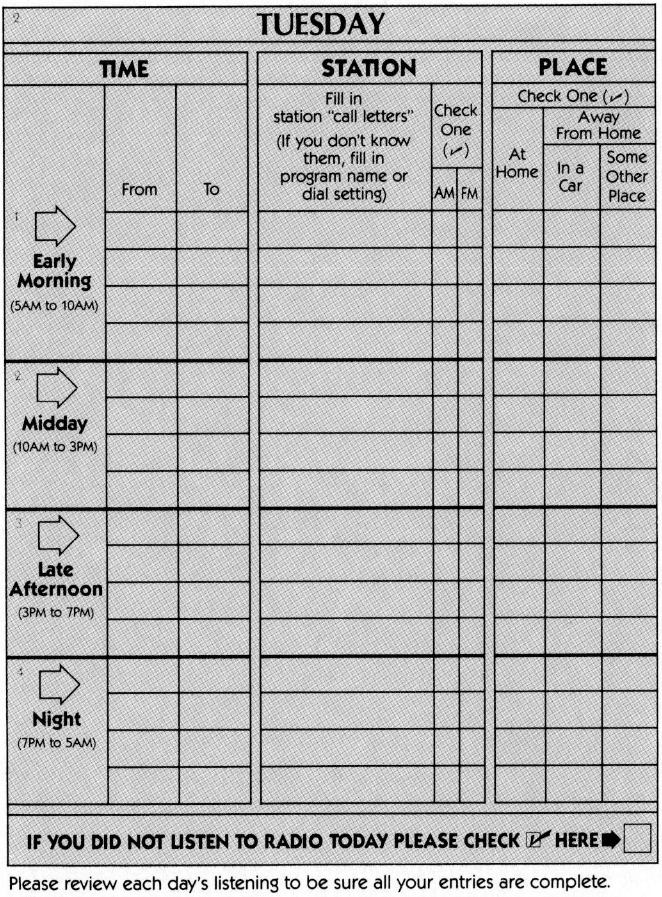
2

TUESDAY

TIME			STATION			PLACE		
				Check One (✓)		Check One (✓)		
			Fill in station "call letters" (If you don't know them, fill in program name or dial setting)			At Home	Away From Home	
	From	To		AM	FM		In a Car	Some Other Place
1 Early Morning (5AM to 10AM)								
2 Midday (10AM to 3PM)								
3 Late Afternoon (3PM to 7PM)								
4 Night (7PM to 5AM)								

IF YOU DID NOT LISTEN TO RADIO TODAY PLEASE CHECK ✓ HERE ➡ ☐

Please review each day's listening to be sure all your entries are complete.

FIGURE 15.7

A sample page from an Arbitron diary. *(Courtesy of © Arbitron Ratings Company)*

radio surveys

An Arbitron-covered area not covered by Nielsen is radio. For many years Arbitron has been the leading company in this area. It measures 250 individual radio markets at least once a year. Larger markets are measured more often; in fact, in several very large markets the measurement is essentially year-round.

MSAs and TSAs

For each market, Arbitron measures two areas, metro survey area (**MSA**) and total survey area (**TSA**). The TSAs are geographically larger than the MSAs so will not receive the lower powered stations as clearly as the MSAs. That is why Arbitron has two areas—the lower powered stations will not compare favorably with the powerhouses in the TSAs, but can hold their own in the MSAs.

samples

For each market, Arbitron determines the number of homes needed for a valid sample. It then selects these homes from a nationwide mailing and telephone list that includes unlisted numbers. Arbitron employees send letters

Specific Audience
MONDAY - FRIDAY 6AM - 10AM

	Persons 12+	Men 18+	Men 18-24	Men 25-34	Men 35-44	Men 45-54	Men 55-64	Women 18+	Women 18-24	Women 25-34	Women 35-44	Women 45-54	Women 55-64	Teens 12-17
+WAAA WRRR METRO	8.6	7.2	1.2	10.1	13.1	5.4	4.3	9.8	11.2	10.4	11.1	3.1	14.9	8.6
WBBB METRO	.9	.7			3.6			.8	1.0				6.4	2.9
WCCC METRO	6.4	5.0		.8	6.0	5.4	14.9	8.5	3.1	3.2	4.0	12.5	12.8	
WDDD METRO	.4							.6	1.0	1.6				1.4
WDDD-FM METRO	8.7	9.8	23.8	12.4	4.8	1.8		6.2	14.3	10.4	1.0	3.1		18.6
TOTAL METRO	9.1	9.8	23.8	12.4	4.8	1.8		6.9	15.3	12.0	1.0	3.1		20.0
WEEE METRO	6.8	5.7	17.9	3.9	3.6		2.1	5.0	10.2	7.2	2.0		6.4	25.7
WFFF METRO	6.6	9.1	4.8	3.9	8.3	16.1	19.1	5.4	1.0	3.2	7.1	7.8	8.5	
WGGG METRO	1.3	2.1		2.3		3.6	6.4	.8		.8	2.0		2.1	
WGGG-FM METRO	20.6	23.6	19.0	25.6	21.4	37.5	21.3	20.6	6.1	19.2	28.3	37.5	25.5	2.9
TOTAL METRO	22.0	25.8	19.0	27.9	21.4	41.1	27.7	21.4	6.1	20.0	30.3	37.5	27.7	2.9
WHHH METRO	3.9	3.8	2.4	8.5			6.4	4.0	8.2	4.8	3.0	3.1		4.3
WIII METRO	11.3	10.7	26.2	8.5	6.0	12.5		10.2	21.4	12.8	4.0	4.7	4.3	22.9
WJJJ METRO	4.3	3.8	2.4	3.9	4.8	1.8	8.5	4.4	4.1	2.4	6.1	3.1	4.3	7.1
WKKK METRO	2.4	3.8	1.2	3.1	11.9	1.8		1.5		3.2	3.0			
WLLL METRO	7.3	4.3		2.3	7.1	5.4	4.3	10.8	7.1	10.4	15.2	9.4	4.3	1.4
WMMM METRO	2.7	.5					2.1	4.8	6.1	5.6	4.0	3.1	6.4	1.4
WZZZ METRO	.5	.7		2.3				.4	1.0	.8				
TOTALS AQH RTG	26.3	26.2	23.5	28.2	27.5	29.0	29.9	28.3	29.0	28.3	31.4	31.5	26.1	17.8

Footnote Symbols: * Audience estimates adjusted for actual broadcast schedule. + Station(s) reported with different call letters in prior surveys - see Page 5B. # Both of the previous footnotes apply.

ARBITRON RATINGS

WINTER 1987

FIGURE 15.8

A sample of the type of report Arbitron furnishes to subscribing radio stations. *(Courtesy of © Arbitron Ratings Company)*

to people asking for their cooperation and then follow these up with phone calls. Diaries are then sent to the homes and each person over the age of twelve is asked to keep a separate diary of radio listening. As with the TV diaries, these are mailed back to the Arbitron headquarters where the details are key-punched and analyzed. Printed reports are then sent to subscribers—primarily radio stations.

Arbitron will provide, for a fee, specialized reports for any customers who want them.[4]

Other Measurement Services

Nielsen and Arbitron are the largest and best known rating services in the United States, but there are many other measurement companies that engage in specialized services or, in other ways, have created their own niches.

One company experimenting in the peoplemeter arena is R. D. Percy and Company, which has developed a metering system that includes a **heat sensor** that searches a room to determine how many people are in it. Like the Nielsen and AGB meters, people have their individual buttons they push to indicate they are watching TV. If the heat sensor finds more or less people in the room than have punched in, it puts a graphic on the TV screen asking "Who is in the room?" All the people present are then to repunch their buttons. Percy and Company says that their type of meter will be able to tell advertisers how many people stay in the room to watch the commercials.[5]

heat sensor meters

Another whole area of measurement is called **pretesting.** Both advertisers and programmers like to know the success probability of an idea before they invest in it heavily. Organizations such as Preview House and Preview Studios gather people into an auditorium and show them both proposed programs and commercials. They elicit the reactions of the audience members by questionnaires, various button-pushing techniques, or electronic techniques that measure perspiration. Sometimes audience members are given products that might be associated with certain programs in order to obtain reactions.

pretesting

Other companies test commercials while the participants remain in their homes. These companies see that the ads are placed in a set period of programming on cable TV, a UHF station, or a willing VHF station. Potential testees are then telephoned and invited to watch the programming of a particular channel. The following day the same people are called again and asked their opinions of the commercials.

commercials

An entirely different kind of measurement is conducted by **TVQ.** It researches attitudes and opinions about specific television programs and personalities. It mails a questionnaire to different families each month to determine the degree to which they are aware of particular shows or people and how much they like the shows or personalities.

TVQ

There are many measurement companies that specialize their services to appeal to particular needs. For example, there is one company that will analyze news programs and another that explores characteristics of the teen-age audience. A few design their services primarily for cable TV. Some of

specialized services

these companies analyze the audiences of the various cable networks while others have methods for determining how many people are watching local origination or public access channels, neither of which are included in Nielsen or Arbitron surveys. Still other companies concentrate on gathering ratings data for public radio or TV stations, medium market radio stations, or low-power TV stations.

Other measurement companies specialize not so much in the customers they serve, but rather in the type of methodology they employ. For example, several companies excel at gathering people together for **focus groups** that discuss advertisements, programs, or whatever the customer wants to have discussed. Other companies do in-depth, one-on-one interviewing. Some specialize in telephone interviewing by calling people and asking them what they are presently listening to or watching (coincidental) or what they have listened to or watched in the past (recall). One company has a special electronic scanning device that records automobile radio listening. It can set its device up on a street corner and count the number of passing cars tuned to each station.

There are also numerous companies that will take ratings data that a station or network has obtained from Nielsen, Arbitron, or other companies and analyze it to find particular strengths that can be pitched to advertisers. They might find, for example, that a station's 4:00 to 6:00 P.M. slot has the second highest ratings if you add together women and children. Often these companies also offer (for a fee) to make suggestions as to how ratings can be improved.

These specialized ratings companies tend to come and go. They must have a service that media people want, and they must offer it at a cost that the media organizations are willing to pay. Many can not do this and, hence, go out of business.[6]

Measurement Calculation

Several different types of statistics are reported by audience research companies. The main one, of course, is the **rating,** which is basically a percentage of the households watching a particular TV program or listening to a particular radio station. Rating percentages take into consideration the total number of households having TV sets or radios.

rating

Assume that the pie in figure 15.9 represents a sample of five hundred television households drawn from 100,000 TV households in the market being surveyed. The rating is the percentage of the total sample. Thus, the rating for station WAAA is 80/500, or 16 percent; the rating for WAAB is 50/500, or 10 percent; and the rating for WAAC is 70/500, or 14 percent. Usually when ratings are reported, the percentage sign is eliminated; thus, WAAA has a rating of 16. Sometimes ratings are reported for certain stations and sometimes for certain programs. If WAAA aired network evening news at the particular time of this rating pie, then this news would have a rating of 16 in this particular city. Of course, national ratings are drawn from a sample of more than just one market. There are approximately 87 million households in the

Regulation and Business

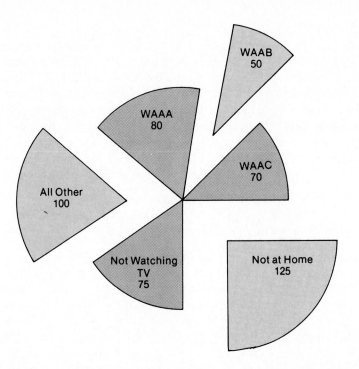

FIGURE 15.9

Ratings pie.

United States, so the number of households watching national programs can be determined by multiplying the rating times 87 million. For example a rating of 20 is 20 percent times 87 million, or 17.4 million.

A different measurement that is based on a universe, defined as all TV households using TV at the time, is called **share** of audience. In the pie shown above, 300 homes had the sets on—80 to WAAA, 50 to WAAB, 70 to WAAC, 100 to all others. In the other 200 homes, either no one was at home or the TV was off, so they do not count in the share of audience total. Therefore, WAAA's share of audience would be 80/300, or 26.7; WAAB's share would be 50/300, or 16.7; and WAAC's share would be 70/300, or 23.4. A share-of-audience calculation will always be higher than a rating unless 100 percent of the people are watching TV—an unlikely phenomenon.

Another calculation involves homes using TV (**HUT**). This is the percentage of TV households that have the set tuned to anything. (For radio, the term used is "persons using radio" or **PUR.**) In the above example, the HUT figure is 60; 300 out of 500 households had the sets tuned to something (300/500 = 60 percent).

Another way to look at these three statistics is by their formulas:

share

HUT

$$\text{Rating} = \frac{\text{homes tuned to station}}{\text{total TV homes}}$$

$$\text{Share} = \frac{\text{homes tuned to station}}{\text{homes using TV}}$$

$$\text{HUT} = \frac{\text{sets turned on}}{\text{total TV homes}}$$

Several other less important measurements are taken from time to time. One is a time-period measurement, which is usually calculated for radio and deals with an average quarter hour (**AQR**). It is a calculation of the average number of persons listening to a particular station for at least five minutes during a fifteen-minute period and is called average quarter-hour persons (**AQH persons**). Another measurement called average quarter-hour rating (**AQR rating**) can be calculated by dividing the AQH persons by the total population being measured. For a radio station, the total population is usually the number of people residing in the Arbitron ADI. This AQR rating is similar to a regular rating in that people tuned to a particular station is divided by total people. The formula is:

$$AQH \text{ rating} = \frac{AQH \text{ persons}}{\text{population}}$$

Another measurement called gross impressions (**GIs**) is the sum of the average quarter-hour persons for all spots in a given schedule. For example, Pete's Car Lot might be interested in knowing the total number of people that heard its KXXX ad on Tuesday at 9:05 A.M., 1:05 P.M., and 5:05 P.M. A ratings company might find that thirty people were tuned in for at least five minutes between 9:00 and 9:15, twenty from 1:00 to 1:15, and forty from 5:00 to 5:15. Therefore, the gross impressions would be ninety. A related statistic is the gross rating points (GRPs), which is the sum of the average quarter-hour ratings for all spots in a given schedule. If the total population that could have heard Pete's Car Lot commercial was one thousand, then the 9:05 rating would be 3, the 1:05 would be 2, and the 5:05 would be 4, adding up to 9 gross rating points.

Another related statistic is the **cume,** which is the number of different persons who tune in to a station over a period of time. For example, Pete might want to know the total number of different people who heard the commercial. The thirty people listening around 9:00 would still constitute a rating of 3, but ten of the twenty people listening at 1:00 might have been listening at 9:00, so for the cume only the ten new people would be counted for an additional rating of 1. At 5:05 there might be four people who heard the commercial at 9:05 and six who heard it at 1:05 and thirty who were hearing it for the first time, so the rating number used would be 3. The cume rating would then be 7 (3+1+3). Cumes are also used by stations to find out what their audience is for a variety of times. For example, a station might want to know the rating based on the number of different people who listen at 7:00 A.M. on Monday, Tuesday, Wednesday, and Thursday. The data might be as follows, based on a sample of one hundred.

Monday: ten people

Tuesday: seven people, three of which are new

Wednesday: six people, two of which are new

Thursday: ten people, one of which is new

The cume would be 10+3+2+1=16

Regulation and Business

Coverage is another term sometimes used in audience measurement. It refers to the number of homes that could be reached if all conditions were ideal. In other words, it is the number of homes that can see or hear the station's signal. Obviously, stations in a large city like New York will have greater coverage than stations in a small town like Cheyenne, Wyoming.

coverage

Another measurement, **circulation,** refers to the number of homes that tune in a station over a set period of time. Usually circulation figures are given in terms of one week and represent a station's drawing power, as opposed to the drawing power of one particular program. Circulation is calculated as the number of homes that tune in a station at least once a week. Therefore, if over the course of a week ninety households out of one thousand tune in radio station KZZZ, its circulation would be 9. This is an arbitrary type of count that came into being because advertisers wanted some way to compare radio station effectiveness with newspapers. It is easy to calculate a newspaper's circulation by counting the number of papers sold. There is no comparable calculation for radio and TV stations, but the number of people tuning in per week is considered to be fairly equitable.

circulation

Another statistic that is very important to advertisers is the cost per thousand, or **CPM** (M is Latin for thousand). This is an indication to the advertiser of how much it is costing to reach one thousand households. For example, if an advertiser pays $30 for a radio spot and the ratings show that spot is reaching five thousand households, then it is costing the advertiser $6 for each one thousand households. Generally advertisers want CPMs close to $5 for general shows such as network programs that reach large audiences. However, with the recent erosion of the audience, network CPMs sometimes reach double digits. An advertiser who buys a national TV ad for $100,000 would be happy with a rating of 20, for that means 17.4 million households are receiving the message. A price of $100,000 for 17.4 million households equals $5.74 as a CPM. Of course, CPMs for subgroups will be much higher. An advertiser wishing to reach working-age men at 2:00 P.M. might find the CPM for that group is about $50 because very few would be tuned to TV.

CPM

A variation on CPM is **cost per rating point.** For this, the cost of the ad is divided by the rating. If an advertiser paid $50,000 for a spot that received a rating of 20, the cost per rating point would be $2500. Since each rating point is worth 870,000 (1 percent times 87 million households), the cost per household would be .00287 cents ($2500 divided by 870,000). Therefore, the cost per thousand households would be $2.87, which is the same as the CPM calculated above. CPM and cost per rating point are just two different ways of looking at the same phenomenon.

cost per rating point

The large number of different types of audience measurement—rating, share, HUT, AQR persons, AQR rating, GI, GRP, cume rating, coverage, circulation, and others—have been developed to meet differing advertising needs. One advertiser might be more interested in the cumulative number of people hearing an ad than the number hearing at any particular time; another advertiser might be interested primarily in finding a station that delivers the largest share of the audience.[7]

Ratings under Fire

Criticism of ratings does not attack their existence. Ratings are needed in order for broadcasters and advertisers to know how widely their programs are being received. Criticisms of ratings stem more from their methodology and interpretation.

samples

All ratings are based on samples, rather than on the entire population. Therefore, it is the viewing and listening tastes of a very small percentage of people who are representing the tastes of the entire audience. At present it is simply too expensive and impractical to survey the whole population; therefore, samples must be used for audience measurement, just as they are for scientific research. A frequent criticism of audience research methodology is that the sample size is not large enough. The placement of peoplemeters in two thousand homes out of a total of 87 million is often decried, even by those who revere the ratings, as a sample size too small to prove anything.

Rating companies provide critics with a wealth of authoritative "proof" that their sample, though small, is statistically sound. These companies also point to their continuing attempt to make their sample representative of the entire population with the same percentages of households with particular demographic characteristics in the sample group as there is in the population of the entire United States. For example, a rating sample would attempt to have the same percentage of households headed by a thirty-year-old high school educated black woman with three children living in the city and earning $12,000 per year as in the population of the whole United States. But just which of these characteristics are important to audience measurement can be avidly debated. Some also debate the extent to which changes in population data should be incorporated into samples. Because of the cost involved in installing meters, only part of the sample is changed each year, leaving rating companies open to the criticism that they do not represent present-moment demographics. Someone who is in the sample as a bachelor could marry and have a child before being taken out of the sample.

gathering samples

Once a company has determined the type of sample it wishes to compose, it can find population percentages from readily available government data. Some companies then gather samples to meet these percentages by contacting people listed in telephone books, a technique criticized because it does not include those with unlisted numbers. Others draw samples by canvasing designated neighborhoods, a procedure criticized because neighborhoods are not necessarily homogeneous in their demographics.

Even after determining the ideal households to include in a sample, the companies are faced with the problem of uncooperative potential samples. About 30 percent of the people contacted refuse to have machines installed on their sets and about 50 percent do not want to keep diaries. Substitutions must be made for uncooperative people, which may bias the sample in two ways—the substitutes may have characteristics that unbalance the sample and uncooperative people, as a group, may have particular traits that bias the sample.

Problems with ratings methodology do not cease once the size and composition of the sample have been decided and the people selected. There is also the problem of receiving accurate information from these people. Each rating method has its drawbacks in this regard.

Any system that employs meters is dealing with a technology that can fail. A certain number of machines do develop mechanical malfunctions, further reducing the sample size in an unscientific manner. The older machines, such as the SIA, recorded only that the set was on; they did not reveal whether anyone was watching it. People could turn on the TV set to entertain the baby— or the dog. This problem has been solved by the peoplemeters, but people can watch TV with the peoplemeters and not bother to push the button. This deflates the ratings. The heat sensitive meters often record large dogs as people, and that certainly confuses things. Some people attempt to appear more intellectual by turning on cultural programs but not actually watching them, or they may react in other ways contrary to their normal behavior simply because they know they are being monitored. machine problems

Diaries are subject to human deceit also, for people can lie about what they actually watched. Some people simply forget to mark down their viewing or listening and then attempt to fill in a week's worth of programming from memory—perhaps aided by *TV Guide*. In addition, only about half the diaries sent out are returned in a form that can be used for analysis. diary problems

The telephone interview techniques can also contain bias—one form being the bias of the interviewer. If the person doing the interviewing has certain preferences toward programs or stations, he or she may influence a person who is unsure of what actually was seen or heard to respond toward the interviewer's preferences. Many interviewers are inexperienced, unskillful, and working only as temporary employees. In addition, interviews can be biased by unanswered phones. Lack of memory or purposeful inaccuracies can affect interviews based on recall, while calls made to determine what a person is viewing or hearing at the moment are unable to supply data involving late night or early morning programs. interview problems

Stations, too, can influence ratings. Those aware of rating periods sometimes engage in the practice called **"hypoing."** They broadcast their most popular programming, hold contests, give away prizes, and generally attempt to increase the size of their audience—usually a temporary measure. hypoing

Another overall criticism of ratings has been that they are too secretive; that the rating companies keep methodology and sampling techniques too close to the vest. This opens the door for corruption and incompetence. secretiveness

Even after rating material has been collected, it is still subject to computer error, printing error, and, even more important, differences in results. Arbitron and Nielsen have often differed as much as several rating points on certain programs or time periods. Even within one service, there are discrepancies. For example, Nielsen's overnights from machines often differed from its later diary reports. With the advent of the peoplemeter, the differences have been even more glaring. Ratings among children and teens are significantly lower for the peoplemeters as compared to diaries and other meters, discrepancies ostensibly because young people do not push the buttons. On the other hand, sports programming rates higher with peoplemeters. The speculation here is that men, watching a sports game intently, will push the buttons while in the past most women kept the family diaries and did not always record the husband's sports viewing. At any rate, the discrepancies among services makes skeptics question the accuracy of any of the ratings.

Although rating company methodology is often criticized, management interpretation is the area most criticized. Rating companies publish results and really cannot be held responsible for how they are used. This area of error is the domain of broadcasting and advertising executives.

emphasis

The main criticism is that too much emphasis is placed on ratings. Even though rating companies themselves acknowledge that their sampling techniques and methodology yield imperfect results (usually not accurate to one percentage point), programs are sometimes removed from the air when they slip one or two rating points. Actors and actresses whose careers have been stunted by such action harbor resentment. Trade journals will headline the ratings lead of one network over the others when that lead, for all programs totaled, may be only two or three points. Ratings should be an indication of comparative size and nothing more, but in reality their shadow extends much further.

quantity only

The overdependence on ratings often leads to programming concepts deplored by the critics. In a popularity contest designed to gain the highest numbers, stations and networks neglect programming for special interest groups. All programming tends to become similar, geared toward the audience that will deliver the largest numbers. Programmers place emphasis on viewer quantity, often to the neglect of creativity, station image, public access, flexibility, availability to the community, and station services to advertisers.

imitation

Dependence on ratings tends to perpetuate the imitative quality of programming. When one show receives a high rating, many similar shows are spawned. Ratings indicate what people liked in the past but give no clue as to what people will like in the future.

qualitative factors

Advertisers are also guilty of being slaves to quantity. Often a small but select audience might be best for a specific purpose—for example, estate planning insurance might be better advertised on a classical music station with a low rating than a Top 40 station with a high rating. Audience measurement barely touches on qualitative factors such as opinions and attitudes of people toward programs and products, brand name recognition in relation to program identification and purchasing decisions, and level of audience attention to selected material such as commercials. Advertisers are not prone to demand such research. They, like broadcasters, seem content to assume that quantity is the primary goal and that cost-per-thousand—regardless of the composition, attitude, or attention of the thousand—will move goods from the shelf. With the present audience measurement structure, ratings should be used as an aid, not an end. All too often this is not the case.[8]

Research on Research

Because of the numerous loopholes that can be found in audience measurement procedures, various groups have been formed to test the reliability of the rating company data. Most of this activity grew out of a House of Representatives committee hearing on ratings held in 1963. The Federal Trade Commission had previously issued cease and desist orders to several rating companies telling them to stop misrepresenting the accuracy and reliability of their reports. The FTC charged the rating companies with relying on hearsay information, making false claims about the nature of their sample populations, improperly combining and reporting data, failing to account for nonresponding sample members, and making arbitrary changes in the rating figures. The House picked up on these charges for its hearing.[9]

House inquiry

While the House inquiry was in progress, the broadcasters themselves formed the Broadcast Rating Council to monitor, audit, and accredit the various rating companies. This organization, which later changed its name to **Electronic Media Rating Council,** serves as a watchdog that oversees the rating companies. Among its duties are checking the sample design; checking the implementation of the samples and the extent to which the predesignated sample is achieved; checking the interviewers and the controls placed on them; verifying the fieldwork of interviewers; checking the procedures for handling diaries from the time of receipt through final data processing; and checking published reports and procedures to ascertain the reliability of printed output. The EMRC has developed minimum standards in each of these areas and awards its own accreditation to the companies that meet or exceed the standards. Submitting to accreditation is voluntary, but most rating companies do it because broadcasters are their major customers.[10]

EMRC

In 1963 the National Association of Broadcasters and the networks set up the Committee on Nationwide Television Audience Measurement (**CONTAM**), which undertook several basic research projects. One dealt with the effect of sample size on accuracy in rating estimates. For the study, CONTAM used 56,000 completed diaries that had already been processed so that ratings results were known. CONTAM selected various sized samples from this 56,000, ranging from one hundred samples of fifty each to one hundred samples of 2,500 each, to see if sample size affected the probability of the sample having the same rating as the 56,000 total. CONTAM concluded that the larger the sample, the greater was the probability of accuracy, but; however, at best, samples could only be an indication of the true value. Another study investigated the significance of the differences in rating methods and found that Nielsen and Arbitron agreed in their program rankings 94 percent of the time—a fairly large percentage of agreement.

CONTAM

An investigation of the effect of no-answer phone calls revealed significant differences. CONTAM, through determined follow-up calls, found that many people who did not answer their phones were in fact home and watching TV but for various reasons did not answer the phone. The difference was enough to change the viewing from 52.5 to 57.5. CONTAM also found that interviewer technique can bias ratings.[11]

unlisted numbers

Another industry group investigated people who had refused to cooperate in ratings surveys and compared them with people who had cooperated. It found that the cooperators generally watched more TV, had larger families, and were younger and better educated than the noncooperators. This is a factor that could tend to inflate all ratings, making it appear that broadcasting permeates society to a greater extent than it really does.

This cooperation factor was borne out in 1966 by the **Politz** research organization, which telephoned 12,000 households to check their viewing habits. It found 41 percent of households using TV at night, while Nielsen found 55 percent during the same time period. This would indicate that ratings favor broadcasters over advertisers, for they artificially increase the thousands, thus increasing the CPM.[12]

The All-Radio Methodology Study (**ARMS**) was formed by the NAB and the Radio Advertising Bureau in 1963 to perform several studies to determine the best procedures for gathering radio rating data. This group found that the coincidental telephone method was the most accurate and that diaries tended to be more accurate if they were collected in person than if they were mailed.[13]

NBC network radio undertook a three-year study of Cumulative Radio Audience Method (**CRAM**), which paid particular attention to the effect of sample nonresponse. NBC augmented the coincidental telephone technique by calling back at a later time and by interviewing everyone in a household over thirteen years of age rather than just one person. It found by these methods that radio listening was usually higher than that reported by rating companies. Another part of the CRAM project utilized daily phone calls to the same sample for an entire week. This method realized about 13 percent higher cooperation rate than the diary method.[14]

These various studies shed light on many of the controversial aspects of ratings. They also serve to keep the audience measurement companies from slipping into slovenly methodology.

In Defense of Ratings

The intangibles associated with broadcast listening and viewing are enormous. Unlike newspapers, the consumption of which is spread out over time, broadcasting vanishes as soon as it has been presented. In addition, radio waves refuse to obey political or geographic boundaries, making the concept of a market a muddled one at best. Networks and stations cannot survey their own audience totals, for the numbers would be suspect. Then, too, the desired result is thrice removed from the original stimulus. Programs are produced to attract a large audience to watch the commercials, which theoretically will induce people to buy products. No wonder, then, that supplying meaningful ratings is difficult. And yet something is needed because millions of dollars are involved. Given these parameters, ratings are the best method devised as yet.

Rating companies can defend their methodology by declaring that conscious manipulation of the truth by sample participants will average out. While one person is inflating his diary to improve the rating of his favorite rock station, another is exaggerating to an interviewer the number of listening hours for a country-western station.

To the critics of sampling procedure, the rating companies can reply, "All right, you come up with a better idea." No two people in the country are exactly alike, so sampling procedures must do the best they can. In general, the methods used by rating companies are as good as any yet devised. samples

The size of the samples is such that no one claims they are accurate to more than three points. Larger, more refined sample sizes could be easily accommodated if subscribers were willing to pay the cost.

Likewise, more qualitative data could be gathered, interviewing techniques could be improved, education about ratings could be more widespread—but someone must pay. There has been no demand for these improvements because subscribers—broadcasters and advertisers—have not been willing to foot the bill. qualitative data

Hypoing can affect ratings, but it is outlawed by the FCC. The FCC has never prosecuted any stations for this violation, but on occasion, rating companies have left stations out of the report because of blatant hypoing. hypoing

Rating companies claim that they must be somewhat secretive, particularly in regard to identification of the sample households. If any significant number of households were bribed, this could severely endanger rating validity. The formation of the Electronic Media Rating Council serves to keep rating companies honest. secretiveness

As to the fact that some ratings in all probability inflate the overall use of radio and TV, the rating companies can state that this may be true but because it is done uniformly, no one suffers. Broadcasting is a choice advertising medium, and if the numbers watching or listening were reported as slightly smaller, all stations could simply increase their CPMs and advertisers would wind up paying the same total dollars. Besides, there are other ratings that deflate the use of radio and television. inflation

The effects that ratings have on program content are simply the result of the democratic process. Audience members get what they vote for. If stations were to use another criterion, say creativity, as the basis for advertising rates, the situation would be far more unjust than is the present quantitative rating system. Creativity is an abstract that really has not been defined, let alone counted. democratic process

Perhaps the peoplemeters will bring refinement of techniques and more intelligent exercise of interpretation. Perhaps they will make everything worse. Regardless, audience measurement is presently and no doubt will continue to be a vital part of the telecommunications process.

Conclusion

Audience measurement has become an important part of the telecommunications business because it enables advertisers and broadcasting executives to estimate how many households are viewing their programs or commercials.

Many different methodologies have been used or are being used to determine audience size and characteristics, but they all have flaws. Fan mail and prizes were used in early days, although not very scientifically. Telephone recall was used by the Crossleys but suffers from the flaw that people do not always remember what they heard or saw in the past. Coincidental telephone technique, such as that used by Hooper, solves the problem of recall but has the additional problems of people not answering their phones or people lying about what they are doing. With face-to-face interviewing, the interviewer can have biases that influence the answers of the interviewee. Mechanical devices, such as the SIA, can break down and also do not indicate if anyone is actually watching a set that is turned on. Peoplemeters are dependent on people pushing buttons. Diary methods, such as those that have been used by Arbitron and Nielsen for their station indexes, are only as accurate as the memories of the people who fill them out and are subject to both intentional and unintentional human inaccuracies. Even the buttons used for pretesting can be sabotaged by someone who is tired or does not want to tell the truth. Obviously, no perfect method of audience measurement has yet been devised.

The two largest audience measurement companies are Nielsen and Arbitron, which, working with ADIs or DMAs, publish numerous reports for advertisers, networks, and stations. Nielsen only reports on television, while Arbitron handles both radio and television. The fact that the two services do not always agree creates skepticism of the measurement system. A European company, AGB, is involved with peoplemeters.

The figures the companies collect can be calculated in many ways: ratings, shares, HUTs, AQR persons, AQR ratings, GIs, GRPs, cume ratings, coverage, and circulation. By using these different calculations creatively, stations can make themselves more attractive to advertisers than they might be if only ratings were calculated.

Because of all the questions that have been raised regarding sample size, representativeness, and secrecy on the part of the rating companies, various government and broadcast-related organizations, such as EMRC, CONTAM, ARMS, and CRAM, have run experimental studies on measurement methodology. Although these studies have shed light on the subject, they have not devised a model for major improvements of the system.

Measurement information, with all its flaws, is still better than no information at all. Ratings may improve as both measurement and media change.

Thought Questions

1. To what extent should ratings affect whether or not a program is taken off the air?
2. Should rating companies and broadcasters be more concerned with qualitative information or quantitative data? Why?
3. Could broadcasting exist without ratings? Explain your answer.

Personnel

Overview

People are, of course, the main element in the radio and television fields. Because these fields are perceived as glamorous, there are many more people who want jobs in telecommunications than there are jobs. Nevertheless, jobs are available in both the traditional media and the newer media.

Many telecommunications entities organize their personnel into four categories or divisions: programming, engineering, sales, and business. This is because many of the reports that the FCC has requested from time to time ask for information regarding these categories. At most facilities such a breakdown represents the main ongoing activities.

This chapter is organized according to these four areas and will cover the following topics:

Jobs in the programming area—program manager, disc jockey, music librarian, reporter, access coordinator, talent, writer, broadcast standards, producer, director, unit manager, stage manager, stage hand, makeup, and scenery

Jobs in engineering—chief engineer, radio board operator, camera operator, audio engineer, technical director, character generator and computer graphics operators, projectionist, video tape recorder operator, video shader, lighting director, cable designer, and maintenance employees

Duties in sales—selling commercial time, coordinating with station reps and ad agencies, traffic, public relations, research, grant application writing, and selling to consumers

Types of work in the business area—accounting, bookkeeping, office procedures, purchasing, legal, personnel, and payroll

Typical organization charts

Types of jobs allied with telecommunications—program and commercial producers and distributors, brokers, consultants, lawyers, equipment manufacturers, and teachers

Pros and cons of having an agent

Major telecommunications unions and guilds

Preparing for a job in telecommunications

Rewards and pitfalls of radio and TV employment

I've been fired twice, canceled three times, won some prizes, owned my own company, and made more money in a single year than the President of the United States does. I have also stood behind the white line waiting for my unemployment check. Through television I met my wife, traveled from the Pacific to the Soviet Union, and worked with everyone from President John F. Kennedy and Bertrand Russell to Miss Nude America and a guy who played "Melancholy Baby" by beating his head. With it all, I never lost my fascination for television nor exhausted my frustration.

Bob Shanks,
when vice-president of ABC

Programming

radio program managers

Programs are the product of radio and TV stations or networks. Therefore, the programming departments are roughly equivalent to the manufacturing departments of most companies and, like these counterparts, are the largest departments. They are headed by **directors of programming** who oversee the total programming concept and are responsible for both locally originated material and programs obtained from outside sources.

For most radio stations this is not an overwhelming job; production and programming are very similar because most of the programming is live disc jockey shows. At a typical radio station the **disc jockeys** report to the program director, as does the **music librarian,** who is responsible for cataloging new records and filing old ones. If the station is strong on news and public affairs, the program director also oversees managers of those areas and their reporters.

Most radio network program heads are quite concerned with news, although music and features figure predominantly in some networks.

TV program managers

Handling all of a TV entity's programming and production is not a job for one ordinary mortal, so most program directors have staffs. This is true for commercial, cable, public stations and networks, and even some low-power TV and MMDS operations. These staffs include **assistant program directors** who help to acquire programming material and a **studio production coordinator** who schedules studios and production personnel. Cable TV systems that are involved in access or local origination programming may have one or more people whose job is to coordinate community program ideas and productions.

talent

Talent usually falls under the province of the program director. A station, network, or system that produces many news programs, talk shows, documentaries, children's programs, entertainment shows, and so on will have contracted talent on its payroll. This is usually supplemented by other talent hired to perform on a limited number of programs on a **free-lance** basis. Sometimes the program director finds the people who are needed through **talent agencies.** TV entities also employ announcers whose main duty is to go on the air during a station break and read whatever announcements are necessary. In many stations this material is being prerecorded and announcers are being phased out.

music librarian

A TV station may or may not have a music librarian, depending on the quantity of music it uses. Generally the only canned music needed is for theme music, since background music for dramas and comedies is created by free-lance composers.

writers

Writers are handled in a manner similar to actors. A station or network may have several writers on staff to handle regular copy and will hire free-lance writers to script programs.

Much of the programming of stations and networks is produced by independent production companies, so the talent, musicians, and writers work for them. People at the networks and stations have the job of selecting which of the independently produced programs they wish to air.

FIGURE 16.1

Larry Stewart
attempting to add to
his knowledge about
blindness before
writing a PBS
documentary, "Out of
the Shadows . . . Into
the Sun." Larry spent
several days
blindfolded with only a
guide dog for eyes.
Here he is pictured
with guide dog senior
instructor Russ Post.
(Courtesy of Larry Stewart)

Members of the **broadcast standards** department are sometimes part of the programming department or at least work with it. These people check all scripts, programs, and commercials to be aired to make sure that they are acceptable to the local community. This department also often makes sure that copyrighted materials used on programs have received proper clearance.

broadcast standards

The production area, which is sometimes under programming and sometimes separate from it, includes producers, directors, unit managers, stage managers, stagehands, makeup, graphics, and scenery. **Producers** are key people because each is responsible for bringing together all the elements necessary for taping a program or series of programs. This includes scripts, graphics, sets, props, and music. Producers are also responsible for seeing that the show does not go over budget. Producers for a complicated show may have assistants; for example, a game show producer may assign to an associate producer the task of obtaining all the prizes. Sometimes several producers will report to an executive producer who handles interrelated problems and usually has important input into the budget.

producers

TV **directors** have primary authority while the program is being taped. They give directions to the crew members during rehearsal and taping and have responsibility for the artistic composition of the pictures and sound. They are, in essence, the "boss" during productions. Sometimes they will have associates, perhaps to operate the timers and handle time cues.

directors

Some stations and networks have **unit managers** who are responsible for seeing that all the necessary equipment is readily available for each taping session. For example, a producer might tell the unit manager that an upcoming show will require three audio tape recorders. If the station has only

unit managers

(a)

(b)

FIGURE 16.2

(a) Producer Melinda Cotton (*center*) conferring with director Gary Greene prior to a news show broadcast. Anchor persons Wendy Wetzel and Arline Radilo look on. (b) Director Harry Ratner calling for a camera cut. (c) Stage manager Richard Jansen making some script notations. *(Courtesy of KOCE-TV, Huntington Beach, California)*

(c)

three audio tape recorders, and one is usually used for over-the-air station breaks, then the unit manager will have to try to convince the producer to change the script to require only two recorders or else convince the people involved with the air schedule to use a different means of station identification while this particular show is being taped.

stage managers

A **stage manager** (also called a floor manager or floor director) is responsible for what happens in the studio while the director is in the control room. This person quiets the studio when the taping is to begin, cues the talent when he or she is to start talking, and gives the talent time signals to show how much time is left in the program.

others

Stagehands set up scenery and props before and during a show and strike them afterwards; **makeup artists** apply basic makeup for most performers and also create character makeup when needed; and the **scenery department** includes people who design, build, and paint sets.

396

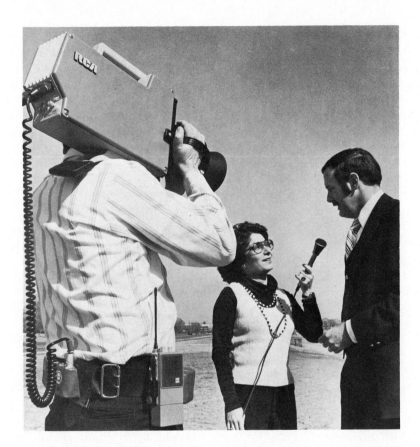

FIGURE 16.3

These TV news department staff members gather a story at a remote location. *(Courtesy of RCA)*

Stations that emphasize local news often have their own separate **news departments** with reporters, anchorpeople, producers, and technicians who work only on news. If a company owns both a radio and TV station in the same city, it may have one combined news department for both.

There are many other variations within stations, cable systems, and networks. In some stations, programming and production are separate; the production manager has the primary responsibility, leaving the program manager with only informal responsibility for production. Frequently broadcast standards will report directly to the general manager. Of course, titles vary all over the lot—what one system calls a manager, another may call a coordinator, director, or vice-president.

news departments

variations

Engineering

Heads of engineering departments are usually called **chief engineers.** Reporting to them are engineers and technicians who operate, design, install, and maintain the equipment. Small radio stations often operate with just one engineer, making him or her the chief engineer. Sometimes they do not have an engineer at all but hire a consultant who performs maintenance as needed.

chief engineers

Disc jockeys may then operate the turntables, audio boards, and tape recorders as they produce their shows. Other radio stations have engineers playing the material for the disc jockeys, as well as planning for new equipment, repairing old equipment, and deciding on equipment setup for ease of operation.

In television stations or networks it is possible that the chief engineers seldom touch the equipment. They have so many engineers under them that their total time is taken up with scheduling and supervising. Sometimes this makes for unhappy chief engineers because their background and interests lie with schematics, levers, buttons, and knobs but they must deal with paper and people.

The nature of television, as distinct from radio, makes it impossible for the talent to operate equipment as can a disc jockey. Numerous engineers and technicians work on any particular TV program—camera operators, audio engineers, technical directors, character generator operators, computer graphics operators, projectionists, video tape recorder operators, video shaders, and the lighting director and crew. Certain people may develop specialties, but all engineers should be able to handle most positions. For example, the people who run cameras should know enough about audio to be able to set up microphones and operate the audio board.

The **technical directors** of shows operate the switcher to change the picture being aired from one source to another. They execute the aesthetic decisions of the director, regarding what should be sent over the air when and in what manner, by operating the appropriate controls. The technical directors are responsible for the overall technical standards of a show and are the ones to decide whether or not a taping session should be stopped because of a minor problem with the video tape recorder. They are in charge of the technical crew of a particular show and work with the director to make sure the crew receives appropriate break time and mealtime.

The technical directors' crew includes the **camera operators,** who obviously are responsible for obtaining the shots the directors want; the **audio engineers,** who place microphones in the studio for best possible pickup, move boom microphones during the show, and operate the audio console, tape recorders, and turntables in order to achieve the sound that the directors specify; the **character generator** and **computer graphics operators,** who type in the appropriate material and see that the right pages are ready to be incorporated into the program at the right times; the **projectionists,** who operate the film chain if film or slides are to be inserted into the program; the **video tape recorder operators,** who record the program as it is being performed and also edit material; and the **video shaders,** who operate remotely located camera controls to obtain the best picture possible from the cameras.

The **lighting directors** and their crews set up lights before the show and handle any lighting changes during a production session. In some stations these people report to the production manager rather than to the chief engineer so that they can be in on production meetings and contribute their aesthetic expertise about lighting effects that create time or mood. In fact, in some organizations all the people who operate equipment during a shoot report to the production manager rather than the chief engineer.

Technical director Toby Baker at the switcher. *(Courtesy of KOCE-TV, Huntington Beach, California)*

Engineer Jeri Shepard performing master control switching during a station break. *(Courtesy of KOCE-TV, Huntington Beach, California)*

The engineers also go on remote shoots where they operate independent camera-audio-VTR setups or cameras and control room equipment placed in a remote truck or van. When the material is brought back from the field, engineers edit it.

The engineers not only help produce programs, they also keep the station on the air. Video tape recorder operators see that the proper tapes of programs and commercials are aired at the scheduled times. As with radio, engineers

other duties

FIGURE 16.6

A maintenance
engineer checking a
camera's electronics.
*(Courtesy of General
Telephone Company)*

design and repair equipment and keep abreast of the latest technological advances so that they can advise management on equipment purchases.

Within the cable TV industry, most engineers are involved with planning and stringing the cable and only a few work with production equipment. Likewise, in the more esoteric areas such as direct broadcast satellite, engineers are employed primarily to design systems rather than to operate or maintain equipment.

Sales

sales reps

In commercial radio and TV stations and networks the sales department is mainly responsible for selling commercial time. At a small radio station the general manager and/or disc jockeys may double as salespeople. Larger stations will have salespeople and perhaps a national sales manager to coordinate with the station representative or advertising agency. Obviously, the networks deal only with national advertising.

traffic

Another responsibility frequently placed in a station's sales department is **traffic.** This responsibility includes listing all the programs and commercials to be aired each day in a station **log.** Traffic is often part of the sales department because the most important coordinating job involved with the log is scheduling the various commercials. As time sales are made, someone must make sure that the ads are actually scheduled to air under the conditions stipulated in the sales contract. However, traffic also involves the scheduling of program content, so sometimes it is part of the programming department and sometimes it is even part of the business area.

Public relations, promotion, and research are all sales-oriented aspects PR, promotion, and research of telecommunications that are extremely important—in fact, so important that some chief executives have these functions report directly rather than through a sales department. The **public relations** function of a station or network is to try to build general goodwill between the broadcaster or cablecaster and the general public, a service particularly essential at license renewal or franchise renewal time. **Promotion** involves all methods used to encourage people to listen to a particular station, network, or program, such as special contests, billboards on buses, ads in newspapers, and feature stories in magazines. Those who work in **research** mainly coordinate with the rating services and analyze data for their own purposes, but sometimes they conduct station- or network-generated research projects.

Public broadcasting stations do not have sales departments as such. fund raising However, they usually have people in charge of obtaining grants, community contributions, and other funding, which in many ways resembles a sales function.

At some telecommunications entities, sales departments are involved obtaining subscribers primarily with obtaining subscribers. Such entities include cable TV, MMDS, SMATV, and the telephone-computer data base services. Many cable networks and some cable systems are now selling commercial time in a manner similar to commercial TV stations and networks. In addition, cable networks have salespeople, usually called **affiliate relations representatives,** to convince cable system operators to carry their particular network on one of the cable system's channels.

Business

Telecommunications, like any other business, must distribute paychecks, sort mail, balance the checkbook, and sweep the floor. Such housekeeping functions are essential to smooth operation and generally reside in the domain of a business manager who reports to the chief executive, such as the station manager or president.

Accounting and **bookkeeping** functions fall under this business manager accounting or sometimes are in a separate finance department. The duties include keeping records of payroll expenditures, equipment and supply purchases, inventories of supplies, cash receipts, cash expenditures, and the like. Companies regularly compile balance sheets, which list all assets and liabilities of a station or network at a specific point in time, and profit and loss statements, which show how much the company received and spent over a period of time so it can tell whether its expenses are exceeding its income.

At large, sophisticated operations, accountants may keep track of unused facilities, idle time of personnel, budgeted costs of specific programs, and efficiency factors of various departments. The job of the business department is to keep all records needed for FCC reports, income tax returns, Social Security reports, and insurance claims. In recent times the computer has been widely utilized for these functions.

General office management, also in the business area, may include overseeing the janitorial services, distributing work to a secretarial pool, making sure the mail is handled efficiently, arranging for public tours of the facility, and establishing security regulations. Purchasing and stockroom procedures constitute another function of the business office. Legal matters involving advice to keep a company out of trouble, as well as actual lawsuits, are handled under the business department. Personnel matters such as hiring, firing, promoting, reprimanding, mediating, and negotiating with unions are also handled by the business department. Of greatest importance to most employees is the payroll department, which issues the paychecks.

How all these functions are distributed depends on station or network size, market, format, success, and top-management philosophy. Very few radio stations have personnel departments; this would be unjustified for a total staff of ten to twenty people. TV networks, on the other hand, have eight to ten people whose full-time job is to interview job applicants. A TV network will have several staff lawyers, who may even report directly to the president if a problem seems particularly sticky. A station, however, will probably hire a lawyer from a law firm as needed. Many small stations have a policy of no visitors, simply because there is no physical space for the extra bodies; hence, they have neither public tour nor security needs. Networks, on the other hand, have guards at every gate and regular formal tours conducted by hired pages. Many stations and cable systems are lucky to have one secretary, let alone a secretarial pool. In some instances, the station manager is also the janitor.

Organization Charts

No two broadcasting entities are organized in exactly the same way, but theoretical organization charts of a TV station and a medium-sized radio station might be like figures 16.7 and 16.8.

Figures 16.9a–c are actual organization charts from various types and sizes of organizations.

Allied Organizations

In addition to people who work directly for commercial broadcasting stations and networks, public broadcasting stations and networks, cable TV systems and networks, and other electronic media, there are many other people who spend time engaged in telecommunications work: television program producers and distributors; radio program producers and distributors; television commercial producers; radio commercial and jingle producers; advertising agencies; station representatives; audience measurement companies; brokers; station finance companies; management consultant firms; public relations, publicity, and promotion services; talent agencies; employment services; communication attorneys; equipment manufacturers; consulting engineers; music licensing groups; unions; news services; and radio and television departments of colleges and universities.[1]

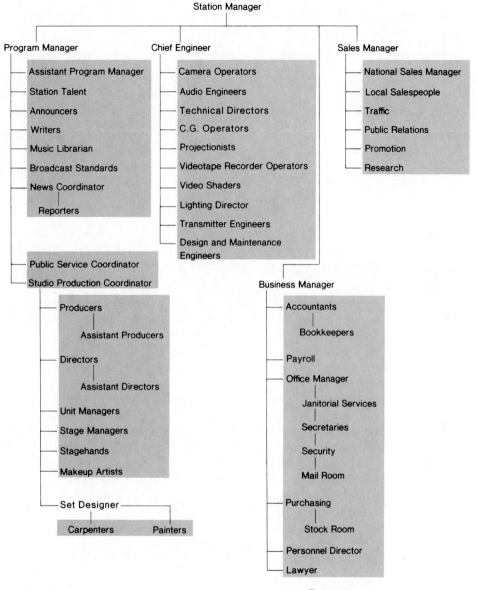

Station Manager

Program Manager
- Assistant Program Manager
- Station Talent
- Announcers
- Writers
- Music Librarian
- Broadcast Standards
- News Coordinator
  - Reporters
- Public Service Coordinator
- Studio Production Coordinator
  - Producers
    - Assistant Producers
  - Directors
    - Assistant Directors
  - Unit Managers
  - Stage Managers
  - Stagehands
  - Makeup Artists
  - Set Designer
    - Carpenters
    - Painters

Chief Engineer
- Camera Operators
- Audio Engineers
- Technical Directors
- C.G. Operators
- Projectionists
- Videotape Recorder Operators
- Video Shaders
- Lighting Director
- Transmitter Engineers
- Design and Maintenance Engineers

Sales Manager
- National Sales Manager
- Local Salespeople
- Traffic
- Public Relations
- Promotion
- Research

Business Manager
- Accountants
  - Bookkeepers
- Payroll
- Office Manager
  - Janitorial Services
  - Secretaries
  - Security
  - Mail Room
- Purchasing
  - Stock Room
- Personnel Director
- Lawyer

FIGURE 16.7

Television station theoretical organization chart.

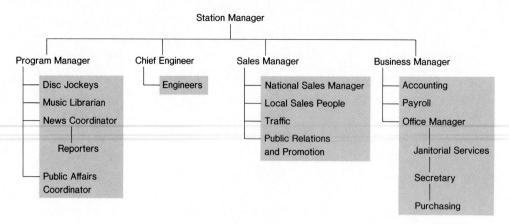

Station Manager

| Program Manager | Chief Engineer | Sales Manager | Business Manager |

Program Manager
- Disc Jockeys
- Music Librarian
- News Coordinator
  - Reporters
- Public Affairs Coordinator

Chief Engineer
- Engineers

Sales Manager
- National Sales Manager
- Local Sales People
- Traffic
- Public Relations and Promotion

Business Manager
- Accounting
- Payroll
- Office Manager
  - Janitorial Services
  - Secretary
  - Purchasing

FIGURE 16.8

Medium-sized radio station theoretical organization chart.

mixed jobs

production companies

brokers

Of course, the activities of these people often extend beyond telecommunications. For example, advertising agencies are concerned with print media as well as broadcast media; their job is to find an advertising mix for their customers, not just to arrange commercials for radio or television. Nevertheless, it is interesting to note the extent to which telecommunications, itself a service, is served by other organizations.

Many companies produce radio or TV programs or commercials. The number of jobs available in this area, however, is somewhat deceptive. A few very successful production companies—MTM, Lorimar, Universal, and Columbia Pictures Television—supply large numbers of programs to the TV networks. However, most production companies are small groups who hope to hit on something that will bring financial reward. Many of them never make it and collapse, although they may reappear some years later with a new name and the same people involved. For the most part, independent production companies cannot afford to own their own facilities, so they rent regular station or network facilities and personnel in order to produce their programs. Therefore, the camera operator who works on the channel 5 noon news may at one o'clock begin working on an entertainment show to be aired on CBS. Independent production, therefore, does not necessarily create a large number of jobs for new people.

Most of the allied organizations are dealt with in other sections of this book, but a few bear describing in slightly more detail. **Brokers** aid in the buying and selling of radio and television stations and cable systems in much the same way that real estate agents do for homes. Individuals or organizations who wish to sell a facility may list it with a broker, who then tries to find a suitable buyer under the guarantee of a commission on the station sale price. Of course, stations can be sold without the aid of a broker.

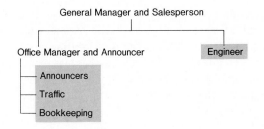

(a)

FIGURE 16.9

(*a*) Small-market radio station organization chart. (*b*) Medium-sized cable TV company organization chart. (*c*) Large-market TV station organization chart.

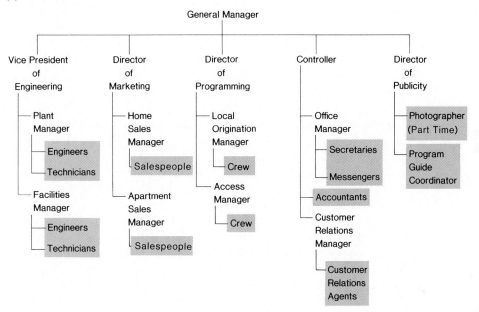

(b)

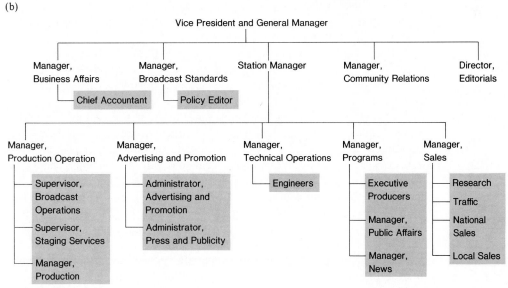

(c)

Consulting engineers, management consultants, promotion services, and attorneys are usually hired by stations on a short-term basis. A new cable system may hire a consulting engineer to determine the best location for a headend. A radio station that has been given permission to increase its power may hire a consultant so that all the technical procedures are properly accomplished. A station having personnel problems may hire a management consultant to suggest a new organization, or a station wishing to execute a spectacular two-week promotion for a particular program might hire a promotion service. When a legal problem surfaces, an attorney is hired to handle the case.

Agents

A word should be said about **agents;** they are generally controversial in a "needed by all, resented by all" role. Their function is to find work for creative people in return for a percentage—usually 10 percent—of the amount earned.

The largest talent agency is **William Morris,** founded in 1898 to represent talent in the legitimate theater and vaudeville. It has grown steadily, adding representation for motion pictures, radio, phonograph records, and television. In its earlier days it was all but eclipsed by **MCA,** which was founded in 1924 as the Music Corporation of America to represent dance bands and orchestras. MCA grew rapidly, representing not only musicians, but actors, writers, producers, directors, composers, and others in the creative fields. It also began packaging shows with its own talent (for which it collected fees for the entire shows) and it acquired **Universal Studios** as its production house. In 1962 this situation came to the attention of the Justice Department, which invoked the **antitrust act** on the grounds that one company should not represent talent on one hand and produce programs with that talent on the other. Because MCA made more money collecting on its shows than on its individual stars, it kept the Universal production unit and gave up the talent agency business, thus leaving William Morris as the largest agency.[2]

It is virtually impossible for an actor or actress or a free-lance writer, director, producer, or other creative person to become successful without an agent because networks and most production companies deal only through agents and not directly with talent.

In some ways this can be advantageous to the person. An agent can describe a client in glowing terms that the client could not say about herself or himself. An agent is constantly out in the field representing people and possesses a wealth of information about what jobs will be available. The agent can also negotiate salary and contract, leaving the talent "above it all." Many agents can give sound professional advice to help a person improve. On the

other hand, agents who are so inclined can abuse their power. Critics of agents sometimes contend that the agent receives 10 percent of everything the talent earns and sometimes does nothing to deserve it. Because an agent represents a large number of people, he or she will not always have one particular client's interest at heart and sometimes will even sell out a client or make an undesirable deal for the client in order to continue doing business with a particular production company. Sometimes an agent will get a client involved in a disastrous project simply to obtain the 10 percent commission.

Networks and production companies also have ambivalent feelings toward agents. They would rather deal with agents than with hordes of aspiring actors, but once an actor becomes known and desired by the production houses, they would rather the agent disappeared so that they could deal directly with the person.

Unions

Another broadcasting phenomenon very important to the production scene is unions. Like agents, unions are much more important in the large cities than in small towns. Unions operate only where they are voted in by the employees—usually where there is a concentration of activity. At present about 33 percent of the TV stations, both commercial and public, and 8 percent of the radio stations deal with one or more unions. Of course, the major networks are unionized. The unions are involved with cable satellite programming, but local cable systems rarely deal with unions. Unions negotiate wages and working conditions. They also prosecute members of management who violate provisions of the union contract and union members who work for less than union scale or otherwise violate contract agreements.

One of the most prominent of the broadcasting unions is the American Federation of Television and Radio Artists (**AFTRA**), a union for performers in live and videotaped programs. AFTRA was begun in 1936 as a union of radio artists in New York, Chicago, and Los Angeles and was called the American Federation of Radio Artists (AFRA). Today its jurisdiction extends to all fifty states and covers radio transcriptions, phonograph records, and slide-film recordings, as well as live and taped radio and TV programs and commercials.

Another performance union, the Screen Actors Guild (**SAG**), was organized in 1933 by actors discontent with the poor pay of motion picture studios. SAG and AFTRA have had several jurisdictional disputes over the years, the most bitter occurring when videotape was introduced. SAG felt it should have the jurisdiction because the material was being recorded to be kept much as film is kept. AFTRA felt it should have control because taping was accomplished in a TV facility. AFTRA won the battle, but at present SAG has jurisdiction over anything produced at one of the traditional film studios, even though the medium used is videotape.

Another acting union is the Screen Extras Guild (**SEG**), which tries to ensure that people with walk-on parts receive adequate pay. At various times plans to merge SEG, SAG, and/or AFTRA have been proposed, but none of these plans have made it through the unions' political process.

Historically, acting unions have had an extremely high unemployment rate. It is estimated that 80 percent of the members earn less than $2,000 a year from acting.[3] Most of these people find other types of employment elsewhere, many in menial jobs that will allow them the freedom to look for acting work.

If a station or network is a signatory to a performer's union, it must pay at least the minimum wage negotiated by the union to all performers. Complicated formulas specify that a principal performer is to be paid more than someone who has less than five lines, who in turn is paid more than an extra. And, of course, the few stars of the business can demand far more than the minimum wage. The union also negotiates such matters as meal and rest periods, credits, wardrobe, use of stand-ins, travel requirements, dressing room facilities, and residuals. This last is a thorny issue involving the formula for paying performers when shows or commercials are rerun. Performers feel the residual percentage should be high because reruns eliminate jobs. Networks and stations want to keep the residuals low so that rerun costs are low.

AFM

Another performance union is the American Federation of Musicians (**AFM**), which has jurisdiction over musicians who perform live or taped on radio or TV. AFM is not as important to broadcasting now as it was when radio networks had their own orchestras. During the late 1930s and early 1940s AFM forced stations to employ musicians even though they had no work for them by making the stations sign contracts stating that they would hire a certain quota of musicians based on the station's annual revenue. These contracts were declared illegal in 1940 and were not renewed, but the AFM threatened networks and record companies with strikes if they supplied music to stations that did not employ musicians, so many of the quotas continued. In 1946 Congress specifically outlawed the use of threats to require broadcasting stations to employ individuals in excess of the number needed to do the job.

IBEW

The first established union for broadcasting engineers was the International Brotherhood of Electrical Workers (**IBEW**), which had originally been formed in the late 1880s by telephone linemen. This union successfully struck a St. Louis radio station in 1926 and later obtained network contracts. In 1953 NBC technicians formed their own union, which became known as the National Association of Broadcast Employees and Technicians (**NABET**). Now NABET and IBEW compete, sometimes vigorously, to convince employees of stations to vote them into power.

NABET

A third technical union, International Alliance of Theatrical Stage Employees and Moving Picture Machine Operators of the United States and Canada (**IATSE**), moved into television from motion pictures. In some stations this union only has jurisdiction over stagehands and in other stations it includes stagehands and some equipment operators.

IATSE

For all of these unions, members must pay an initiation fee and then regular monthly dues, usually an amount related to income earned. One of the points of frustration to people trying to obtain a first job in broadcasting is that one must have a job to join a union but must join the union in order to get a job. Breaking into that vicious circle is often difficult.

guilds

There are several broadcasting guilds that do not operate in exactly the same way as unions, but they are involved with setting pay rates and negotiating. Three of the most important are Directors Guild of America (**DGA**), Writers Guild of America (**WGA**), and Producers Guild of America (**PGA**).

In addition to the major telecommunications unions and guilds—
AFTRA, SAG, AFM, IBEW, NABET, IATSE, DGA, WGA, and PGA—
there are many smaller unions for groups such as costumers, model makers,
set designers, makeup artists, and carpenters.[4]

Job Preparation

Gaining employment in telecommunications is not easy—there are many more
people who would like to be so employed than there are jobs. A recent survey
conducted by the Broadcast Education Association showed that over 300 colleges have programs leading to degrees in radio and television. These schools
graduate approximately 12,000 students a year, ostensibly into an industry
that directly employs only slightly over 240,000 people.[5]

openings

 Obviously, however, the persevering can be employed in broadcasting.
Although obtaining a college education is not essential for most of the entry-
level jobs, it is generally a wise course to take. Stations, networks, and cable
systems are interested in hiring people with promotion potential and generally
feel that a college education makes people more promotable.

 Enrolling in one of the three hundred colleges with programs leading to
a broadcasting degree is advisable, but equally important is having broad
knowledge in other fields. Someone wishing to enter the news field would be
almost valueless if he or she were an outstanding video tape recorder operator
but knew nothing about national or international affairs. Political science, history, writing, and journalism courses should be a must for reporters. Likewise,
accountants and salespeople should have proper business courses; engineers
should know electronics; writers should emphasize composition and literature;
directors should be knowledgeable about drama, music, and psychology; and
graphics personnel should know art and photography.

college courses

 A student, while in college, should make every attempt to obtain experience in the industry through part-time jobs, **internships,** or nonpaid fill-in
work. Many telecommunications openings have an "experience required" tag
attached to them, and people who have at least made the attempt to rub elbows
with the industry can sometimes get a foot in the door for these jobs.[6]

experience

 Students would be well advised to join telecommunications organizations in order to meet people in the field. The old saying "its not what you
know but who you know" is often what enables you to obtain employment.
Many of these organizations have special reduced rates for students.

organizations

 Students should also begin reading the trade journals of the industry.
Broadcasting magazine, a weekly published in Washington, D.C., is the "bible"
of the industry and covers the gamut of telecommunications industry news.
Once a year the same organization that publishes *Broadcasting* publishes
Broadcasting Yearbook. This supplement includes names, addresses, and facts
about all radio and TV stations in the country and all cable systems, advertising agencies, networks, program suppliers, equipment distributors, and other
telecommunications-related groups. Other trade publications of value include
Variety, Hollywood Reporter, and the myriad of specialized journals put out
by media organizations.

trades

The first job is the hardest to obtain and usually involves a great deal of letter writing and pavement pounding. Many of the trades contain want ad sections that provide excellent leads. Letters of introduction and resumes can be sent to stations, networks, cable systems and allied organizations. *Broadcasting Yearbook* can serve as an excellent source for names and addresses for these potential employment sources. Phone calls or visits to as many facilities as time and money will permit can also lead to employment.

The nature of the first job is not nearly so important as getting in the door. Most facilities promote from within, making it much easier to obtain the desired job from within than without. It is also easier to move from one facility to another if some experience has been gained, regardless of what the experience was. Patience tempered with aggressiveness and geographic flexibility are traits that are likely to lead to advancement.

Career Compensation

On the average, broadcasting is not an especially high-paying industry. True, there are the superstars who make millions, but they are few and far between. Others earn a decent but not sensational living. There is no standard pattern on fringe benefits but most employers provide medical insurance and life insurance, as well as paid sick leave and vacations for full-time employees.

What undoubtedly attracts most people to the telecommunications industry is the glamour, excitement, and power of the profession. All of these are present (albeit to a lesser degree than most people think) and they do make for a richer, fuller, more rewarding life than many people experience in other day-to-day occupations.

One by-product of this glamour, excitement, and power, however, is extreme insecurity. Except for those in a few highly unionized jobs, no one is shielded from the pink slip. Last year's superstar can be this year's forgotten person—and, worse yet, have no means of support. Many people who "make it" in the business tend to spend like millionaires in order to maintain an image. Then, when the image is only a fad of the past, they have no nest egg to comfort them. The industry encompasses many free-lancers who are constantly in the state of being fired and hired and whose outrageously high salaries for one week's work are balanced by two years of unemployment. It is a hard field to break into and a hard field to stay in, but the rewards are worth it to those who endure.[7]

Conclusion

Typical opportunities and requirements in the telecommunications personnel categories of programming, engineering, sales, and business can be seen in an examination of people who want to be announcers, producers, camera operators, salespeople, and accountants.

Would-be announcers may seek employment as radio disc jockeys or TV announcers, the former jobs being more exciting than the latter. They may also work on a free-lance basis for companies that produce commercials or jingles. Within a studio structure, announcers report to the program manager and work closely with the music librarian, the director, and the engineers. For success in the field, announcers should have an agent and join AFTRA. College training in theater arts and radio-TV is wise. Starting pay is relatively low, but big-time announcers can earn salaries of six figures. The secure jobs do not have the excitement or variety of the insecure free-lance jobs.

Producers have more opportunities in TV than in radio and have more prestige at networks than at stations or cable systems. They can also work free-lance or, occasionally, permanently for program or commercial production companies. They report to executive producers or program managers and must have particularly strong relationships with directors—directors bring producers' concepts and elements together in the studio or at the remote setting. Producers often observe production, but directors decide the moves of the production crew members and the engineers. Producers may or may not join PGA, which is not as strong as the actual unions, and they may or may not have an agent, depending on the amount of free-lance work they wish to undertake. A liberal arts background with emphasis on TV production and a pleasant but aggressive personality should equip them for the role. Top producers can earn top dollars, but most earn significantly less than their stars.

People who want to become camera operators may find work in commercial, public, or cable TV and, to some extent, in independent production companies. In most companies camera operators are in the engineering department and report to the chief engineer; however, in small operations they may be in programming and double as producers or directors. They should be familiar with other engineering areas in addition to camera, such as audio, VTR, switcher, and lighting. During production they follow the commands of the director and also take some direction from the technical director, who is in charge of the engineering aspects of a particular show. In large cities they need to join one of the engineering unions and should try to gain membership in a union whenever the opportunity presents itself. Camera operators can earn good consistent pay, but it rarely reaches the upper limits that top performers and producers make.

Salespeople can sell time for commercial radio or TV stations or networks and, to some degree, for cable entities. They also can be employed in selling hookups to consumers for cable or other media. Many salespeople are employed by station representatives and equipment manufacturing companies. They report to the sales manager or, if the company is small, to the general manager. They must coordinate time sales with the people in traffic so that commercials are properly placed on the log. They also work with people in the business department concerning the overall financial health of the station or network. A background in business in general and the peculiarities of the telecommunications industry in particular will help them with entry and continuing jobs. Effective salespeople can earn a great deal of money if they work on commission. Sales is also an excellent avenue to general management.

Most accountants in the telecommunications field did not necessarily intend to join that industry. They have backgrounds in business that enable them to move from one industry to another. They can be employed at stations, networks, cable systems, production or distribution companies, ad agencies, station reps, audience measurement companies, brokerage firms, consulting firms, talent agencies, equipment manufacturers, and just about anywhere else in the industry. They report to the business manager and work with executives in all departments. Their jobs are more stable than most others in the telecommunications field, but accountants do not obtain very high salaries and their jobs do not have the excitement of production or engineering.

Thought Questions

1. Explain why you think all radio and TV stations should or should not be unionized.
2. Do you feel that, overall, an agent is an asset or deterrent to a performer? Why?
3. Should the telecommunications departments in colleges and universities cut back on the number of students they admit so that fewer people will be prepared to enter the job market? Why or why not?

Other Telecommunications Forms

*M*ost of this book so far has been concerned with uses of telecommunications in the United States by the public at large. Needless to say, the United States does not hold a monopoly on any of the forms of electronic media. They are evident throughout the world, sometimes in forms that vary significantly from what Americans see and hear on a daily basis.

Likewise, not all forms of telecommunications are produced and distributed with the idea that anyone might want to watch or listen to them. Some are produced in such a way that the masses can not even have access to them. These are intended for employees of one particular company or even members of one particular family.

Both the corporate and international telecommunications forms have unusual and interesting aspects.

Corporate, Government, and Personal Uses of Telecommunications

17

I believe in the future . . . a man in one part of the country may communicate with another distant place.

Alexander Graham Bell, 1878

Overview

Many uses are made of telecommunications by such entities as schools, businesses, hospitals, government offices, libraries, the military, and individual people. Although these uses may not be as "glamorous" as traditional broadcasting, they are often quite complex and quite abundant. Some of the special uses made of telecommunications that will be discussed in this chapter include:

The history of nonmass telecommunications

Improvements in phone systems

The introduction and growth of cellular radiotelephones

The numerous data banks available to stock brokers, broadcasters, lawyers, libraries, individuals, and other elements of society

Possible future directions of data banks

The multitude of uses made of computers and videodiscs for typing, financial management, information, and training

Procedures for one-shot and regular teleconferences

Advantages and disadvantages of teleconferences

History and varied uses of ITFS

Methods by which video programs are produced for industry, education, and personal use

The Development of Nonmass Communications

Most of what this book has dealt with so far has been electronic media that are intended for the mass audience. Just about anyone can turn on NBC news, listen to music on radio, or go to the local video store to rent a cassette. But only a select few can see a video about a new airplane design by Boeing, or access the TRW credit information data file, or work with a videodisc that trains fighter pilots, or even see the home videotape of Suzie's first birthday party. small group use

All of these are specialized uses of telecommunications intended for small groups who have something in common and have a specific need for the information.

With the exception of the old-fashioned telephone, these types of uses of telecommunications began in the late 1960s and early 1970s. At first, video history equipment was so bulky and expensive that only TV stations and networks could afford and house it. After **helical** recording was developed and smaller and cheaper cameras and VTRs were developed, this technology became available to other organizations.

Education was one of the first areas to become involved with the use of school use this equipment. Schools bought cameras and other video equipment to teach students to become broadcasters and to enhance classroom and learning activities. Industry, too, took to video to improve training of personnel and to communicate through **teleconferencing.**

Use by educational, corporate, government, medical, and military organizations became so widespread that in 1971 two groups were formed to aid communication among these various video users. These groups were called the Industrial Television Society and the National Industrial Television Association. Eventually these two groups merged into an organization that is now called the **International Television Association** (ITA).[1] ITA

Another development of the late 1960s and early 1970s that led to use of telecommunications by the corporate and government sector was the growth of the **computer** industry. The computer revolutionized the modern office and, computers eventually, the methods by which people obtained information. Hundreds of data banks were set up to deliver information to the modern library, hospital, corporation, and even home. Schools used the interactive capabilities of computers to help students learn. Sometimes the computer was connected to a **videodisc** that gave it even greater capability.

Audio was not left out of all this. It, too, became more of a tool for busi- audio ness, government, and individuals through improvements in the old-fashioned phone and developments in new telephone techniques.

POTS (Plain Old Telephone Service)

Plain old telephone service isn't so plain anymore, but the term is used to designate those parts of telephone service that deal with the traditional function of getting a person's voice from one place to another. It does not include data services, which generally come under PANS (pretty amazing new services).

phone provisions

Voice transmission has been important to all areas of business and to individuals for almost a century. Recent advances have enhanced its value even more. Starting with the modest hold button, telephones now have provisions for redialing numbers all by themselves, telling when another call is waiting, allowing a number of people to talk to each other all at once, and storing selected numbers. How many people do you know who don't own an answering machine? There are 800 numbers that allow for toll-free calls, and, of course, there are the controversial 976 dial-a-porn numbers. Even airplanes have telephones so business people can call the office from thirty thousand feet up.

All of these advances have aided the corporate-government sectors by enabling them to conduct business more efficiently. They also make life more convenient for individuals. They have also spawned quite a few new businesses, such as answering machine manufacture and sales, the design and sale of telephones themselves, and various dial-a-message services. The dial-a-porn services, which receive several million calls a day, cost about $3 million a year to operate and earn about $9 million.[2]

Cellular Radiotelephones

older mobile phones

Another innovation in telephones involves cellular radiotelephones. These are mobile telephones used primarily in automobiles. There had been a **mobile telephone service** available for years, but it involved relatively high-powered transmitters that covered a large area. Because of the high power, only a limited number of phones could operate in any particular area, totaling about 160,000 mobile phones in the entire country. There were always long waiting lists to obtain a mobile phone, so they did not add a great deal to business efficiency.

cells

With the advent of cellular phones, however, an almost unlimited number of mobile phones is possible. The system operates by dividing a particular geographic area into rather small, discrete parts called **"cells."** Low-powered transmitters and receivers serve each cell—so low-powered, in fact, that several phones can transmit on the same frequency in the same cell without causing interference to each other. A person making a phone call from an automobile does so with a low-power transmitter whose signal is received at a central cell complex. This cell complex is tied by phone lines to the regular local and national phone system so the caller in the automobile can reach anywhere in the country. Similarly, the automobile caller can receive calls from anywhere in the country. The call goes by phone lines to the central cell location and then goes from a cell low-power transmitter to a receiver in an automobile.

switching system

The limited area of a cell might appear to be a disadvantage for cellular radio, but the real beauty of it is that as traffic moves from one cell to another, the cellular system has the ability to hand off signals from one cell to the next. This is accomplished by a switching system that determines when an in-progress call is at too low a signal level and, therefore, switches it to a closer cell. The switching process takes only a few milliseconds, so the caller is not even aware of changing from one cell to another.[3]

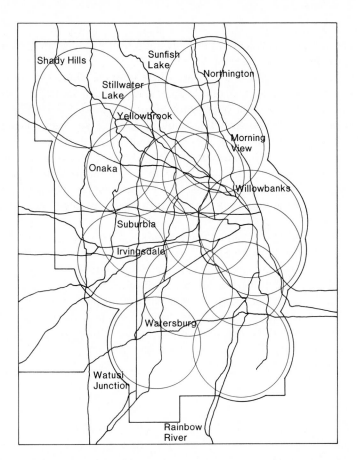

FIGURE 17.1

A drawing showing
how the cells of a
cellular radio-telephone
system overlap.

The technology for cellular radio was in development at AT&T's Bell
Telephone Laboratories as early as the 1960s. In 1971 Bell presented the FCC
with a detailed study of how the system would work. Three years later the
FCC authorized the concept and several years after that it allowed several
experimental systems to be developed. One was a ten-cell system in Chicago
serving two thousand customers and operated by what was then AT&T's Illinois Bell Telephone Company. The other experimental system consisted of
seven cells and two hundred customers in the Baltimore-Washington area operated by American Radio-Telephone Service, Inc.—a radio common carrier
involved in mobile radio. Response to both systems from the customers was
overwhelmingly positive. In April of 1981 the FCC authorized cellular radio
and invited applications to apply for franchises to operate the various areas.

    The structure for cellular radiotelephone was designed to foster healthy
competition by following a model that allowed both phone companies and other
companies to be involved. Up to two systems can operate in any area. One of
these systems automatically goes to the local phone company if that entity
desires it, but the other can be operated as competition to the phone company

early experimentation

ownership

by some other corporation. Most of the "other" corporations who have applied for and received franchises are companies traditionally in the commercial radio business or the paging business.

At first the FCC decided to award the cellular franchises by comparative hearing, whereby it would choose the most qualified candidate for the second system. However, it was so inundated with applications that it went instead to the **lottery** system. This caused some of the applicants to merge so that they would be involved in at least part of the financial remuneration.[4]

Cellular radio has been a good business. Once established within an area, the sign-up rate and sale of equipment is high. Fortunately the FCC established one technology for cellular radio so that, although the phones and other equipment are manufactured by numerous companies, they are compatible with each other.

Not all areas of the country have cellular phones as yet, and the concept is not very effective in rural areas. Therefore, some people have two car telephones, a cellular phone, and a mobile phone of the old-fashioned variety.

Mobile phones are used primarily for business. A salesperson can make phone calls while driving between sales calls. Someone stuck in traffic can answer the phone messages that have piled up on his or her desk. A doctor can keep in touch with what is happening at the hospital. Of course, they can be used for personal calls, too—confirming (or breaking) a date or reminding someone to pick up the laundry.[5]

Data Banks

One of the new uses for telephones during the last several decades involves data banks. These are essentially computers filled with information that can be accessed by customers. The growth of this industry was aided by the invention of the **modem,** a device that connects a computer to a telephone. Bits and bites stored in a computer could be sent through a modem, which, in essence, translated them into telephone language and sent them over phone lines to an individual telephone connected to a modem. This second modem translated them back into computer language and sent them to the individual's computer where they could be viewed and manipulated.

Once this process became available, many companies developed data banks of information with the idea of selling this information to others. The most common setup was that a person or company paid an initial fee to join the data bank and then paid for the amount of time it spent accessing the information. For example, a data bank might charge $300 as an initial fee and for that the users would receive a special password enabling them to access the bank. Then they would be charged $6 for each hour of time they spent retrieving information. Some of the hourly fee would go to the company that had set up the data bank and some would go to the phone company to cover line charges, in a similar manner to the way everyone is charged for phone calls.

Other Telecommunications Forms

One of the first groups to be targeted for data bank services was stock brokers. They are in need of instant information that is constantly updated. This is an ideal use for data banks. Several companies such as Quotron Systems and Bunker Ramo set up these financial services in the 1960s.

stock brokers

Another fairly early service was provided to lawyers. They need to keep up to date on the myriad of legal decisions that are made each day. Lexis set up such a service and earmarked it for law libraries and individual law offices.[6]

lawyers

Data banks are also ideal for librarians. The card catalogue for the Library of Congress is on a data bank, as are most of the indexing services such as Readers Guide and Psychological Abstracts. By using these services, someone can come up with a list of articles or books on a certain subject much more quickly than the old-fashioned way of looking through drawers of cards and stacks of books. There are also services such as Nexis (a brother of Lexis in that both are owned by Mead Data, which is a subsidiary of the Mead Paper Company), which offer on-line access to entire articles from hundreds of publications.[7]

librarians

Over the years thousands of data bases have been established; at present there are close to twenty-five hundred.[8] Many have come and gone because they have not been able to attract enough users to cover costs; however, there are large industries that can support numerous banks.

The medical industry, for example, needs current information, which is supplied by a number of different data services. Travel agents use all the airline schedules. The defense industry is provided with information regarding who has received which government contracts for how much. Even the broadcasting industry has a number of services aimed at it. Audience measurement figures are available on data bases, as is weather information and "morgues" (news stories from the past).

other services

Other information available in various data banks includes the current status of any congressional legislation, hundreds of specialized newsletters that deal with such information as tobacco exports, organized religion, ski conditions, Japanese baseball scores, book and movie reviews, home decorating, and trivia.[9]

Obviously, some of this material is intended more for individuals than companies. Several companies, most notably The Source and CompuServe, have established services aimed at individuals in their homes. They also serve business because those customers can pay more. Generally there is one rate for accessing information during normal working hours and a lower rate for accessing information at night.

individual use

One of the problems with various data banks is that each has its own system for entry and for obtaining information. This makes it difficult for someone to learn how to access all the information that might be needed. At first the only people who accessed information were trained librarians. This meant that the inducer had to work through an intermediary. To circumvent this, a few companies were set up that combined various data services into one and made the access and retrieval the same for all the information. Dialog is one such company, consisting of over three hundred data bases.[10]

retrieval problems

Videotext is, in a way, another attempt at simplifying data bases. The companies that experimented with this took information from data banks and added some of their own trying to design a package of information that would best suit the at-home customers they were trying to serve.[11]

On an even grander scale, there is an international group working to establish integrated services digital network (**ISDN**). This would establish an international standard for data banks and similar services so that worldwide there would be a model network that would provide for capturing, storing, processing, and transporting most of the information that society needs.[12]

Another problem with data bases is security. Obviously, the companies that compile and then sell their information do not want people obtaining it for free. Other data base information, such as credit ratings, is confidential and should not be given out unless there is a bona fide need to know. Computer "hackers" have broken into various bases, so tighter security measures have been devised and have been at least partially successful.

Another problem is what is referred to as **"keyboarding."** In order for information to get into a data base, someone has to actually type it. For newer information, this is not a great problem. Most books and magazines are now typeset using something that is computer based, so this material can be dumped into a data base without retyping it. Therefore, many of the indexing and magazine article services only go back to the 1970s when this technology was adopted. Getting earlier information into data bases is a huge job. Only a fraction of the books in the Library of Congress are in data bases because of the huge task involved with totally retyping them.

Several technologies are developing that can solve this problem. One is **electronic scan** and the other is **voice-activated computers.** With electronic scan, a machine similar to a Xerox machine scans the page of a book or magazine and translates that information into the digital matter needed for the computer. This means that, instead of retyping material, it can simply be copied. With voice-activated computers, someone can read the information and the computer translates the voice into bits and bytes.

At present most of the data base material is sent over the telephone. However, another technology—the videodisc—lends itself well to storage of information. One videodisc can hold the entire Encyclopedia Britannica. Accessing the disc is cheaper because it does not involve phone line charges. The disadvantage is that it cannot be updated frequently as the phone-based data systems can. For information such as stock market prices and weather, videodiscs are inadequate, but they are an excellent method for storing books and magazines from the past.[13]

All of the data base actions discussed so far involve retrieving information. There are also services called **transactional services** or interactive video that are incorporated within data bases. These services allow the inducer to take some action, such as buying a ticket, making an airline reservation, or buying a shirt. The banking industry has involved itself with this service, setting up systems by which people can transfer money from one bank to another and pay bills without writing checks.

Other Telecommunications Forms

Another interactional service in development is **electronic mail.** Through this service, companies or individuals can send letters, advertisements, or other information to another person, company, or to a whole group. Each entity on an electronic mail network has an "address" to which others can send material. The person looks at this mail by going to a computer and typing in the code needed to retrieve the messages. This method has advantages over the present mail system because of its speed. It also saves trees because very little of it needs to be committed to paper.[14]

electronic mail

That which does need to appear on paper can be printed out from data banks by connecting the computer to a printer. This, of course, raises questions of **copyright,** especially if articles or books are copied.

Computer buffs have set up their own interactive networks, some of which are connected to formalized data banks and others of which are ad hoc, just to communicate with each other in much the same way that ham radio operators do. One of the interactive services that CompuServe provides is called Litforum. It is designed primarily for writers so they can show each other material such as scripts, poems, and stories and then critique and interact regarding them.[15]

Data banks are still a growing unstable business. They have achieved popularity, but not to the degree that some people predicted.[16]

Other Computer/Video Applications

Computers have become ubiquitous at the workplace and in the home. They are used extensively for **word processing** and accounting by all sectors of the economy, including broadcasting.

Some station and network newsrooms are now computerized so that stories typed into a computer by a reporter can be retrieved at another computer by a producer, who can then rewrite or integrate the story into a newscast.[17] Stations use computers to bill customers, keep program logs, analyze ratings, keep track of income and outgo, and handle a myriad of other functions.

broadcast uses

A few companies have instituted **telecommuting.** People work at their homes on personal computers and then send the work to the company by modem and phone lines. This is particularly apt for handicapped people and people who must take care of small children.

telecommuting

Many of the computer/video applications deal with information and training in an interactive way. **Kiosks** exist in hotels, shopping malls, airports, hospitals, and other places that contain information for the people visiting there. For example, a hotel may list on a video screen the organizations holding meetings there. By touching a particular part of the screen, someone can find out the meeting rooms of a particular organization and then the session occurring in a particular room. Likewise, someone in a shopping mall can touch a video screen to find out the names of shoe stores in that mall and can even be shown on a map how to walk to a certain shoe store.[18]

kiosks

FIGURE 17.2

A computerized newsroom. *(Courtesy of Dynatech Newstar)*

FIGURE 17.3

A computer/video directory to help visitors find their way around. *(Courtesy of Nelson and Harkins, Chicago)*

pilot training

Corporate and government training is now often enhanced with computers and video. Sometimes a videodisc machine is involved because of its huge storage capacity. Military pilots can be trained by putting them in a huge dome where three-dimensional visual images, created by a network of computers, can simulate aerial combat. These images change to respond to the actions of the pilot in training. A mistake on the pilot's part results in nothing more harmful or expensive than a simulated crash.[19]

Workers repairing equipment can consult a computer-disc configuration if they run into a difficult repair problem they have not encountered before or cannot remember how to handle. The screen can lead them step by step through the method of repair.[20]

Schools use computers for training also. Most children are fascinated by computers and programs designed for them can be excellent teaching tools. Children can proceed at their own pace and branch into additional information when they do not understand something. These programs can have the same fascination for children as video games do.[21]

Video games are certainly an example of a computer/video craze that many individuals were caught up in for a period of time. In the early 1980s arcades were abuzz with video games and the home TV set was often used more for the games than for watching TV. Companies, especially Atari, had huge profits for several years. Then, as quickly as the video game craze came, it vanished, a victim of the fickle public.[22]

educational uses

video games

Teleconferencing

Teleconferencing (also known as **videoconferencing**) is a fairly new method of communication for corporations or other organizations. It involves renting satellite time and using it to transmit video information from place to place. Usually this video information consists of people making speeches to people in other parts of the country or the world. Teleconferencing usually has an interactive telephone element so that viewers can phone questions to the speakers.

One of the early teleconferences was held in 1979 and involved the American Soybean Association. This conference linked together speakers in Tokyo, Rio de Janeiro, London, and Atlanta through a satellite conferencing setup. In this way, approximately fifteen hundred soybean growers were able to hear experts from four continents discuss market outlooks for their crops. Television cameras in the four cities were connected to uplinks that enabled the remarks made at each location to be fed to satellites that transmitted the messages to a number of meeting halls.[23]

early teleconferences

Similar teleconferences have been held for many other organizations—medical associations, universities, business affiliates. These are of a one-shot nature and are usually handled by companies that have arisen to arrange teleconferences. These companies provide the rooms, the television equipment, and the satellite time. Several hotel chains, such as Holiday Inn and Hilton, have equipped many of their hotels with satellite dishes and rooms that are permanently prepared for teleconferencing. In this way, conferences can be held in many cities at the same time, all using the same hotel chain.[24]

one-shot teleconferences

Sometimes teleconferencing is conducted on a regular basis as opposed to a one-shot conference. For example, J.C. Penny Company regularly uses teleconferencing to link its product managers with buyers at the various stores. This greatly cuts down on the amount of time required to get the product through the system. Similarly, automobile companies use satellites to introduce their new lines to their distributors. The Army uses teleconferencing for technical training.[25]

regular teleconferences

Some corporations with branches in several cities arrange for their top executives to "meet" through teleconferencing. Cameras and microphones at each location pick up what executives there say and beam it to the other locations. In this way, the executives can interact in a manner that is similar to what they would do if they were all in one room.

disadvantages

One disadvantage of teleconferencing, though, is that it is not exactly like meeting in the same room. The spontaneity, camaraderie, and body language that occur in regular meetings are often lacking in teleconferencing. Some executives don't come across well on TV and feel very awkward in a teleconferencing situation. In addition, sometimes there are technical problems that flaw the interaction.[26]

advantages

One of the advantages of teleconferencing is that it cuts down on expenses. The costs of teleconferencing, which run about $300 an hour for the room and $300 an hour for the satellite time, are less than the cost of transportation, lodging, and meals for all the executives that would need to be moved around the country. It also saves on the time those executives would spend wending their way through airports and sitting on airplanes.[27]

Teleconferencing has not grown as quickly as predicted and certainly has not destroyed the airline business, one of the early worries. Several companies use it frequently and have their own dedicated systems. Others use it infrequently and rent the facilities when needed. Most government and corporate entities have yet to try it.

Instructional Television Fixed Service (ITFS)

Instructional Television Fixed Service is a system of broadcasting that is at a higher frequency than regular commercial broadcasting; it cannot be received on a regular household TV set. As a result, it is used for specialized programming. Most of this programming is of an educational nature because educationally oriented institutions have control of ITFS. However, they often use it in ways that service business and industry.

channels

The FCC created ITFS in 1971 when it set aside frequencies between 2500 and 2690 megahertz solely for educational use. This band is divided into twenty-eight channels, each having a bandwidth of six megahertz. The channels are organized into seven groups of four channels each and are given the designations of A, B, C, D, E, F, and G. They have been given out in groups of four so that one educational institution might be given the four A channels in an area and another might be given the four C channels. The channels have fairly short range; in general, their coverage area is about twenty-five miles.

process

In order to operate the channels, the group receiving the license has to put up a **transmitter** that is connected to a studio or playback facility. It must then install an **antenna** and a **downconverter** on the roof of each site that is to receive the ITFS signal. This downconverter changes the frequency to one that can be accommodated by a regular TV set. Therefore, the people using ITFS tune the set to regular channels, such as 3 or 5 or 7, and watch what is being sent out from the ITFS transmitter. Often there is a telephone link from the receiving area to the ITFS studio so that people receiving program material can ask questions.[28]

owners

The main operators of ITFS channels are colleges and universities, school districts (both private and parochial), public TV stations, and government agencies. There was no great rush to obtain ITFS frequencies after they were set aside in 1971. A modest number of organizations have set up ITFS systems and use the channels in constructive, but limited, ways.

uses

School districts use ITFS systems to send programs that teachers can use in the classroom. Sometimes the districts produce their own programs (e.g., a show to demonstrate what the local fire department does); sometimes they acquire them from production companies (e.g., a series to teach Spanish to third graders); and sometimes they merely retransmit what is on the local public TV station at times that are more convenient for the classroom schedule (e.g., "Sesame Street" from 11:20 to 11:40 A.M.). After school is out, the ITFS system might send in-service training programs that teachers can watch in their own schools.

Colleges and universities sometimes use ITFS to exchange courses or to set up extension centers. If colleges are within twenty-five miles of each other, a professor from one college can teach a course in statistics that can be received by the other college, which might not have a statistics teacher. Or the statistics course might be beamed to a local grade school and people from the community could receive the course as though it were an extension course offered in their local community. With the aid of the specially provided telephone, they can ask questions.

A number of universities have also set up cooperative arrangements with local businesses. These businesses often want to upgrade training for their employees (perhaps a course in marketing or economics). A professor at the university can offer the course, which is then sent through the ITFS transmitter to the antenna and downconverter located at the business. The course can be scheduled so that employees can watch it during their lunch hour or right after work. Hospitals are also big customers for these types of courses because busy doctors and nurses need constant updating.

Public TV stations own ITFS mainly so they can retransmit their in-school programs several times a day to meet the scheduling needs of the schools.

Municipal authorities use ITFS primarily for communication. The police chief might talk to police at all the local stations before they start their shifts. Fire fighters at each fire station might watch training material. Sometimes these groups use ITFS to conduct miniteleconferences whereby people in various parts of a city interact.[29]

The use of ITFS grew slowly during the 1970s, but by the 1980s many of the channels still had not been used. The FCC began considering turning over some of the channels to multichannel multipoint distribution service (**MMDS**) so they could be used for commercial purposes. Educators did not want to lose their channel allocations so fought this change. One plan that grew out of this was that the Public Broadcasting Service apply for channels

MMDS entry

throughout the country to set up a **National Narrowcast Service** that would carry educational, training, and informational programs to businesses, industries, and colleges. In 1984 PBS was granted eighty-two channels[30] and has recently begun this service.[31]

The FCC did end up giving some ITFS channels to MMDS operators, but less than originally proposed. The original proposal called for allocating eleven channels to MMDS. In the end only the E and F groups (eight channels) were given. The final proposal also allowed ITFS systems to rent their channel capacity to MMDS; some of them have done that, particularly in the evening hours.[32]

ITFS has survived and stabilized. It is not an overly significant force in the corporate-government-education world, but it does provide some valuable services.

Video Production

Nonbroadcast video intended for use within the corporate, educational, medical, and government structure has become quite important to many entities. Most of it began in the late 1960s when helical recorders became available, although the military used training tapes during the 1950s.[33]

training

Nonbroadcast video has a multitude of uses, most of which can be lumped under training or communication. Within the training area it is used to teach people concepts or skills. For example, a manufacturing company may produce tapes showing the work performed at each station on an assembly line. When new employees are hired, they are shown the tape relevant to their job. In this way, a trainer does not need to teach the same material over and over to each new employee. Within an educational institution, a teacher may tape a guest speaker and then show the tape to future classes so that the guest does not need to come back each time.

Another way tape is used to train is to tape people performing a certain task and then play back the tape so they can see things on which they could improve. For example, simulated sales calls can be set up and taped to see how a new salesperson would handle the situation. The tape is then played back and critiqued by an experienced salesperson. Many athletes use tapes in this way to improve their skills. The slow motion and still frame characteristics of tape machines aid this training technique.

communication

One of the most common ways tape programs are used to communicate involves employee orientation. A tape will be made that covers the history of the organization, its functions, and its benefits. This is shown to new employees so they can better understand the organization they have joined. Some companies have regularly scheduled news programs about company progress and employee happenings. A number of general managers tape regular reports that are shown to all employees. Programs are also produced that cover the affairs of the entire industry of which the company is a part. Sometimes tapes are made about the features of new products, which are sent to all the companies that will be distributing the product.[34]

Other Telecommunications Forms

FIGURE 17.4

Taping an industrial TV program to instruct telephone operators in safety procedures. *(Courtesy of General Telephone Company)*

Some of the tapes produced are intended to communicate with the outside public rather than employees. Hospitals, for example, tape programs about their procedures or about certain medical conditions, which are shown in waiting rooms or hospital rooms. An executive may want a tape about his or her company to show at a conference.[35]

There are several different ways that these programs are produced. Often a company will have an **in-house staff** of about three people who will have at their disposal a studio and television equipment. They will be responsible for producing the company programs. They may initiate ideas or other people within the company may come to them with ideas. They are responsible for seeing that all the elements needed—performers, equipment, and graphics—come together at the appropriate time. They then act as crew during the taping, sometimes augmented with **free-lancers.** Some of the programs are taped in a studio, but many are taped at remote locations, particularly around the plant.

production methods

This group of people is housed in different places in different organizations. Most often it will be housed under public relations (often called corporate communications) or under training. Sometimes it is in the general administrative area and reports to the chief executive.

Another way programs are produced is to have them taped **out-of-house** by a production company. Often both in-house and out-of-house production methods are used. Simple productions, which can be handled by a small crew and limited equipment, are done in-house; more complex productions are taped out-of-house. Someone within the company must find the proper production group and oversee the program so that the end result is what the organization wants.

Some programs can be bought **off-the-shelf,** but generally corporate video is so company-specific that preproduced programs do not work. Something like a general health and safety program might be useful, however.

Once programs are produced, they must be distributed to the intended audience. At one time schools, hospitals, and companies had **closed-circuit** television systems. Wires were run from a master control area to various rooms

distribution

that had TV sets. These sets were plugged into special connections that delivered the programming from the master control area. At a certain time particular programs would be played, and if the TV set was turned on, it would show the program. During the 1960s and 1970s a fair number of schools had closed-circuit systems so programs could be shown in classrooms. Companies had assembly areas wired so news programs could be shown there and hospitals wired patient rooms.

The problem with closed-circuit TV was that everyone had to be ready to watch a particular program at a particular time. If a teacher had not finished the reading lesson, he or she had to stop anyhow because the science lesson was about to start on TV. Another drawback was that programs could not be stopped in the middle and discussed.

In general, **videocassette recorders** have replaced closed-circuit TV. These are inexpensive and portable enough to be wheeled to particular places where people want to watch programs. A teacher, trainer, or individual can operate the machine and even stop the tape and discuss it if desired.[36]

individual uses

The coming of the cassette recorder also made TV producers out of many individuals. People who had filmed family events on super 8 switched to small-format videotape, and people who had never even taken very many photographs started using the easy-to-operate cameras and VCRs.

Video equipment has made an important inroad into both the private and public sectors of American business. It is used more widely than teleconferencing and ITFS, but not as widely as computers.

Conclusion

Audio, video, and data play an important part in today's business world. The nonmass uses of telecommunications began in the late 1960s and have progressed ever since.

In the audio area, the telephone has added many features including call waiting, conferencing, and answering machines. Mobile telephones were greatly aided by the invention of the cellular concept. The old mobile phones were so high power that very few of them could operate in one area. The cell concept allows phones to be switched from one transmitter to another while in transit. This allows for an almost infinite number of phones.

In the video area, companies, schools, hospitals, and the like use television primarily for training and communication. Sometimes this is intended for internal distribution and sometimes it is meant for the outside world. Videotaping provides plentiful opportunity to train people in a nonrepetitive manner that is interesting and nonthreatening. Both in-house and out-of-house videos can prove effective. The most common means of distribution is the videocassette player, which has essentially replaced closed-circuit systems.

Teleconferencing can bring people together from around the country or around the world through the use of video equipment, satellites, and phone interconnects. This system has its interpersonal drawbacks, but it can save on transportation cost and inconvenience. ITFS is used primarily by education, but can also serve as a bridge between town and gown when the educational facilities transmit material that is useful to corporations. It can also be used to deliver programs to classrooms, train teachers, and provide for the exchange of college courses. ITFS has lost some of its frequencies to MMDS, but many of the rest are spoken for by PBS for its National Narrowcast System.

Data banks that utilize computers, modems, and telephones have grown rapidly in recent years. They contain up-to-the-minute information helpful to stock brokers, doctors, lawyers, and others, as well as more long-term information pertinent for library research. Systems such as Dialog, videotext, and ISDN attempt to standardize data banks and make them more user friendly. Problems associated with security and keyboarding may be solved by videodiscs, electronic scanning, and voice-activated computers. Also on the horizon are more transactional services and electronic mail. Computers are also used for word processing, accounting, and, to a limited degree, telecommuting. Computers can give out information at kiosks and train people through simulated setups and branching programs. Although business is the biggest user of data banks and computers, they have also become popular with individuals, including the temporary video game frenzy.

Thought Questions

1. Should more of the ITFS channels be given to MMDS? Why or why not?
2. How could your school make use of video production in ways that it presently does not?
3. How do you think data banks will be used ten years from now?

International Telecommunications

18

The orbiting satellites herald a new day in world communications. For telephone, message, data, and television, new pathways in the sky are being developed. They are sky trails to progress in commerce, business, trade, and in relationships and understanding among peoples. Understanding among peoples is a precondition for a better and more peaceful world. The objectives of the United States are to provide orbital messengers, not only of words, speech, and pictures, but of thought and hope.

Lyndon B. Johnson

Overview

Generally, what has been covered in this book has been concerned with telecommunications in the United States. However, most of the world's population now has access to radio and TV, and the procedures used by the various countries differ significantly from each other and from those used in the United States. Countries also influence each other through the interweaving and exchange of electronic media products. This chapter will give an overview of various aspects of the international broadcasting scene by discussing the following topics:

Early radio in foreign countries

Varying technical standards around the world, including 525 and 625 lines, NTSC, PAL, and SECAM

Government, private, and mixed systems of ownership

Financing by taxes, licensing, and advertising

Content and creation of programming in various parts of the world

Controversies that exist concerning the sale of foreign programming (especially American) to other countries

Ways in which people in various countries are trained to enter the broadcasting business

The extent to which nations have or have not embraced the newer media

The history, organization, and programming of the BBC and IBA

The history, organization, and programming of the Soviet Union

Peculiar quirks of broadcast systems in various countries

International broadcasting that countries aim at each other, including the activities of Voice of America

The AFRTS programming to service people

The functions of the International Telecommunication Union

The organization and activities of COMSAT and INTELSAT

Aspects of International Telecommunications

The concept of international telecommunications encompasses a great deal. On one hand, there is the broadcasting that occurs internally within particular countries. This is intended for the national audiences, although on occasion programs are sold to other countries. In order to operate a radio-television system, countries have set up differing types of mechanisms to handle programming, distribution, finances, regulation, and ownership.

Different countries also lay different claims to the various "firsts" associated with broadcasting. For example, the Russians observe May 7 as Radio Day, claiming that the beginning of the technology was on May 7, 1895, when the Russian, Alexander Popov performed a radio demonstration. The Dutch and the Canadians began broadcast services in 1919, prior to the beginning of the United States station KDKA. The Germans made a TV system available to the public in 1935, and the British Broadcasting Corporation started regularly broadcast television in 1936.[1]

An entirely different aspect of international telecommunications involves the transmission of material by one country to another country or countries. This is often propaganda based. The intent of the transmitting country is to show itself in the best light to other countries of the world. Closely aligned with this are the negotiations that occur between countries regarding the use of spectrum space by each national entity and the uses these countries make of satellite transmission.

These various aspects of international telecommunications will be covered in this chapter, beginning with the systems of electronic media used within individual countries, particularly Britain and the USSR. "Different strokes for different folks" is a truism that definitely applies here. Organizations vary as boundary lines are crossed, and yet there are common elements that glue the world's telecommunications systems into a unified whole.

Technical Standards

One of the irritating technical problems of international telecommunications is that various countries have different systems for taping and transmitting electronic information. The United States operates on a television system that scans **525 lines** on the face of the TV tube and employs a color coding system for color TV that was approved by a committee of engineers called the National Television Systems Committee. This color coding system is referred to as **NTSC.** The Americans hoped it would be adopted throughout the world, but European engineers quipped that NTSC stood for "Never Twice the Same Color" and developed their own system called **SECAM** (Sequence Couleur a Memoire), which American engineers in turn labeled "Something Essentially Contrary to the American Method." Another system, **PAL** (Phase Alternate Line), was developed in hope that it would yield "peace at last." This was not the case. At present approximately half the world uses PAL and the other half is split almost evenly between NTSC and SECAM. Most of the SECAM and PAL systems scan at **625 lines,** making the United States among the minority that scan at 525.[2]

HDTV

There is hope that the entire world will develop one system for **high-definition television,** which would greatly improve the exchange of programming from one country to another. Now when programs are exchanged they must be dubbed from one system to another, which causes loss of picture quality. Another option when exchanging programs is to distribute them on film, which does have a worldwide standard in both 16mm and 35mm.[3]

Ownership

The different systems that developed adopted different methods of ownership. In the United States, electronic media facilities are generally owned by private companies, which are usually responsible to stockholders. This form is rare in the rest of the world, especially outside the western hemisphere.

government ownership

In many countries, particularly those with totalitarian governments, the governments own and control all of broadcasting. In other countries, private corporations have actual ownership, but are closely tied to the government.

charters

The government in these countries issues **charters** outlining provisions that the corporation must follow. Usually the government department overseeing broadcasting is the same one that overseas the mail, telegraph, and telephone services. Because of this, there have been some historically close ties between the telephone and radio/TV.

As time has progressed, ownership in some countries has evolved into a mixed system of both government and private ownership.[4]

Financing

mixture

A mixture has also evolved in many countries regarding financing. Originally, most government-owned systems were financed by general government funds collected through taxes. Charter systems were financed by license fees collected from people who owned radio and TV sets. Private systems were funded by advertising.

commercials

All these systems worked fairly well for radio, but television was a much more expensive proposition and many countries found they could not support it through general taxes or licenses. As a result they allowed commercials, but usually on a much lower-key basis than in the United States. Foreign country commercials are much more likely to be clustered at times that do not interrupt programming than are those in the United States, and they are less likely to be flamboyant and hard sell. The United States even has somewhat of a mixed system because public broadcasting is largely noncommercial and does receive government funding.

unique methods

Some governments have come up with unique methods of funding broadcasting, such as publishing program guides that include advertisements. Egypt's president, Gamal Abdel Nasser, instituted a surcharge on all electrical bills because his government was having so much trouble collecting license fees from people.[5]

Because advertising is not an overwhelming concern in most countries, **audience measurement** does not take on the significance that it does in the United States. Many countries engage in some sort of research regarding how many people are watching or listening to various programs, but it is usually not at all secretive and is intended primarily to help alter the programming so that it will be more effective.

Programming

Programming is, of course, a great concern to all countries. The emphasis the United States places on entertainment programming is not predominant worldwide. Although most countries program over 50 percent entertainment, they place much more emphasis on educationally oriented material of both a formal and informal nature than does the United States. Religious programming is also high, particularly in the Arab world where church and state are closely intertwined.

In the United States, programming is created primarily by privately owned groups—stations, networks, and production companies. The government has only limited control over the content; controversial ideas and criticisms of government procedures appear frequently. Outside the United States, such a hands-off attitude by the government is rare. In most totalitarian countries, programming is created within the government, usually by a department of education or propaganda, and material is often censored by a committee. Many of the programs are paternalistic and often espouse the virtues of the government. The electronic media become essentially a public relations arm of the government, so government policies are not questioned. In less totalitarian countries, programming is produced by private companies, but the content is implicitly or explicitly scrutinized by the government.

As with ownership, some countries have evolved into a mixed system wherein some of the programming is produced by the government, some is produced by private companies, and some is imported from other countries.[6]

Foreign Exchange

The amount of foreign programming on a national service often becomes very controversial within particular countries. On one hand, they want to maintain their national identity, but on the other hand, they want quality programming without high cost. Radio was inexpensive enough that most countries could handle their own programming, even when they first began the service. Television was another matter, however, and programming available through international syndication could be obtained must less expensively than similar programming produced locally. As a result, a large market built up for the exchange of foreign programming.

Many countries sell or exchange their programs throughout the world. Mexico provides much of the programming for South America, and the Soviet Union does likewise for the iron curtain countries. Britain sells a great deal of its programming, primarily to countries that used to be part of its commonwealth. Canada, France, Germany, Brazil, and Australia all make international sales. The granddaddy of them all, however, is the United States, which sells more programming than any other country.[7]

American programming in foreign countries, especially the developing nations, has been controversial. Politically, this programming is often unwelcome because many of the cultural modes it displays are different from those of the nation in which the program is aired. News and information programming supplied by American radio and TV is often accused of having a Western bias, and entertainment programming is criticized for decadent morals.

Usually as countries develop their TV facilities, they buy less foreign programming and depend more on their own production. Many of them set **quotas** regarding the amount of foreign programming that can be shown. Different countries have differing attitudes about this, however. New Zealand, which feels isolated from the rest of the world, will program up to 60 percent foreign-produced shows so it can feel in touch with the world. Canada, which is always fearful of being absorbed by American culture, is constantly fighting to keep down the amount of American programming that crosses the border.[8] In the radio area, countries may produce their own programs, but the music aired on a large number of the stations represents America's ubiquitous contribution to the world.

Another element of foreign exchange of programming involves the number of broadcasting outlets in a country. If a country has just one or two government-run networks, it may not purchase a great deal of outside material. But as more and more countries are developing private commercial networks and stations, the demand for product goes up. So the cutback in the purchase of foreign programming, which most countries went through as they became developed, has in some ways been offset by the increase in broadcasting facilities.[9]

This is welcome news for the countries that sell programming because they have come to depend on the foreign market. Many United States program producers do not turn a profit on productions until they are sold overseas. Increasingly, this is becoming true for British programs.

The international market is filled with financial vicissitudes. What one nation is willing to purchase for $40,000, another may only be willing to purchase for $400. The annual program market, held in **Cannes,** France, is where most international buying takes place; selling patterns can vary widely from year to year. For awhile only dramas sold well overseas, but now comedies are hitting their stride. Once the programs are sold, there are often difficulties in collecting the money, particularly from some of the more volatile Latin American countries. In addition, **piracy** has become a big problem because many nations are not sympathetic to the United States copyright laws.[10]

Of course, the country that imports the least programming of all is the United States. Generally it imports only 2 percent of programming and most of that is British material seen or heard on public broadcasting.[11]

American programming

quotas

increased outlets

selling practices

Personnel Training

In many countries, radio and television are not taught within the formal educational structure of high schools and universities. Instead, there are special institutes that teach the subject matter, usually to people who are already employed in the industry. Sometimes these institutes are operated by separate companies, but more often they are operated by the broadcasting entity in the country.

institutes

A common pattern is that a person is hired by the government broadcasting organization on the basis of potential and personal characteristics. Often this hiring occurs at about age eighteen or after what would be the equivalent of an American high school. Sometimes college graduates with general liberal arts types of degrees are hired. These people then receive up to a year of training in radio and television, usually at the facility where they will be working. This consists of formal instruction plus **"internships"**—short periods of time spent observing in the various departments of the organization. Once this is completed, the person is placed in a job that tries to take into account the needs of the company and the talents of the person.

on-the-job training

There are numerous variations on this system. For example, sometimes people are sent to foreign countries with well developed broadcasting, such as England, and are trained there. Other times they receive only a few days of training, but then receive frequent in-service instruction throughout their careers.[12]

Newer Media

In general, foreign countries have not been as quick to adopt newer media as the United States has. Cable TV has penetrated Europe, but not to the degree that it has in the United States and Canada.[13] Very little of the Orient is wired. Australia voted down installing cable TV, in part because it knew it could not fill all the channels with Australian-produced programming and it did not want a huge influx of American programming.[14]

cable TV

Direct broadcast satellite may become more successful in some other parts of the world than it has in the United States. India has actually been using it for years to beam educational programs to people in rural areas who come to the nearest village to watch a TV.[15] DBS also shows signs of success in Europe where several companies are actively pursuing the idea. One of the thorny problems in this area is that satellite signals transmitted by one country can easily be picked up in others, given the geography of Europe and Asia. This has made some countries uneasy. Poland, for example, would not be happy having its residents receive programming from Britain, just as France does not particularly want to be inundated with German programming.[16]

DBS

Videocassette recorders are rapidly becoming popular around the world, even in third world countries. Of course, they are particularly popular in Japan.[17]

videocassettes

There are several countries that have been more successful with teletext and telephone delivered data services than has the United States. The most shining example is France, where the government underwrites much of the cost to the people.[18]

Overall, many countries have influenced each other regarding telecommunications. Television, in particular, has grown up in an era when international communication is quite common. Although American programming is the most watched foreign programming, the British organization for radio and television has been most imitated throughout the globe, mainly because Britain had so many colonies for a long period of time. The Russian system is, of course, an excellent example of broadcasting in a totalitarian state. Its model has been followed in most of the countries influenced by the Russian form of government.

The British System

British broadcasting emanates from two different organizations— the British Broadcasting Corporation (**BBC**) and the Independent Broadcasting Authority (**IBA**).

As in the United States, various experimental radio programs were broadcast during the 1920s. At first the government banned them because they were interfering with other types of wireless communication, but in 1922 the British Broadcasting Company was formed under a license from the Post Office. Five years later, in 1927, this organization became the British Broadcasting Corporation under a Royal Charter.[19]

This charter gave the BBC a monopoly on all radio broadcasting. It created a Board of Governors consisting of twelve members appointed by the monarch for five-year terms. This was to be the policy-deciding group and the day to day operations of the BBC were to be carried out by a **Director-General** and his/her staff. The charter also stated that there could be no advertisements on the BBC, that it had to broadcast daily impartial accounts of the proceedings in Parliament, and that it could not give its own opinion on current affairs. Most of the provisions of this charter are still in effect. The major changes are that the charter now encompasses television as well as radio, and the BBC no longer has a monopoly.[20]

Because there were no advertisements, the funding for the BBC came from money people paid to own a radio. This money was collected annually by the Post Office as a license fee. When TV was added, people paid licenses on both radios and TV sets. Eventually the license fee was dropped on radios, so now people pay a license only on their TV sets and this supports both BBC radio and BBC TV.[21]

Originally the programming on BBC radio was very paternalistic. Designed to upgrade tastes, it was often referred to as programming from "Auntie BBC." The original organization consisted of three program services, each designed to lead the listener to the next for a higher cultural level. At the lowest level was the **Light Programme,** which consisted of quiz shows, audience participation programs, light music, children's adventure, and serials. Level two was the **Home Service,** which included good music and drama, school broadcasts, information about government, and news. The highest level, called the **Third Programme,** was classical music, literature, talk, drama, and poetry. All these services were national and were the only radio in Britain.

This programming was an easy target for outside competition. When Radio Luxembourg began broadcasting popular music of the 1950s and 1960s, many of the British people began tuning in; British companies even began buying ads on Luxembourg stations.

"pirate radio"

In 1963 **"pirate ships"** anchored off the coast of England began broadcasting rock music on frequencies that could be picked up with ordinary radios. This programming became so popular that when the government planned to suppress it, there was such public outcry that the BBC had to change its programming before these pirate ships could be eliminated.[22]

four program services

The result was a new, renamed, four-level radio setup. **Radio 1** programs rock and pop music. **Radio 2** resembles the old Light Programme with panel and quiz games, comedy shows, and light music. **Radio 3** is similar to the old Third Programme, broadcasting primarily serious music. **Radio 4** is the main talk service with news, current affairs, parliamentary reports, and other programs similar to the old Home Service. In addition, the BBC established local radio stations. Over thirty of these stations are now on the air throughout the country programming primarily information of interest to local listeners.[23]

early BBC TV

BBC television began in 1936 and in 1937 televised the coronation of George VI—quite a feat for that time. The operation was stopped in 1939 because of the war and then resumed in 1946 utilizing a 405-line scanning system. In 1953 it telecast the coronation of Queen Elizabeth II, an event that greatly increased the sale of television sets and brought television to the attention of the British people. When color was introduced in 1967, the BBC converted to a 625-line PAL system.[24]

two program services

In 1964 the second BBC TV network was established and the two were called BBC I and BBC II. Unlike radio, the overall program content of both services was similar, but the individual programming hours were different. For example, at 8:00, BBC I might have sports and BBC II a drama; then at 9:00, BBC I might have a drama and BBC II a documentary. The original intent of this programming was to upgrade appreciation.

formation of IBA

This paternalistic programming, like its radio counterpart, came under fire and the result was the establishment of the Independent Broadcasting Authority. The IBA was originally formed in 1954 as the Independent Television Authority (**ITA**) after a heated debate in Parliament regarding the quality and role of television. In 1972 its responsibilities were extended to include the setting up of independent radio stations, over five hundred of which are now on the air. Along with these extended responsibilities, its name was changed to Independent Broadcasting Authority.[25]

FIGURE 18.1

The London complex that serves as headquarters and production facility for BBC-TV. *(Courtesy of Barbara and Dave Huemer)*

IBA radio

The organization of the independent radio stations is fairly simple. These are local stations similar to the BBC local stations except that they receive their money from advertisements rather than government allocation.

IBA TV

The organization of independent television is quite complex, however, and very different from the BBC. Although the IBA is overseen by an eleven-member Board appointed by the Home Secretary and led by a Director-General, it is actually a regionally based national network. It does not produce any programs itself. All programming comes from fifteen independent companies, called **ITVs,** selected and appointed by the IBA. Each of these companies serves a different area of the country and programs local material, as well as the national fare. Each contributes a certain number of programs to the national feed. The companies in large cities generally contribute more than the companies in outlying areas. The closest parallel to this in the United States is public broadcasting, where local stations produce programs for PBS.[26]

structure

IBA is not noncommercial. In fact, it and the independent companies receive all their funding from commercials. The ITVs sell commercial time within the programs they produce. Regulations for these ads have been set up by the IBA in accordance with the Broadcasting Act that established it. These regulations include provisions that limit ads to an average of six minutes an hour, state they have to come at natural program breaks, and prohibit advertisers from sponsoring program content (i.e., they cannot buy programs, they can only buy spots).[27]

The independent companies receive the money from the ads, but they then pay IBA to rent the national television transmitters. In this way the IBA receives the money it needs for its operation.

FIGURE 18.2

The IBA headquarters in London.

To make matters a little more confusing, there is also a shared network, **ITN,** which provides news to all the regional companies. Since 1983 there has also been a nationwide breakfast program produced by a separate company called TV-am.[28]

In 1982 a new independent network, **Channel 4** (C4), was authorized. It is funded by additional subscription fees paid by the ITVs, but they, in turn, collect the advertising for Channel 4 and provide it with many of its programs. Other programs come from other countries and production companies that are not affiliated with either the ITVs or the BBC. C4 pays IBA for the use of the transmitters in much the same way that the ITVs do. One of the reasons C4 was established was to provide more serious education-oriented programming than that served on ITV.[29]

Channel 4

To some extent this has happened, but overall British programming has evened out. With the coming of IBA, BBC I lightened its programming so that it could maintain its audience. BBC II appeals to narrower interests, but it, too, has some popular programs. By American standards, all the British programming services are heavy on educational programming. During the day there are many programs, both radio and TV, for children in schools. At other times there are continuing education programs. One unique partnership in which the BBC participates is the **Open University.** Established in 1969, this department produces courses for university credit that are funded by the government through its Department of Education and Science, rather than being funded by the license fees.[30]

Open University

FIGURE 18.3

A production scene from "Coronation Street," one of IBA's most popular programs. Its cast of characters often interacts in the pub, Rovers Return Inn. *(Courtesy of Granada Television)*

controversies

Independent TV has a larger budget than the BBC. This is because advertising generates more money than people are willing to contribute through license fees. In fact, the hue and cry from the public against the ever-increasing license fees and the complaining from the BBC about shrinking budgets led the House of Commons to introduce a bill that would allow advertising on the BBC. This bill caused a high degree of emotion and was defeated in 1985. [31]

Another element causing much consternation in England is the status of the new technologies. Cable TV was not introduced into the country until 1982, and its introduction was greatly feared by the established broadcasting services. Direct broadcast satellite was authorized around the same time and caused even more fear, but neither has made much headway. The BBC and IBA have been fairly undaunted by the new technologies, except perhaps for videocassettes, which have become quite popular.[32]

teletext

One area that is usually considered new technology in the United States is actually old technology in England—teletext and videotext. BBC engineers began experimenting as early as 1966 with ways of sending information in the vertical blanking intervals of the TV signal. By 1969 they had developed a system to use for captioning for the deaf and foreign language translation. In 1972 both the BBC and the IBA inaugurated teletext systems using the same technology, but the IBA allowed for advertising space. The BBC's system is called **Ceefax** (see facts) and the IBA's is called **Oracle** (an acronym for Optical Reception of Announcements by Coded Line Electronics). People have to adapt their TV sets with decoders and keypads if they wish to receive the teletext at all times, but the pages are also shown when regular TV is not being broadcast. In 1977 the British Post Office began offering a videotext service called **Prestel** (short for press and tell), which is delivered over phone lines.[33]

The British experience has been imitated in many parts of the world, both intentionally and unintentionally. The government-run charter system followed by more private commercial systems is the pattern in many countries. It is not common, however, in dictatorships.

The Russian System

Dictatorships tend to structure their broadcasting in a manner similar to that in the Union of Soviet Socialist Republics. Here the only systems are those that are state owned and state controlled.

Russian radio did not start out that way when it was first developed experimentally by Alexander Popov in the early 1900s. Some of the first radio stations were run by labor unions and educational organizations, and some were run by the government. Gradually the government took over the unions and educators and broadcasting was, in reality, government run.

Vladimir Lenin saw radio as a significant force to enlighten the illiterate masses and made radio development one of his primary goals after the 1917 revolution. Programming was primarily educational, including such information as sanitation techniques and medical care. Because the Soviet Union is such a large country, the transmitters in Moscow were high powered (40 kilowatts in 1927). Short wave was also used to transmit radio. A fair portion of the nation was wired, with radio programs being delivered to loud speakers situated in the main squares of large cities.

By 1936 there were five state-run radio networks, each serving a different part of the country, but each carrying essentially the same programming. There also were some local networks to serve regional needs. All of them were, and are, supported by general government funds.

During World War II, 1200 kilowatt stations emanated from Moscow and Leningrad, mainly broadcasting information for the troops.

After the war, radio went back to an organization that included central and local state-operated systems. In 1949 it was placed under the Council of Ministers of the Communist Party, the highest party organization in the USSR.

Television was started in the 1950s using the 625-line SECAM system. In 1957 the operation of both radio and television was given to a newly created committee, the State Committee for Radio and TV. This is still the governing body for broadcasting. The people on the committee are all Communist Party members and, as a group, report to the Council of Ministers.

Today there are five radio networks programmed in Moscow and a number of local stations. The five networks include: the **First Program,** which is the most important network devoted half to socio-political messages, including news, and half to the arts; the **Second Program,** which consists of music and information bulletins; the **Third Program,** which is literary and musical and includes such material as operas and plays; the **Fourth Program,** which is devoted to symphony music; and the **Fifth Program,** which is music and information.

early radio

national and local

early TV

national and local

The development of television became a national objective in the 1950s. A huge production facility was built in the outskirts of Moscow. One of the big televised events was the World Youth and Student Festival held in Moscow in 1957. Programming emphasized news and education, but there was also variety, drama, sports, and a quiz show called "Funny Questions Evening."

One of the government's major problems was the delivery of the programming to all parts of the country, which includes eleven time zones and over sixty languages and dialects. The solution was satellite transmission. In 1965 the government launched its first communications satellite and delivered the programming at many different times of day and dubbed into different languages.

Four national TV networks and a sprinkling of local stations are currently in operation. First Channel is the main network and contains news, culture, movies, and concerts. It is particularly known for its 9:00 P.M. news "Time," which regularly reaches 100 million people. Second Channel is primarily for Moscow and features information and art. Third Channel is educational programming material for school children, college courses, and programs to teach the Russian language and other languages. Fourth Channel consists of sports, arts, and politics. Commercials are not aired, but there are programs that tell consumers the values of various goods.

Cable TV was started in the USSR, but the government decided not to make it a priority and it withered. Of late, videocassette recorders have become popular.

According to the Communist Party, broadcasting can be open as long as there are no slanderous fabrications that defame the Soviet state and social system. For the most part, programs are taped, so the content can be carefully controlled. Occasionally live programs present problems. Recently a woman reading her poetry live ended by supporting Andrei Sakharov, the dissident physicist.

Overall, broadcasting in the Soviet Union is looked at primarily as a tool to educate the masses.[34]

Other Systems

Most of the world's broadcasting systems have borrowed in some ways from the American, British, or Russian systems. Each country has its own peculiarities, however, which give it color and interest.

Most of the European countries have government-operated systems similar to those in Britain. However, they have been quicker to place limited advertisements on TV than has the BBC. They are also starting to develop private systems to compete with the government ones.[35]

Canadian broadcasting originally consisted entirely of privately owned stations supported by advertising. These stations tended to be in large cities because stations in sparsely populated areas could not attract enough advertising to make money. As a result, the Canadian Broadcasting Corporation (**CBC**) was formed by the government to provide programs to these sparsely

Margin notes: production facility · satellites · four program networks · cable TV · government stipulations · Europe · Canada

populated areas, as well as to the cities. The CBC gets most of its money from the government, but it also has limited advertisements. Two radio and two television networks are operated by the CBC, one in English and the other in French. Canada has a quota that requires 60 percent of programs to be Canadian produced. Canada also has a well-developed cable TV system.[36]

Australia

Australia has an expansive geography similar to Canada and, therefore, has only one nationwide service, the **Australian Broadcasting Corporation** run by the government. There are several private networks, but they operate only in populated areas. The country used to have a quota system requiring at least 50 percent of programming to be nationally produced. Now, however, they have gone to a point system whereby homegrown products receive more points than foreign imports. The point system also governs types of programs. For example, a drama gets more points than a game show. To keep licenses, stations are supposed to average at least one point per hour.[37]

Japan

Japan has an advanced system of broadcasting that includes the national service **NHK,** which is similar to the BBC but has more autonomy. There are also private radio and TV stations that are more or less local in nature. The production of programs (and commercials) is very high tech.[38]

China

Chinese radio started in the 1920s in the American mold then switched to the Russian model in the 1950s. During the 1950s the Russians began helping the Chinese to establish TV. When the two countries severed relations the Russian advisors left. During the Cultural Revolution, much of the broadcasting structure was closed down and what remained programmed endless excerpts from Mao's *Little Red Book* and several operas chosen by Mao's wife. More recently, Chinese broadcasting has started down the road toward modernization. This includes commercials and the airing of some American programming.[39]

Cuba

At one point, Cuba was the bastion of commercialization. Radio stations would program one-minute programs that consisted of thirty seconds of news, twenty-five seconds of commercials, and a five-second time check. The commercials were loud, hard sell, and repetitive. When Castro came into power this stopped and the broadcasting system took on characteristics similar to those of the USSR.[40]

Brazil

Brazil has five television networks, but one of them, **Rede Globo,** dominates. It claims to be the fourth largest network in the world, chosen each evening by 70 percent of the Brazilian TV viewers. Its most popular programs are novellas, which are closely akin to soap operas. Many of these are sold to other countries.[41]

Africa

Most of the African countries have systems patterned after Britain or France because they were colonies of those countries. When they obtained their independence, many Africans closed down radio and TV for awhile, but then opened them up again in a style similar to what they had been. The Union of South Africa was one of the last countries in the world to initiate television and has had separate services for blacks and whites.[42]

Saudi Arabia

Saudi Arabia now has very modern broadcasting, but for many years an organization of Muslim religious leaders prevented its introduction. Radio won its way into the country because a demonstration transmission of readings from

FIGURE 18.4

A set used for the taping of religious programs in Malaysia.

the Holy Quran went off smoothly. The devil cannot read the Quran, so the religious leaders accepted that radio was not the work of the devil. TV had a harder time because the Quran forbids creating images of living people. King Faisal thought TV was needed for nationalistic purposes, however, and had a system built that went on the air in 1965.[43]

Malaysia

Malaysia modeled its broadcasting after the BBC during the years when it was a British colony. It began with two state-run networks and then added a private network, owned primarily by the leading newspaper and controlled rather closely by the government. Malaysia is primarily a Muslim country, so broadcasting has a strong religious content, including daily readings of the Quran. The Prime Minister and cabinet are somewhat authoritarian, so the government has a fairly strong hand in what is programmed.[44]

Thailand

Perhaps one of the most intriguing broadcasting systems is in Thailand, where there really is no system at all. There are no channel assignments and no technical standards. No one knows exactly how many radio and TV stations there are because they come and go and even change frequency and call letters during a single day. Everyone involved just seems to have a lot of fun.[45]

International Broadcasting

Another entirely different type of broadcasting that many countries engage in involves programming that is not intended for the national audience, but for other countries. Countries generally do this out of nationalistic interests. They want to make sure that people in other parts of the world are exposed to events as they view them. They also do it out of altruism—a desire to see that less fortunate people are exposed to culture through radio.

FIGURE 18.5

Voice of America's
master control board.
Two technicians on
around-the-clock duty
feed programs to
VOA's U.S.
transmitters and switch
all channels at every
station break. *(Courtesy
of United States Information
Agency)*

The totalitarian countries lead in this type of broadcasting. The USSR programs over 2200 hours a week and China airs almost 1500 hours. By contrast, the United States broadcasts slightly over 1200 hours per week and the United Kingdom close to 1000.[46]

hours of broadcast

Obviously, each of these services broadcasts many programs simultaneously in many different languages—forty-two for the United States and thirty-seven for England.[47]

The American international broadcasting is prepared by the Voice of America (**VOA**), which is part of the United States Information Agency. It is headquartered in Washington, D. C., where it has thirty-three studios. The programs are transmitted around the world by a complicated system that includes satellites, microwave, short wave relay, telephone land lines, and transmitters.

VOA

VOA started during World War II and was part of the Office of War Information. In 1948 Congress enacted legislation "to promote better understanding of the United States in other countries" and made VOA a permanent part of foreign policy housed in the State Department. In 1953 the United States Information Agency was formed and the VOA became part of it.[48]

The programming emphasizes carefully prepared newscasts, but music, sports, and features are also broadcast.[49] American music is generally the lure that gets people to listen. Sometimes programs are shipped on tape to regular radio stations in various countries and aired that way. Most people, however, hear VOA over short wave. It is impossible to know how many people listen, but the VOA received half a million letters from overseas in 1986 and the BBC overseas service estimates that its regular audience is 120 million.[50]

International broadcasting is not without its conflicts. Countries often try to jam other countries' signals.[51] When the VOA started Radio Marti to broadcast to Cuba in 1985, Florida stations were fearful that Fidel Castro would retaliate by using supertransmitters to jam them, but this did not happen.[52] During periods when the international situation appears to be somewhat peaceful, funds are often cut from this type of transmission.[53] The need for and effects of international broadcasting can most assuredly be debated, but it seems to be something that the major powers feel is necessary.

Armed Forces Radio and Television Service

Another type of overseas broadcasting is undertaken by the American armed forces for service people who are in locations where English language broadcasting is not available.

This, too, was started during World War II by some enterprising servicemen in Alaska who set up a transmitter and wrote to some Hollywood stars asking for radio programs. The stars couldn't send the programs because of security regulations, so the servicemen contacted the War Department in Washington. The result was the establishment of Armed Forces Radio (**AFRS**). Its mission was to give servicemen a touch of home and to combat Axis Sally and Tokyo Rose, dulcet-voiced girls broadcasting appeals intended to demoralize American soldiers.

Some of the broadcasting was done by short wave and some was done by troop-operated stations. While most of the stations were in fixed locations where the troops were stationed, a number of the them actually moved along with the advancing armies. Studios were set up in Los Angeles so that big name stars could perform on programs such as "Command Performance" and "Mail Call," which were sent to the AFRS facilities.

In 1945 there were about three hundred AFRS stations around the world, but this shrunk to sixty after the war. With the advent of the Korean War and the Vietnam conflict, the service once again expanded, this time including television. In 1954 AFRS became the Armed Forces Radio and Television Service (**AFRTS**).

At present, AFRTS is under the Department of Defense and operates over seven hundred outlets on land and on ships at sea. It programs some material specially for the military, but most of the programming is regular commercial and public radio and TV with the commercials deleted.[54]

International Telecommunication Union

Radio waves do not stop at national boundaries. If Mexico were free to allocate radio spectrum in any way it wanted, its stations might very well interfere with those of the United States. Even more to the point, just about any country in Europe can affect the broadcasting of any other European country. Some agency is needed to establish international guidelines and resolve disputes.

FIGURE 18.6

Arthur Shields (*left*) and Gary Cooper perform for AFRS's "Mail Call" about 1943. *(Courtesy of True Broadman)*

ITU

That body is the International Telecommunication Union, an arm of the United Nations. This organization periodically arranges a **WARC** (World Administrative Radio Conference) or a **RARC** (Regional Administrative Radio Conference). Numerous countries, including the United States, send delegates to these conferences to determine how the electromagnetic spectrum will be used.[55]

ITU delegates have dealt with problems such as increasing the size of the AM band and shrinking the bandwidth of individual radio stations.[56] The thorniest problem they have had to deal with of late is the allocation of satellite orbital slots to various countries. The third world nations want to make sure that slots are available to them when and if they develop the technology to launch their own satellites. The United States and other developed countries feel the precious slots should not lie fallow and should be used, at least temporarily, by the countries that can make use of them. A compromise has been worked out in part that would guarantee each country one orbital slot, but details of this have yet to be stabilized.[57]

COMSAT and INTELSAT

COMSAT and INTELSAT are both organizations that provide satellite services of an international nature. Their organization, duties, and interrelationships are complex and confusing and subject to constant change.

COMSAT (Communications Satellite Corporation) was the first to be formed. In the early 1960s the United States Congress felt the need to have an organization that would represent the United States in matters dealing with

COMSAT

satellites. As a result, it passed the Communications Satellite Act of 1962, which set up COMSAT. The provisions of the act called for COMSAT to be a private company regulated by the FCC. Half the stock was to be held by the public at large and half was to be held by companies such as AT&T and RCA, which might be engaged in satellite business. At first, the fact that the communication companies that would use COMSAT were its owners looked a bit like the inmates running the asylum. Over the years, these companies sold most of their stock to the general public and now own less than 1 percent of COMSAT.[58]

INTELSAT

INTELSAT (International Telecommunications Satellite Organization) was formed in 1964, largely at the urging of the United States. It was set up as an organization of member nations to deal with international satellite transmission. At first COMSAT was the manager of INTELSAT on behalf of these member nations. In 1973 INTELSAT became an independent organization owned by a consortium of countries, each of which pays a yearly fee based on its use of satellite communications. COMSAT represents the United States in INTELSAT and owns the biggest share of it (about 25 percent). Over 110 different countries are part of INTELSAT, which is headquartered in Washington, D.C.

INTELSAT has launched many satellites over the years until today it owns and operates more satellites than any other entity in the world. It has sixteen satellites, which carry two-thirds of international telephone traffic and almost all international television feeds. When a nation wants to use a satellite for some event, such as telecasting the Olympics or a World Soccer Championship, it rents satellite time from INTELSAT, which then makes all the arrangements for the satellite feed.[59]

Ever since its establishment, INTELSAT has had what amounts to a monopoly on providing satellite services for the world at large. However, after much debate at both the national and international levels, it was decided in the late 1980s that other companies should be allowed to compete with INTELSAT.[60]

uses

COMSAT deals with satellite feeds relevant to the United States. It rents satellite space from INTELSAT just like other countries do, but it rents more time and most of what it contracts for is used in the United States. As far back as 1964 it arranged for the Olympic ceremonies in Tokyo to be sent live to the United States. Since that time it has been responsible for many other Olympic coverages, as well as coverages of space flights, Nixon's visit to China, the meeting in outer space between United States and USSR astronauts, and presidential inaugurations. Among its other transmissions are interconnects between the United States mainland and Hawaii and Alaska, the "Hot-Line" between Moscow and Washington, the transmission of several newspapers from one part of the country to another, emergency communication during disasters, teleconferencing, and an international mail service between London and New York called INTELPOST.[61]

Other Telecommunications Forms

FIGURE 18.7

The lobby of the INTELSAT building in Washington, D.C., displaying some of the satellites INTELSAT has launched.

COMSAT owns its own satellites called COMSTARS, but it rents these primarily to AT&T, MCI, Sprint, and others for domestic telephone service. COMSAT becomes involved in many other operations, some of which have caused it financial problems. For example, Satellite Television Corporation, the company that proposed direct broadcast satellite, was a subsidiary of COMSAT. COMSAT is also the United States signatory to INMARISAT, which offers satellite communications services to ships at sea. COMSAT frequently acts as consultant when parts of the world want to set up regional satellite communications systems, such as the Arab world establishing AR-ABSAT.[62]

Both COMSAT and INTELSAT have been embroiled in many controversies ranging from improper international influence to misappropriations of funds.[63] The satellite world is still young and changing and the organizations dealing with it find they must change, too.

Conclusion

International telecommunications consists of broadcasting within countries and without. Internally, most countries have adopted characteristics that can be seen in the American, British, or Russian systems. Technically, countries generally use 525-line NTSC (American), 625-line PAL (British), or 625-line SECAM (Russian) for scanning and color coding. Ownership is usually private (American), charter (British), or government (Russian). Financing comes from advertising (American), licenses (British), or taxes (Russian). The purpose of programming is to entertain (American), uplift (British), or educate (Russian). In reality, most countries have a mixture of these systems; this is true also of America, Britain, and the USSR.

Controversies often arise regarding how much imported programming should be allowed on a country's broadcasting system and how much emphasis should be placed on newer media.

Two totalitarian countries, Russia and China, are the leaders in external broadcasting, followed by the United States and Britain. This type of programming is usually at least somewhat propagandistic. Programming for service people is also beamed abroad, as evidenced by the Armed Forces Radio and Television Service. Because radio waves do not obey national boundaries, conferences sponsored by the International Telecommunication Union of the United Nations are often held to plan spectrum use. Satellites have become so essential to most countries that various SAT organizations have been formed to handle international uses of the birds.

The international sphere is a fascinating one that needs constant fine tuning as world conditions change.

Thought Questions

1. What are the major advantages and disadvantages of the American, British, and Russian systems of broadcasting?
2. Should the networks in the United States show more foreign programming than they do? Why or why not?
3. Should the United States place more or less emphasis than it does on Voice of America broadcasts? Defend your answer.
4. If you could start broadcasting from scratch in some newly formed country, how would you organize it?

1

1. Eric Barnouw, *A Tower in Babel: A History of Broadcasting in the United States to 1933,* (New York: Oxford University Press, 1966), 7.

2. *Broadcasting/Cablecasting Yearbook,* 1987, (Washington, D.C.: Broadcasting Publications, Inc., 1987), A–2 and D–3.

3. "Valenti Has Good/Bad News for NATO," *Variety* (October 22, 1986): 16.

4. "Where Things Stand," *Broadcasting* (April 6, 1987): 40.

5. *Serving the Nation, Enriching Our Communities: 1986 Annual Report,* (Washington, D.C.: National Assocation of Broadcasters, 1986), 26.

6. "Cable Penetration Nears 50 Percent in U.S.," *Variety* (April 2, 1987): 6.

7. "VCR Penetration in U.S. Hits 49 Percent," *Variety* (May 11, 1987): 4.

8. *Serving the Nation,* 26.

9. James Traub, "The World According to Nielsen," *Channels* (January-February 1985): 26–32.

10. *Serving the Nation,* 26.

11. "Public Gives TV News High Rating," *Broadcasting* (May 13, 1985): 58.

12. David Handler, "TV Finished First, Friends Second, Helping Others Third," *TV Guide* (April 18, 1987): 20–21.

13. *Public Attitudes Toward Television and Other Media in a Time of Change,* (New York: Television Information Office, 1985), 5; and *America's Watching,* (New York: Television Information Office, 1987), 5.

14. Ibid., 24 and 22.

15. Bert E. Bradley, *Fundamentals of Speech Communication,* (Dubuque, Iowa: William C. Brown Publishers, 1984), 7.

16. W. A. Belson, *"Facts and Figures": The Comprehensibility of Two Programmes in the Facts and Figures Series,* (London: British Broadcasting Corporation, March 17, 1955).

17. P. H. Tannenbaum, *Instruction Through Television: A Comparative Study,* (Urbana, Ill.: University of Illinois, June 1956).

18. Wilbur Schramm, "Learning From Instructional Television," *Review of Educational Research* (April 1962): 156–67.

19. Thomas James Aylward, "A Study of the Effect of Production Techniques on a Televised Lecture," (Ph.D. dissertation, Syracuse University, 1958).

20. One such summary is Lynne S. Gross, *Experimental Research in Educational Television,* (Burlingame, Calif.: California Advisory Council on Educational Research of the California Teachers Association, January 1968).

21. Ronald Drabman and Margaret Thomas, "Does TV Violence Breed Indifference?" *Journal of Communication* (Autumn 1975): 86–89.

22. Melvin S. Heller and Samuel Polsky, "Responses of Emotionally Vulnerable Children to Televised Violence," *Studies in Violence and Television* (New York: American Broadcasting Co., 1976), 1–52.

23. Bromley H. Kniveton, "Social Learning and Imitation in Relation to Television," in Ray Brown, ed., *Children and Television* (Beverly Hills, Calif.: Sage, 1976), 237–66.

24. Tom Link, "Saving What We've Seen," *Emmy* (Winter 1980): 39–41.

25. Joseph R. Dominick, "New Technologies: Implications for Research," *Feedback* (November 1980): 3–4.

26. Rebecca B. Rubin and Alan M. Rubin, *Communication Research: Strategies and Sources,* (Belmont, Calif.: Wadsworth Publishing Company, 1986).

27. *Statistical Abstracts of the United States, 1988* (Washington, D.C.: U.S. Department of Commerce, Bureau of Census, 1988), 408.

NOTES

28. Ibid., 311 and 753.

29. "$510 Million's the Mark to Beat Now," *Broadcasting* (May 20, 1985): 39.

2

1. For a more detailed account of Maxwell, see Orrin E. Dunlap, Jr., *Radio's 100 Men of Science* (New York: Harper and Brothers Publishers, 1944), 65–68.

2. For a more detailed account of Hertz, see ibid., 113–17.

3. For a more detailed account of Marconi, see Degna Marconi, *My Father Marconi* (New York: McGraw-Hill, 1962).

4. For a more detailed account of Fleming, see Dunlap, *Radio's 100 Men of Science,* 90–94.

5. For a more detailed account of Fessenden, see Helen M. Fessenden, *Fessenden: Builder of Tomorrows* (New York: Coward-McCann, 1940).

6. For a more detailed account of De Forest, see *Lee De Forest, Father of Radio: The Autobiography of Lee De Forest* (Chicago: Wilcox and Follett, 1950).

7. For a more detailed account of Sarnoff, see Eugene Lyon, *David Sarnoff: A Biography* (New York: Harper, 1966).

8. Erik Barnouw, *A Tower in Babel: A History of Broadcasting in the United States to 1933* (New York: Oxford University Press, 1966), 57–61; and Kenneth Bilby, *The General: David Sarnoff and the Rise of the Communications Industry* (New York: Harper and Row, 1986).

9. David Sarnoff, *Looking Ahead: The Papers of David Sarnoff* (New York: McGraw-Hill Book Company, 1968), 31–33.

10. Barnouw, *A Tower in Babel,* 39–41.

11. Ibid., 61–64.

12. For a more detailed account of Conrad, see Dunlap, *Radio's 100 Men of Science,* 180–83.

13. Robert E. Summers and Harrison B. Summers, *Broadcasting and the Public* (Belmont, Calif.: Wadsworth Publishing Company, 1966), 34.

14. Barnouw, *A Tower in Babel,* 85.

15. Ibid., 86.

16. Ibid., 100.

17. Ibid., 102.

18. William P. Banning, *Commercial Broadcasting Pioneer: WEAF Experiment, 1922–1926* (Cambridge, Mass.: Harvard University Press, 1946), 150.

19. Ibid., 155.

20. "The First 60 Years of NBC," *Broadcasting* (June 9, 1986): 49–64.

21. For a critique of CBS, see Robert Metz, *CBS: Reflections in a Bloodshot Eye* (Chicago: Playboy Press, 1975); and "The Winning Ways of William S. Paley," *Broadcasting* (May 31, 1976): 25–45.

22. For more information, see "The Silver Has Turned to Gold," *Broadcasting* (February 13, 1978): 34–46; and Sterling Quinlan, *Inside ABC: American Broadcasting Company's Rise to Power* (New York: Hastings House, 1979).

23. "After 50 Years the Feeling Is Still Mutual," *Broadcasting* (September 10, 1984): 43–50.

24. Edwin L. Glick, "The Life and Death of the Liberty Broadcasting System," *Journal of Broadcasting* (Spring 1979): 117–36.

25. Joan Huff Wilson, *Herbert Hoover: Forgotten Progressive* (Boston: Little, Brown, and Company, 1975), 112–13.

26. Donald G. Godfrey, "Senator Dill and the 1927 Radio Act," *Journal of Broadcasting* (Fall 1979): 477–490.

27. Summers and Summers, *Broadcasting and the Public,* 45–50.

28. For a first-hand account of "Amos 'n' Andy," see Charles J. Correll and Freeman F. Gosden, *All About Amos 'n' Andy* (New York: Rand McNally, 1929); and Bert Andrews and Ahrgus Julliard, *Holy Mackerel: The Amos 'n' Andy Story* (New York: Dutton, 1986).

29. For more information on radio programming, see Irving Settel, *A Pictorial History of Radio* (New York: Grossett and Dunlap, 1967); Frank Buxton and Bill Owen, *The Big Broadcast, 1920–1950* (New York: Viking, 1972); J. Fred McDonald, *Don't Touch That Dial: Radio Programming in American Life from 1920 to 1960* (Chicago: Nelson Hall, 1979); Mary C. O'Connell, *Connections: Reflections on Sixty Years of Broadcasting* (New York: National Broadcasting Company, 1986); and Vincent Terrace, *Radio's Golden Years: The Encyclopedia of Radio Programs 1930–1960* (San Diego, Calif.: A. S. Barnes and Co., 1981).

30. Erik Barnouw, *The Golden Web: A History of Broadcasting in the United States 1933–1953* (New York: Oxford University Press, 1968), 8–18.

31. Giraud Chester, Garnet R. Garrison, and Edgar E. Willis, *Television and Radio* (New York: Appleton-Century-Crofts, 1971), 35.

32. James A. Brown, "Selling Airtime For Controversy: NAB Self Regulation and Father Coughlin," *Journal of Broadcasting* (Spring 1980): 199–224.

33. "Genesis of Radio News: The Press-Radio War," *Broadcasting* (January 5, 1976): 95.

34. For a more detailed account of radio during World War II, see Paul White, *News on the Air* (New York: Harcourt, Brace, 1947).

35. Barnouw, *The Golden Web,* 46.

36. Summers and Summers, *Broadcasting and the Public,* 46.

37. Ibid., 70.

38. Federal Communications Commission v. Sanders Brothers Radio Station, 309 U.S. 470, March 25, 1940.

39. *Broadcasting Yearbook, 1975* (Washington, D.C.: Broadcasting Publications, Inc., 1975), C–289.

40. Ward L. Quaal and James A. Brown, *Broadcast Management* (New York: Hastings House, 1976), 292.

41. Interview with Dave Morehead, general manager of KMET, Los Angeles, March, 1978.

42. For more insight into the rise of DJs, see Arnold Passman, *The DJ's* (New York: Macmillan, 1971).

43. For a more detailed account of Armstrong, see Lawrence Lessing, *Man of High Fidelity: Edwin Howard Armstrong* (New York: Lippincott, 1956).

44. "FCC Ends Curb on Simultaneous AM, FM Programs," *Variety* (April 1, 1986): 1.

45. "Broadcasters Jubilant as FCC Votes to Lift Most of Regulations Covering Radio," *Variety* (January 15, 1981): 1.

46. "First with Eight," *Broadcasting* (September 17, 1984): 105.

47. *Broadcasting/Cablecasting Yearbook, 1987* (Washington, D.C.: Broadcasting Publications, Inc., 1987), H–72.

48. "ABC Opens New Network Radio Broadcast Center," *Broadcasting* (November 5, 1984): 62.

49. "Sounds of Success," *Newsweek* (December 3, 1979): 132–34.

50. "New Radio Networks Spring Up," *New York Times* (January 11, 1980): III,22:6.

51. "Those 24–Hour Satellite Networks: Update on Radio's Latest," *Broadcasting* (December 14, 1981): 39.

52. *Broadcasting/Cablecasting Yearbook, 1987*, F–53 and F–58.

53. "Westwood One to Buy NBC Radio Networks for $50 Mil," *Variety* (July 21, 1987): 1.

54. Chris Miller, "AM Stereo: After All These Years, Is the Marketplace Ready?" *Feedback* (Winter 1984): 14–21; and "The AM Stereo Fight Continues," *Broadcasting* (April 26, 1986): 68.

55. "High Hopes for 80–90," *Broadcasting* (March 11, 1985): 31.

56. "1986: The Second 50 Years of the Fifth Estate," *Broadcasting* (December 29, 1986): 50.

3

1. The best overall chronicle of the history of television can be found in the three-volume history of broadcasting by Erik Barnouw published in New York by Oxford University Press. The titles and dates are: *A Tower in Babel: A History of Broadcasting in the United States to 1933* (1966); *The Golden Web: A History of Broadcasting in the United States 1933–1953* (1968); and *The Image Empire: A History of Broadcasting in the United States from 1953* (1970). This work will be cited as a major source of information because it gives the most back-up detail. It is also scholarly and thoroughly researched through innumerable interviews, unpublished papers, and oral history collections, as well as the more traditional books, articles, and documents. Barnouw has condensed and updated much of the television material into a one-volume book, *Tube of Plenty: The Development of American Television* (1975). Another excellent one-volume history is Christopher H. Sterling and John M. Kittross, *Stay Tuned: A Concise History of American Broadcasting* (Belmont, Calif.: Wadsworth Publishing Company, 1978).

2. Barnouw, *A Tower in Babel,* 210, 231. These early mechanical systems are best described in A. A. Dinsdale, *First Principles of Television* (New York: John Wiley, 1932).

3. Barnouw, *The Golden Web,* 42, 127, 145. See also "Portrait," *Business Week* (June 7, 1952): 106; and "Portrait," *Fortune* (February 1954): 47.

4. Barnouw, *A Tower in Babel,* 210, and *The Golden Web,* 39–40, 42, 283. See also George Everson, *The Story of Television: The Life of Philo T. Farnsworth* (New York: Norton, 1949); and Stephen F. Hofer, "Philo Farnsworth: Television Pioneer," *Journal of Broadcasting* (Spring 1979): 153–66.

5. Barnouw, *A Tower in Babel,* 66, 154, 210. For Zworykin's own view of the electronics of television, see Vladimir Zworykin, *The Electronics of Image Transmission in Color and Monochrome* (New York: John Wiley, 1954).

6. Barnouw, *The Golden Web,* 126.

7. Ibid., 126–30.

8. Ibid., 242–44.

9. See footnotes 20–23 of chapter 2, "Commercial Radio."

10. Barnouw, *The Golden Web,* 285–90.

11. Jack Slater, "The Amos 'n' Andy Show," *Emmy* (January-February 1985): 48–49.

12. A large number of excellent, highly pictorial works have been published that deal with TV programming through the years. These include Linda Beech, *TV Favorites* (New York: Scholastic Book Services, 1971); Arthur Schulman and Roger Youman, *The Television Years* (New York: Popular Library, 1973); Irving Settl, *A Pictorial History of Television* (New York: Frederick Ungar Publishing Company, 1983): Fred Goldstein and Stan Goldstein, *Prime-Time Television: A Pictorial History from Milton Berle to "Falcon Crest"* (New York: Crown Publisher, 1983); and Alex McNeil, *Total Television: A Comprehensive Guide to Programming from 1948 to the Present* (New York: Viking Penguin, 1984).

13. Barnouw, *The Golden Web,* 295.

14. Most of this information was gained from a 1976 interview with True Boardman, one of the writers blacklisted during the fifties. Three sources interesting to peruse are John Cogley, *Report on Blacklisting* (Washington, D.C.: The Fund for the Republic, 1956); *Red Channels: The Report on Communists in Radio and Television* (New York: Counterattack, 1950): and Larry Ceplair, "Hollywood Blacklist," *Emmy* (Summer 1981): 30–32.

15. Barnouw, *The Image Empire,* 22–23.

16. See note 12.

17. Tony Peyser, "Pat Weaver: Visionary or Dilettante?" *Emmy* (Fall 1979): 32–34.

18. See note 12.

19. Daniel E. Garvey, "Introducing Color Television: The Audience and Programming Problems," *Journal of Broadcasting* (Fall 1980): 515–26.

20. Larry Ceplair, "The End of the Trail," *Emmy* (September/October 1984): 50–54.

21. See note 12.

22. "Dress Rehearsals Complete with Answers?" *U.S. News* (October 19, 1959): 60–62; "Is It Just TV or Most of Us?" *New York Times Magazine* (November 15, 1959): 15+; "Out of the Backwash of the TV Scandals," *Newsweek* (November 16, 1959): 66–68; "Quiz Probe May Change TV," *Business Week* (November 7, 1959): 72–74+; and "Van Doren on Van Doren," *Newsweek* (November 9, 1959): 69–70.

23. See note 12.

24. Barnouw, *The Image Empire,* 160–70; P. M. Stern, "Debates in Retrospect," *New Republic* (November 21, 1960): 18–19; and "TV Debate Backstage: Did the Cameras Lie?" *Newsweek* (October 10, 1960): 25.

25. "Covering the Tragedy: President Kennedy's Assassination," *Time* (November 29, 1963): 84; "Did Press Pressure Kill Oswald?"

U.S. News (April 6, 1964): 78–79; "President's Rites Viewed Throughout the World," *Science Newsletter* (December 7, 1963): 355; and J. H. Winchester, "TV's Four Days of History," *Readers Digest* (April 1964): 204I–204J+.

26. "Satellites," *Broadcasting* (July 20, 1987): 33–44.

27. For Friendly's analysis of this whole situation, see Fred Friendly, *Due to Circumstances Beyond Our Control . . .* (New York: Random House, 1967).

28. For more information on TV's relationship to Vietnam, see Michael J. Arlen, *Living Room War* (New York: The Viking Press, 1969); Oscar Patterson, "An Analysis of Television Coverage of the Vietnam War," *Journal of Broadcasting* (Fall 1984): 397–404; and Edward Fouhy, "Looking Back at 'The Living Room War,'" *RTNDA Communicator* (March 1987): 12–13.

29. Barnouw, *The Image Empire,* 197.

30. Fred Bronson, "Keep on Trekkin'," *Emmy* (December 1986): 42–48.

31. See note 12.

32. Barnouw, *The Image Empire,* 251–52.

33. Elizabeth J. Heighton and Don R. Cunningham, *Advertising in the Broadcast Media* (Belmont, Calif.: Wadsworth Publishing Company, 1976), 268–71.

34. Sydney W. Head, *Broadcasting in America* (Boston: Houghton Mifflin Company, 1976), 215–18.

35. " 'Family Hour' OK but Not by Coercion," *Variety* (November 5, 1976): 1.

36. "FCC Opens Box on Financial Interest," *Broadcasting* (June 28, 1982): 30.

37. "Now an Open Season on All Newsmen," *Broadcasting* (November 24, 1969): 44.

38. See note 12.

39. "FCC Eliminates Number of Regs for TV Stations," *Variety* (June 28, 1984): 1.

40. "FCC Strikes the Flag on TV Ownership Rules," *Broadcasting* (August 13, l984): 35.

41. "Hi-Fi Meets Television," *Newsweek* (June 24, 1985): 78; and "Momentum Builds for TV Stereo," *Broadcasting* (September 9, 1985): 117.

42. "Ratings Down 10 Percent, Shares Off Four Points," *Broadcasting* (November 9, 1987): 35.

43. Lynne Schafer Gross, "Celebrating Independents," *Emmy* (January/February 1987): 68–71.

44. "Capcities + ABC," *Broadcasting* (March 25, 1985): 31.

45. "General Electric Will Buy RCA for $6.28 Billion," *Los Angeles Times* (December 12, 1985): I, 1.

46. "Welch Names Robert Wright to NBC Presidency," *Broadcasting* (September 1, 1986): 34.

47. "NBC's Farewell to Fred," *Newsweek* (July 13, 1981): 69.

48. L. J. Davis, "How Tinker Turned It Around," *Channels* (January/February 1986): 30–39.

49. "CBS Gleam in Ted Turner's Eye," *Broadcasting* (March 4, 1985): 35.

50. "1986 Shakeups in Broadcasting Biz Put Novices in Driver's Seat," *Variety* (January 5, 1987): 14.

51. "Metromedia Vid as Fox Subsid," *Variety* (May 7, 1985): 1.

52. "Murdoch Holds Full Fox Hand," *Variety* (September 24, 1985): 1.

53. "Fox Broadcasting Company: The Birth of a Network," *Broadcasting* (April 6, 1987): 88.

54. "Turner Gets Financial OK to Buy MGM," *Electronic Media,* (August 19, 1985): 1.

55. "Here's a Brief Look at 1985's Media Mergers," *Electronic Media* (January 13, 1986): 88.

56. Stratford P. Sherman, "Are Media Mergers Smart Business," *Fortune* (June 24, 1985): 98–103.

57. D. K. Knapp, "The New Golden Years," *Emmy* (August 1986): 50–55.

58. Suzy Kalter, *The Complete Book of M*A*S*H* (New York: Abrams, 1984).

59. "Satellite Newsgathering: Growing Presence at NAB," *Broadcasting* (April 6, 1987): 71.

60. Fred Bronson, "Live Aid," *Emmy* (August 1986): 76–83.

61. "Home Shopping Network's Shopping Spree," *Broadcasting* (March 9, 1987): 55.

4

1. Erik Barnouw, *A Tower in Babel: A History of Broadcasting in the United States to 1933* (New York: Oxford University Press, 1966), 61.

2. Werner J. Severin, "Commercial vs. Non-Commercial Radio During Broadcasting's Early Days," *Journal of Broadcasting* (Summer 1981): 295–302.

3. Barnouw, *A Tower in Babel,* 215.

4. John Witherspoon and Roselle Kovitz, *The History of Public Broadcasting* (Washington, D.C.: Current, 1987), 8–9.

5. Donald N. Wood and Donald G. Wylie, *Educational Telecommunications* (Belmont, Calif.: Wadsworth Publishing Company, 1977), 32.

6. *Broadcasting/Cablecasting Yearbook, 1987* (Washington, D.C., Broadcasting Publications, Inc., 1987), A-2.

7. Frederic A. Leigh, "College Radio: Effects of FCC Class D Rules Changes," *Feedback* (Fall 1984): 18–21.

8. An excellent history of the NAEB can be found in Harold E. Hill, *The National Association of Educational Broadcasters: A History* (Urbana, Ill.: National Association of Educational Broadcasters, 1954).

9. *National Public Radio Fact Sheet* (Washington, D.C.: National Public Radio, n.d.), 1.

10. Susan Stamberg, *Every Night at Five: Susan Stamberg's "All Things Considered" Book* (New York: Pantheon, 1982).

11. *National Public Radio Fact Sheet* (Washington, D.C.: National Public Radio, 1987), 1.

12. For more information on NPR, see Robert K. Avery and Robert Pepper, "Balancing the Equation: Public Radio Comes of Age," *Public Telecommunications Review* (March/April 1979): 19–30; Joseph Brady Kirkish, "A Descriptive History of America's First Public Radio Network: National Public Radio, 1970–1974" (Ph.D Dissertation, The University of Michigan, 1980); John Kenneth Garry, "The History of National Public Radio, 1974–1977" (Ph.D. Dissertation, Southern Illinois University at Carbondale, 1982); and "NPR Board Moves Toward New Program Sales Plan," *Current* (July 6, 1987): 6.

13. Ibid.

14. *American Public Radio* (St. Paul, Minnesota: American Public Radio, n.d.): 1–7.

15. *Programming Notes,* (St. Paul, Minnesota: American Public Radio, 1987).

16. "A Home Companion Bids Farewell," *Broadcasting* (June 22, 1987): 39.

17. *American Public Radio,* 2.

18. Frederic A. Leigh, "College Radio: Effects of FCC Class D Rule Changes," *Feedback* (Fall, 1984): 18–21.

19. "Last Minute Salvation of NPR," *Broadcasting* (August 1, 1983): 21.

20. "Public Stations Approve NPR Reorganization," *Electronic Media* (May 30, 1985): 3.

21. "What Happened to Mankiewicz?" *Broadcasting* (April 25, 1983): 28.

22. Donald P. Mullally, "Radio: The Other Public Medium," *Journal of Communication* (Summer 1980): 189–97.

23. Lynne S. Gross, "A Study of College Credit Literature Courses Offered on Open Circuit Television," *NAEB Journal* (November/December 1962): 87–88.

24. Robert K. Avery and Robert Pepper, "An Institutional History of Public Broadcasting," *Journal of Communication* (Summer 1980): 126–38.

25. *Broadcasting Yearbook, 1975* (Washington, D.C.: Broadcasting Publications, Inc., 1975), A-7.

26. Clifford G. Erickson and Hyman M. Chausow, *Chicago's TV College: Final Report of a Three-Year Experiment* (Chicago: Chicago City Junior College, August, 1960).

27. Mary Howard Smith, *"Midwest Program on Airborne Television Instruction," Using Television in the Classroom* (New York: McGraw-Hill, 1961).

28. *Broadcasting Yearbook, 1975,* A-7.

29. Gene R. Stebbins, "PTV Membership Support: Its Beginnings," *Public Telecommunications Review* (July/August 1979): 52–54.

30. Carnegie Commission on Public Television, *Public Television: A Program for Action* (New York: Harper and Row, 1967).

31. The Public Broadcasting Act of 1967, Public Law 90–129, 90th Congress (November 7, 1967).

32. Robert M. Pepper, *The Formation of the Public Broadcasting Service* (New York: Arno Press, 1979).

33. Arthur Shulman and Roger Youman, *The Television Years* (New York: Popular Library, 1973), 270–303.

34. *The Public Television Station Program Cooperative* (Alexandria, Va.: Public Broadcasting Service, 1986), 1–3.

35. "CBS to Drop Its Cultural Cable-TV Service After Failing to Draw Needed Ad Support," *The Wall Street Journal* (September 14, 1983): 7.

36. "Public Broadcasting Rocked by Verdict in Captioning Case," *Broadcasting* (November 2, 1981): 28; and "Pubcasters Get Some Good News from High Court," *Variety* (February 23, 1983): 1.

37. "CPB, PBS Strike Truce in Atlanta," *Broadcasting* (February 14, 1977): 31.

38. "Public Broadcasting's Major Players," *Broadcasting* (May 11, 1987): 70.

39. "Twenty Tumultuous Years for CPB," *Broadcasting* (May 11, 1987): 60–75.

40. *Carnegie Commission of the Future of Public Broadcasting: A Public Trust* (New York: Carnegie Commission, 1979); and "Carnegie II's Revised Blueprint for Public Broadcasting," *Broadcasting* (February 5, 1979): 56–65.

41. Willard D. Rowland, Jr., "Continuing Crisis in Public Broadcasting: A History of Disenfranchisement," *Journal of Broadcasting and Electronic Media* (Summer 1986): 251–274.

42. *Made Possible by a Grant from the Corporation for Public Broadcasting* (Washington, D.C.: Corporation for Public Broadcasting, n.d.).

43. Figures come from *Status Report on Public Broadcasting, 1973* (Washington, D.C.: Corporation for Public Broadcasting, 1974).

44. *PBS: Facts About Us* (Alexandria, Virginia: Public Broadcasting Service, 1987): 7.

45. "New Budget Bill Mixed Blessing for Pubcasters," *Variety* (August 3, 1981): 1.

46. "Annenberg Grant Comes Through," *Current* (March 16, 1981): 1.

47. "Pubcasters Get OK to Air Logos of Contributors," *Variety* (April 24, 1981): 1.

48. "More Subscribers to Public Stations Called Answer to Federal Cuts," *Broadcasting* (June 29, 1981): 57.

49. "'Pledgeless' Pledge Drives: Noncommercial Phenomenon," *Broadcasting* (December 8, 1986): 110.

50. *PBS: Facts About Us,* 4.

51. "CPB on the Prowl for New Ways to Fund the Future," *Broadcasting* (November 9, 1981): 29.

52. "10 Pub TV Stations Granted Go-Ahead to Broadcast Ads," *Variety* (February 2, 1982): 1.

53. "PBS Revises Rules for Underwriting in Wake of FCC Decision, *Broadcasting* (April 9, 1984): 92.

54. "The Commercialization of Public TV," *Electronic Media* (June 23, 1986): 1.

55. John Weisam," Public TV in Crisis," *TV Guide* (August 8, 1987): 25–40.

56. *Made Possible by a Grant From the Corporation for Public Broadcasting,* 3–4.

57. "Furor Over a Death," *Time* (April 21, 1980): 32.

58. "PBS & Controversy: Mostly OK," *Variety* (April 16, 1987): 1.

59. "Editorializing Ban for Public Stations Debated Before Supreme Court," *Broadcasting* (January 23, 1984): 95.

60. *PBS: Facts About Us,* 17.

61. *Made Possible by a Grant From the Corporation for Public Broadcasting,* passim.

62. *20 Reasons for Public Television* (Alexandria, Virginia: Public Broadcasting Service, 1986), 2.

63. *PBS: Facts About Us,* 5.

64. *PBS Marks 15 Years of Service to Nation* (Alexandria, Virginia: Public Broadcasting Service, 1984), 6.

65. *Corporation for Public Broadcasting 1985 Annual Report* (Washington, D.C.: Corporation for Public Broadcasting, 1986), 15.

66. *Who Watches Public Television?* (Alexandria, Virginia: Public Broadcasting Service, 1986), 1.

5

1. David L. Jaffe, "CATV: History and Law," *Educational Broadcasting* (July/August 1974): 15–16; and Thomas F. Baldwin and D. Stevens McVoy, *Cable Communication* (Englewood Cliffs, New Jersey: Prentice-Hall, 1983), 8–10.

2. *Broadcasting/Cable Yearbook, 1981* (Washington, D.C.: Broadcasting Publications, Inc., 1981), G–1.

3. Jaffe, 34.

4. For other chronicles of early cable TV, see Mary Alice Mayer Philips, *CATV: A History of Community Antenna Television* (Evanston, Ill.: Northwestern University Press, 1972); and Albert Warren, "What's 26 Years Old and Still Has Growing Pains?"*TV Guide* (November 27, 1976): 4–8.

5. The relationship between cable and government can be found in *Cable Television and the FCC: A Crisis in Media Control* (Philadelphia: Temple University Press, 1973); Steven R. Rivkin, *Cable Television: A Guide to Federal Regulations* (Santa Monica, Calif.: Rand Corporation, 1973); Martin H. Seiden, *Cable Television USA: An Analysis of Government Policy* (New York: Praeger, 1972); and Stuart N. Brotman, *Communications Policymaking at the Federal Communications Commission* (Washington, D.C.: The Annenberg Washington Program, 1987), 33–46.

6. A great deal about early franchising can be learned from Leland L. Johnson and Michael Botein, *Cable Television: The Process of Franchising* (Santa Monica, Calif.: Rand Corporation, 1973).

7. Jaffe, 17.

8. *Broadcasting/Cable Yearbook, 1981,* G–1.

9. Jaffe, 35.

10. *Broadcasting/Cable Yearbook, 1981,* G–1–G–3.

11. Two sources that deal with early local origination are Ron Merrell, "Origination Compounds Interest with Quality Control," *Video Systems* (November/December 1975): 15–18; and Sloan Commission on Cable Communications, *On the Cable: The Television of Abundance* (New York: McGraw-Hill, 1972).

12. Two sources dealing with early public access are Richard C. Kletter, *Cable Television: Making Public Access Effective* (Santa Monica, Calif.: Rand Corporation, 1973); and Charles Tate, *Cable Television in the Cities: Community Control, Public Access, and Minority Ownership* (Washington, D.C.: The Urban Institute, 1972).

13. "Cable Survey Shows Growth," *Video Systems* (January/February 1976): 6.

14. "Righting Copyright," *Time* (November 1, 1976): 92.

15. Margaret B. Carlson, "Where MGM, the NCAA, and Jerry Falwell Fight for Cash," *Fortune* (January 23, 1984): 171.

16. "Compulsory License Hit Again," *Variety* (April 23, 1981): 1.

17. *HBO Landmarks* (New York: Home Box Office, n.d.), 1.

18. Sheila Mahony, Nick Demartino, and Robert Stengel, *Keeping Pace with the New Television* (New York: VNU Books International, 1980), 61.

19. Don Kowet, "High Hopes for Pay Cable," *TV Guide* (June 10, 1978): 14–18.

20. Sheila Mahony, et al., *Keeping Pace,* 131.

21. Ibid., 95.

22. "The Dimensions of Cable: 1981," *Channels* (April/May 1981): 88.

23. "CATV Stats Leapin': Pretax Net 45 percent; Feevee Revenue 85 percent," *Variety* (December 31, 1980): 16.

24. "Cable Revenues Gain Faster than Profits, Survey Finds," *Broadcasting* (November 30, 1981): 52.

25. "The Dimensions of Cable: 1981," 88.

26. Ibid., 88.

27. "The Top Line: Almost $2 Billion; The Bottom Line: Almost $200 Million," *Broadcasting* (January 5, 1981): 75.

28. "Cities Issue Guidelines for Cable Franchising," *Broadcasting* (March 9, 1981): 148; and Shelia Mahony, et al., *Keeping Pace,* 93.

29. Pat Carson, "Dirty Tricks," *Panorama* (May 1981): 57–59+; "Fight for TV Franchise in New England Town Elicits Big Bribe Offer," *The Wall Street Journal* (December 22, 1981): 1; and Laurence Bergeen, "Cable Fever," *TV Guide* (June 6, 1981): 23–26.

30. For more on franchising, see "City's Ability to Regulate Cable Television Industry Questioned," *Los Angeles Times* (August 17, 1981): II, 1; "Warner Amex Loses Bid to Cablevision for Subscriber TV Franchise in Boston," *The Wall Street Journal* (August 13, 1981): 6; "The Gold Rush of 1980," *Broadcasting* (March 31, 1980): 35–56; "Simmering Cable Franchise Issue Comes to a Boil," *Variety* (July 31, 1981): 1; "New York Today Picks Its Cable-TV Winners for Four Boroughs," *The Wall Street Journal* (November 18, 1981): 1; and many other newspaper articles of the late 1970s and early 1980s.

31. "The Top 50 MSO's: As They Are and Will Be," *Broadcasting* (November 30, 1981): 37; "Entertainment Analysts Find Cable Mom-Pop Days Are Gone; Big Bucks Rule the Day," *Broadcasting* (June 15, 1981): 46; and "FCC Okays Sale of TPT to Westinghouse, the Largest Communications Merger Ever," *Variety* (July 13, 1981): 1.

32. "Cable Television Is Attracting More Ads; Sharply Focused Programs Are One Lure," *The Wall Street Journal* (March 31, 1981): 46; "CBS Cable Signs Kraft as Its First Advertiser," *Broadcasting* (August 3, 1981): 59; "ARTS Snags GM as Underwriter," *Variety* (September 2, 1981): 1; and "Bristol-Myers to Furnish Series to USA Network," *Broadcasting* (February 23, 1981): 61.

33. "Scribes Back at Typewriters," *Variety* (July 16, 1981): 1; and "Pay-TV Settlements Will Hasten Debut of Feevee 1st-Run Prod'n," *Variety* (July 15, 1981): 8.

34. Keith Larson, "All You Ever Wanted to Know About Buying an Earth Station," *TVC* (May 15, 1979): 16–19; "Going Once, Twice, Gone on Satcom IV," *Broadcasting* (November 16, 1981): 27; and "Greenlight for Satcom IV Plan," *Broadcasting* (March 29, 1982): 40.

35. "Viacom Becomes Second Satellite Pay Cable Network," *Broadcasting* (October 31, 1977): 64.

36. "GalaVision Reaches First Anniversary, Free Preview Scheduled for October," *TVC* (October 15, 1980): 33–36; "RCA Pay-Cable Plunge Via 50 percent Slice of RCTA," *Variety* (May 11, 1981): 1; "Times Mirror Sat'lite Programs Join Feevee Networks on May 1," *Variety* (March 26, 1981): 3; and "Disney Previews Pay-TV Channel," *Variety* (April 13, 1983): 1.

37. "Getty Throws Feevee Bombshell," *Variety* (April 23, 1980): 1; and "Premiere Goes Down Feevee Tube," *Variety* (June 5, 1981): 1.

38. Some of the articles summarizing basic cable are Michael O'Daniel, "Basic-Cable Programming: New Land of Opportunity," *Emmy* (Summer 1980): 26–30; "The Cable Numbers According to Broadcasting," *Broadcasting* (November 30, 1981); and Peter W. Bernstein, "The Race to Feed Cable TV's Maw," *Fortune* (May 4, 1981): 308–18.

39. For five different views of access programming see Don Kowet, "They'll Play Bach Backwards, Run for Queen of Holland," *TV Guide* (May 31, 1980): 15–18; Susan Beonarcyzk, "Mid-West Cablers Accent on Access," *Variety* (March 17, 1981): 28; "Cable Interconnects: Making Big Ones Out of Little Ones," *Broadcasting* (March 1, 1982): 59; Ann M. Morrison, "Part-Time Stars of Cable TV," *Fortune* (November 30, 1981): 181–84; and the entire July, 1981, issue of *Community Television Review* published by the National Federation of Local Cable Programmers.

40. "Week One for Warner's Qube," *Broadcasting* (December 12, 1981): 62; and "Second Qube Facility Launched by Warner-Amex Near Cincinnati," *TVC* (October 15, 1980): 58–59.

41. "The Two-Way Tube," *Newsweek* (July 3, 1978): 64.

42. "Warner Amex Launches New Retrieval Service in Columbus," *TVC* (February 1, 1981): 68–69.

43. "Valley Cable Seeks Teen Aud with PPV Shows," *Variety* (May 5, 1983): 1.

44. "Home Security Is a Cable TV, Industry Bets," *The Wall Street Journal* (September 15, 1981): 25.

45. "Pay Cable TV Is Losing Some of Its Sizzle as Viewer Resistance, Disconnects Rise," *The Wall Street Journal* (November 19, 1982): 22.

46. "Cable's Lost Promise," *Newsweek* (October 15, 1984): 103–5.

47. "Warner Amex's Lowering Expectations," *Broadcasting* (March 19, 1984): 37; and "Buyers Study How to Divide Group W Cable," *Electronic Media* (January 6, 1986): 3

48. "Viacom's Rise to Stardom," *Newsweek* (November 25, 1985): 71.

49. "HBO Takes Best Seat in the House while Rivals Still Wait in Line," *Los Angeles Times* (June 5, 1983): V, 1.

50. "Basic Cable Networks," *Channels* (Field Guide, 1987): 73–74; "Pay-Cable Networks," *Channels* (Field Guide, 1987): 81; "RCA Venture for Cable-TV Programs to End," *The Wall Street Journal* (February 23, 1983): 8; and "Spotlight Pay-TV Venture Is Seen Ending with Subscribers Being Shifted to Rivals," *The Wall Street Journal* (September 2, 1983): 9.

51. Mark Frankel, "Can Playboy Save Its Skin?" *Channels* (November 1986): 37–40.

52. "Programming Setback for HBO," *Variety* (December 19, 1983): 1; and "HBO Signs $500 Million Film Deal with Paramount," *Broadcasting* (June 30, 1987): 21.

53. "CBS to Drop Its Cultural Cable-TV Service After Failing to Draw Needed Ad Support," *The Wall Street Journal* (September 14, 1983): 7.

54. "Turner the Victor in Cable News Battle," *Broadcasting* (October 17, 1983): 27.

55. "Viacom Agrees to Buy MTV Networks," *Variety* (August 27, 1985): 1; and "Untimely End for Odyssey," *Broadcasting* (January 6, 1986): 160.

56. "Buyer Sought to Acquire Time's USA Network Interest," *Variety* (April 24, 1987): 1; "ABC Video Ent. Acquires ESPN," *Variety* (May 1, 1984): 1; "Daytime + CHN: Joining Together to Stay Afloat," *Broadcasting* (June 20, 1983): 37; "The Reach of MSOs," *Channels* (May 1987): 49; and "Sorting through the Fallout of Cable Programming," *Broadcasting* (October 17, 1983): 29.

57. "Discovery Channel Sets Sail," *Broadcasting* (June 24, 1985): 53.

58. Gregory Kerr, "Lights, Camera. . .Congress!" *Emmy* (May/June 1987): 41–47.

59. "Nebraska Numbers," *Broadcasting* (October 8, 1984): 20.

60. "Most Will Scramble by 1987," *Electronic Media* (March 17, 1986): C–3; and *The Scramble to Scramble: A Satellite Television Dilemma* (Washington, D.C.: Television Digest, 1986).

61. "City Council Votes to Maintain Status Quo on Cable TV Access," *Variety* (April 29, 1987): 1.

62. "Warner Amex Cable Cuts Interactive Programming Feed in Six Major Cities," *Variety* (January 19, 1984): 1.

63. "Home Is Where The Mart Is," *Channels* (Field Guide, 1987): 77–78; and "Cable Value Network and COMB Restructuring," *Broadcasting* (March 30, 1987): 163.

64. Patricia E. Bauer, "Young and Impulsive," *Channels* (May 1987): 50–51; "Hollywood Tries for a New Hit," *Newsweek* (April 8, 1985): 56–58; "Playboy Ent. Joins Growing Ranks of Pay-per-view Hopefuls," *Variety* (June 5, 1985): 1; and "PPV Service Changes Name: Ready to Go," *Electronic Media* (November 4, 1985): 30.

65. "Two Major Restraints on Cable Television Are Lifted by the FCC," *The Wall Street Journal* (July 23, 1980): 1.

66. "Appeals Court Upholds FCC Repeal of Distant-Signal, Exclusivity Rules," *Broadcasting* (June 22, 1981): 32.

67. "Free at Last: Cable Gets Its Bill," *Broadcasting* (October 15, 1984): 38; and "Cable TV Gets a Deregulation Gift From FCC," *Variety* (April 12, 1985): 1

68. "Court Frees Cable TV Firms of Obligation to Carry Local Channels," *Los Angeles Times* (July 20, 1985): IV–1; "Few Surprises in FCC's Must-Carry Order," *Broadcasting* (December 1, 1986): 43; and "Must Carry Back on Track," *Broadcasting* (March 30, 1987): 55.

69. "Black Tuesday Descends on Cable Industry," *Broadcasting* (May 21, 1983): 61.

70. "Court Tells Cable to up Copyright Ante," *Broadcasting* (January 11, 1988): 40.

71. Thomas W. Hazlett, "The Policy of Exclusive Franchising in Cable Television," *Journal of Broadcasting and Electronic Media* (Winter 1987): 1–20.

72. "Supreme Court Upholds Pole-Rate Regulations," *Variety* (February 26, 1987): 1.

73. "NCTA Study Shows that Post-Dereg Rates up Average 6.7 Percent," *Broadcasting* (November 30, 1987): 86.

74. "Cable-TV Systems, Since Deregulation, Are Selling at Increasingly Higher Prices," *The Wall Street Journal* (May 19, 1987): 12.

75. "Coveting Thy Neighbor's System" *Channels* (Field Guide, 1988): 51.

6

1. Several books that detail newer media are: Lynne Schafer Gross, *The New Television Technologies* (Dubuque, Iowa: Wm. C. Brown Publishers, 1986); Loy A. Singleton, *Telecommunications in the Information Age* (Cambridge, Ma.: Ballinger Publishing Company, 1986); George E. Whitehouse, *Understanding the New Technologies of the Mass Media* (Englewood Cliffs, New Jersey: Prentice-Hall, 1986); and Frederick Williams, *Technology and Communication Behavior* (Belmont, Ca.: Wadsworth Publishing Company, 1987).

2. "First Feevee," *Variety* (December 15, 1986): 95.

3. "After Six Trips to the Firing Line, Satellites Finally Put Pay-Television on the Map," *Variety* (December 9, 1980): 1.

4. David H. Ostroff, "A History of STV, Inc. and the California Vote Against Pay Television," *Journal of Broadcasting* (Fall 1983): 371–386.

5. "STV: Scratching Out Its Place in the New-Video Universe," *Broadcasting* (April 7, 1980): 46.

6. *Broadcasting/Cablecasting Yearbook, 1987* (Washington, D.C.: Broadcasting Publications, Inc., 1987): A–7.

7. "No Holds Barred for STV," *Broadcasting* (June 21, 1982): 23.

8. "Hanging the Crepe for STV," *Broadcasting* (October 15, 1984): 46.

9. "SelecTV Finalizes Its Deal to Buy Up Oak's ON TV Subs," *Variety* (February 5, 1985): 1.

10. "Telstar Corp Acquires SelecTV," *Variety* (January 5, 1987): 3.

11. Leonard Shyles, "The Video Tape Recorder: Crown Prince of Home Video Devices," *Feedback* (Winter 1981): 1–5.

12. Bruce Cook, "High Tech: The New Videocassettes," *Emmy* (Summer 1980): 40–44.

13. David Lachenbruch, "Video-cassette Recorders—Here Comes the Second Generation," *TV Guide* (October 28, 1978): 2–6.

14. Ivan Berger, "Videocassette Recorders: Rising Stars of Home Entertainment," *Popular Electronics* (June 1980): 54.

15. "RCA Predicting $1 Bil Retail Year for the VCR Industry," *Variety* (May 13, 1981): 1.

16. Cook, 40–44.

17. "Betamax Decision Overturned," *Variety* (October 20, 1981): 1.

18. "Hollywood Loses to Betamax," *Variety* (January 18, 1984): 1.

19. "3d-Quarter VCR U.S. Population Pegged at 13 Mil," *Variety* (October 25, 1984): 1.

20. "Sony Adds VHS to VCR Format," *Variety* (January 12, 1988): 1.

21. "Kodak Plans to Unveil Its New Qtr-Inch VCR," *Variety* (December 22, 1983): 1.

22. "Camcorders Zoom into Video Products Picture," *Los Angeles Times* (May 18, 1987): IV-1.

23. "Movie Studios Put More Emphasis on Home-Video Pay-TV Markets," *The Wall Street Journal* (May 1, 1984): 33.

24. "Getting a Fix on How They're Using All Those VCR's," *Broadcasting* (May 7, 1984): 74.

25. "Off-Air Recording Is the Dominant VCR Use; Tape Rentals Grow, Buying Last," *Variety* (August 28, 1984): 1.

26. " '83 Vidpiracy Nears $1 Bil," *Variety* (December 21, 1983): 1.

27. "Video Disks," *Broadcasting* (February 2, 1981): 35.

28. Neil Hickey, "Take the Videodisc, Please," *TV Guide* (December 26, 1981): 12–14.

29. "RCA Reports Big Success for Its Vidisk Launch," *Variety* (May 1, 1981): 1.

30. "RCA Gives Up on Videodisc System," *Los Angeles Times* (April 5, 1984): IV, 1.

31. "Laserdisk Players Expand Capabilities," *Variety* (December 27, 1983): 7.

32. "Videodisks Make a Comeback as Instructors and Sales Tools," *The Wall Street Journal* (February 15, 1985): 25.

33. CD-V Format Given Boost From Music, Homevid Firms," *Variety* (April 24, 1987): 8.

34. "MDS/Wireless Cable—Boon or Bane?" *TVC* (December 15, 1980): 195–201.

35. "FCC Asked to Open Up MDS Band as Challenge to Cable," *Broadcasting* (February 15, 1982): 31.

36. "FCC Reassigns Eight ITFS Channels," *E-ITV* (July 1983): 6.

37. "FCC Gets First MMDS Deluge," *Broadcasting* (October 31, 1983): 55.

38. "Now That We've Won It, What Do We Do With It? *Broadcasting* (October 7, 1985): 47

39. "Movement Afoot on MCTV Front," *Broadcasting* (January 2, 1984): 40.

40. "The Quick Fix That Came Late," *Channels* (Field Guide, 1987): 57.

41. "SMATV: The Medium That's Making Cable Nervous," Broadcasting (June 21, 1982): 33–43.

42. "SMATV Free From All But Federal Regs," *Variety* (November 10, 1983): 1.

43. "SMATV," *Channels* (Field Guide, 1987): 70.

44. "Backyard Satellite Dishes Spread But Stir Fight With Pay-TV Firms," *The Wall Street Journal* (April 2, 1982): 25.

45. "More Shots Fired at Comsat's DBS," *Broadcasting* (March 30, 1981): 54.

46. "Lofty Bid for First DBS System," *Broadcasting* (December 22, 1980): 23.

47. "Good News, Bad News in DBS Spacerush," *Broadcasting* (July 20, 1981): 23.

48. "CBS-DBS-HDTV: Putting Them All Together for the FCC," *Broadcasting* (July 13, 1981): 23.

49. *Enter DBS: The Story of Direct Broadcast Development in 1984* (Washington, D.C.: Television Digest, Inc., 1985).

50. "Public Has an Appetite for DBS," *Variety* (April 27, 1983): 1.

51. Michael Pollan, "How the DBS Kids Stole Comsat's Thunder," *Channels* (July/August 1983): 39–41.

52. "DBS Makes Its Long-Awaited Bow Today," *Variety* (November 15, 1983): 10.

53. "Slow Liftoff for Satellite-To-Home TV," *Fortune* (March 5, 1984): 100.

54. "Thinning Rank of DBS Pioneers Heads for July 17," *Broadcasting* (July 16, 1984): 30.

55. "Another Nail in the DBS Coffin: Comsat Bows Out," *Broadcasting* (December 3, 1984): 36.

56. "Home Is Where the Dish Is," *Broadcasting* (September 10, 1984): 92.

57. "S/TMC Offers Sat'Lite Service Package to Home Dish Owners," *Variety* (March 8, 1987): 35.

58. "TCI Company Beaming to Backyards," *Broadcasting* (October 5, 1987): 38.

59. "LPTV," *Broadcasting* (February 23, 1981): 39–66.

60. "FCC Swamped with Applications for New Low-Power TV Stations," *The Wall Street Journal* (October 30, 1981): 29.

61. "LPTV Gets the FCC Go Ahead," *Broadcasting* (March 8, 1982): 35.

62. "Where Things Stand," *Broadcasting* (April 6, 1987): 40.

63. "They Jes' Keep a-Growin'," *Channels* (Field Guide, 1987): 54.

64. "Newest TV Stations Are Low in Power and High in Color," *The Wall Street Journal* (October 23, 1984): 1; and "Low-Power TV: It Plugs the Gaps," *Los Angeles Times* (December 8, 1984): I, 1.

65. Joseph Roizen, "Teletext—A Service That's Coming of Age," *E/ITV* (September 1982): 39–44.

66. Antone F. Alber, *Videotex/Teletext: Principles and Practices* (New York: McGraw Hill, 1985).

67. Cheryl Rhodes, "The Growth of Videotex in Britain and France," *DataCast* (January/February 1982): 23–35.

68. "KNXT and KCET Begin First L.A. Test of Teletext," *Variety* (April 9, 1981): 1.

69. "Teletext/Videotext Out of Cake in Toronto," *Broadcasting* (May 25, 1981): 26.

70. "The Whys and Wherefores of Text Transmission," *Broadcasting* (July 5, 1982): 32–35.

71. "The British Are Coming (With a Teletext Standard)," *Broadcasting* (May 20, 1981): 28.

72. "Joint Venture Will Provide Teletext Service during Summer Olympics," *Broadcasting* (January 30, 1984): 72–73.

73. Interview with Victoria Borne, Assistant Editor, KSL, Salt Lake City, June 1, 1987.

74. Interview with Hillary Goodall, Managing Editor, Taft Broadcasting, May 29, 1987.

75. Some of the books that give history and major developments of the telephone are: J. Brooks, *Telephone: The First Hundred Years* (New York: Harper and Row, 1975); *Events in Telecommunication History* (New York: AT&T, 1979); John R. Pierce, *Signals: The Telephone and Beyond* (San Francisco: Freeman, 1981); and Gerald W. Brock, *The Telecommunications Industry: The Dynamics of Market Structure* (Cambridge, MA: Harvard University Press, 1981).

76. Some sources on computers are: Gary B. Shelly and Thomas J. Cashman, *Introduction to Computers and Data Processing* (Brea, Ca.: Anaheim Publishing Company, 1980); James W. Morrison, *Principles of Data Processing* (New York: Arco Publishing Company, 1979): and

Fred D'Ignazio, *Messner's Introduction of the Computer* (New York: Julian Messner, 1983).

77. "Ma Bell's Big Breakup," *Newsweek* (January 18, 1982): 58-59.

78. Robert Britt Horwitz, "For Whom the Bell Tolls: Causes and Consequences of the AT&T Divestiture," *Critical Studies in Mass Communication* (June 1986): 119–153.

79. "Teletext Service Launched by Group W Cable, *Broadcasting* (April 30, 1984): 150.

80. "High Rollers with High Hopes," *Channels* (Field Guide, 1987): 86–87.

81. "Is ISDN Worth Waiting For?" *Business Communications Review* (March/April 1986): 13–17.

7

1. Martin Clifford, *Microphones* (Blue Ridge Summit, Pa.: TAB Books, 1986); and Alec Nisbett, *The Use of Microphones* (Boston: Focal Press, 1983).

2. "The LP's Wobbly Future," *Newsweek* (February 9, 1987): 52.

3. Glyn Allin, *Sound Recording and Reproduction* (Woburn, Ma.: Focal Press, 1981); and Bill Tullis, "Applying DAT in Television," *Television Broadcast* (November 1987): 70–73.

4. Stanley R. Alten, *Audio in Media* (Belmont, Ca.: Wadsworth Publishing Company, 1981).

5. Alec Nisbett, *The Technique of the Sound Studio* (Woburn, Ma.: Focal Press, 1979).

6. Glen Pensinger, "Hello, Mr. Chips," *Television Broadcast* (January 1987): 62–67.

7. Jack Mathews, "Enter the Age of Video-Made Movies," *Los Angeles Times* (June 1, 1987): IV, 1.

8. "High Wire Video," *Broadcasting* (December 31, 1984): 108.

9. David Cheshire, *The Book of Movie Photography* (New York: Alfred A. Knopf, 1984).

10. Alan Wurtzel, *Television Production* (New York: McGraw Hill, 1979), 336–340.

11. "Coloring the Black and White Past," *Broadcasting* (May 24, 1986): 92.

12. Neal Weinstock, "CGI: An Introduction to the Technology," *E/ITV* (April 1987): 12–17.

13. "Digital Effects Update," *Television Broadcast* (February 1987): 42–45.

14. "Who Makes It," *Television Broadcast* (August 1986): 90.

15. Thomas D. Burrows and Donald N. Wood, *Television Production* (Dubuque, Iowa: Wm. C. Brown Company, 1986), 179–199.

16. "Half-inch is New Measure of TV News," *Broadcasting* (May 25, 1987): 65–66.

17. "State of the Art," *Broadcasting* (October 8, 1984): 72–74.

18. Herbert Zettl, *Television Production Handbook* (Belmont, Ca.: Wadsworth Publishing Company, 1984), 307–343.

19. Ibid., 20–28.

20. "Small World," *Broadcasting* (April 20, 1987): 7.

8

1. Joseph Kerman, *Listen* (New York: Worth Publishers, Inc., 1972), 13.

2. David Lachenbruch, "A Field Guide to the Airwaves," *Channels* (November/December 1983): 48–49.

3. Milton Kiver, *FM Simplified* (Princeton, N.J.: Van Nostrand, 1960).

4. Stan Prentiss, "AM Stereo Wins FCC Approval," *Popular Electronics* (August 1980): 59.

5. "Lid's Off FM SCA's," *Broadcasting* (April 11, 1983): 35.

6. *Broadcasting/Cablecasting Yearbook, 1987* (Washington, D.C.: Broadcasting Publications Inc., 1987): A-4.

7. Interview with Kevin Gross, electronic technician, Bevens Systems, January 8, 1982.

8. *Broadcasting/Cablecasting Yearbook, 1987*, A-5.

9. K. R. Sturley, *Radio Receiver Design* (New York: Barnes and Noble, 1965).

10. Eugene David, *Television and How It Works* (Englewood Cliffs, N.J.: Prentice-Hall, 1962): 48–53.

11. "CBS Breakthrough on HDTV Compatibility," *Broadcasting* (September 26, 1983): 77.

12. "High-Definition TV: So Close and Yet So Far Away," *Broadcasting* (April 7, 1986): 134.

13. Glen Pessinger, "A Modest Proposal," *Television Broadcast* (December 1987): 50–54; "Broadcasters Ask for HDTV Inquiry," *Broadcasting* (February 23, 1987): 46; and "Terrestial HDTV Broadcasting Debuts in Washington," *Broadcasting* (January 5, 1987): 214.

14. Stan Prentice, *Television: From Analog to Digital* (Blue Ridge Summit, Pa.: TAB Books, 1985).

15. Herbert Shuldiner, "Flat Screen Color TV," *Popular Science* (November 1983): 100–102; "Improvements in LCDs," *E/ITV* (October 1985): 66–67; and David Lachenbruch, "TV, TV On the Wall, Will We Ever See You At All," *TV Guide* (July 27, 1985): 30–33.

16. "3-D Takes Aim at the Videocassette Homefront," *Variety* (December 31, 1981): 1; and David Lachenbruch, "Here Come Deepies—Maybe," *TV Guide* (March 14, 1981): 22–26.

17. David Lachenbruch, "Giant TVs that Pop Up, Tiny TVs You Can Talk To," *TV Guide* (September 5, 1981): 38–41; and Gary H. Arlen, *Tomorrow's TVs* (Washington, D.C.: National Association of Broadcasters, 1987).

18. For some of the technical challenges, see Donald G. Monteith and Nicholas F. Hamilton-Piercy, "Building Plant to 499 MHz and Beyond," *TVC* (December 15, 1980): 132–50; John P. Taylor, "Not Enough Channel Capacity? 'Supercable' to the Rescue," *Cable Age* (May 18, 1981): 21–32; and David L. Willis, "The System Rebuild," *TVC* (December 15, 1980): 154–55.

19. William B. Johnson, "The Coming Glut of Phone Lines," *Fortune* (January 7, 1985): 96–99

20. Christopher Podmore and Denise Faguy, "The Challenge of Optical Fibers," *Telecommunications Policy* (December 1986): 341–351.

21. "Intercity Transmission Capacity," *Communications Week* (February 11, 1985): C-1.

22. "Trans-Atlantic Fiber Optic Link Proposed," *Broadcasting* (October 8, 1984): 89–90.

23. Jay C. Lowndes," "Optical Fiber Hardware Threatens Telecommunications Market Share," *Aviation Week and Space Technology* (October 20, 1986): 115.

24. Chris Bowick and Tim Kearney, *Introduction of Satellite TV* (Indianapolis, In.: Howard W. Sams and Co., 1983).

25. "Communication Satellites: The Birds Are in Full Flight," *Broadcasting* (November 19, 1979): 36–47.

26. "Today and Tomorrow in Domestic Communications Satellites," *Broadcasting* (May 19, 1980): 89.

27. "The Bird Business," *Channels* (Field Guide, 1987): 65.

28. "Communications Satellites," 36–47.

29. William D. Houser, "Satellite Interconnection," *Television Quarterly* (Fall 1976): 78–80.

30. Robert N. Wold, "Satellite Distribution Is No Pie in the Sky," *Broadcast Communications* (December 1979): 49.

31. "Demand Exceeds Supply of Westar Transponders," *Broadcasting* (February 4, 1980): 23.

32. "Keeping Up with the Satellite Universe," *Broadcasting* (August 17, 1981): 32.

33. "Satellite Delivery Set for Two More Group W Programs," *Variety* (October 30, 1981): 1.

34. "Radio Joins Television in Move to Sat'Lite Transmission," *Variety* (April 13, 1981): 1.

35. "ABC Targets Switchover to Satellite Delivery," *Broadcasting* (April 27, 1981): 94.

36. "State of the Art Journalism" *Broadcasting* (December 3, 1984): 47.

37. "ABC Makes It a Full House on Galaxy 1," *Broadcasting* (November 30, 1981): 34.

38. "Costs Come Home to Roost For the Birds," *Channels* (Field Guide, 1987): 64–65.

9

1. *A Short History of Radio Music Licenses* (Washington, D.C.: All-Industry Radio Music Licensing Committee, 1987).

2. "Supreme Court OK's Blanket Music Licenses," *Broadcasting* (February 25, 1985): 38–39.

3. "Where Things Stand," *Broadcasting* (April 6, 1987): 80.

4. "First Low-Power Station Licenses Arriving at BMI," *Variety* (July 25, 1986): 7.

5. "Pornographic Rock Lyrics Issue Gets Airing at Radio Convention as PMRC Calls for Record Warning Labels," *Broadcasting* (September 6, 1985): 38.

6. "Supreme Court Ruling on Programming Formats Big Victory for Broadcasters," *Variety* (March 25, 1981): 1.

7. "The Beat Goes on TV: Music Videos, Sign-On to Sign-Off," *Broadcasting* (June 25, 1984): 50–52.

8. "MTV Changing as Video Novelty Wears Off," *Electronic Media* (June 30, 1986): 26.

9. Specific references to drama include: Linda S. Lichter and S. Robert Lichter, *Prime Time Crime: Criminals and Law Enforcers in TV Entertainment* (Washington, D.C.: Media Institute, 1983); Richard Meyers, *TV Detectives* (San Diego, Ca.: A. S. Barnes and Company, 1981); Richard Levinson and William Link, *Stay Tuned...An Inside Look at the Making of Prime-Time Television* (New York: St. Martin's Press, 1981); Dave Kaufman, "10 Best Programs in Television History," *Variety* (43rd Anniversary Issue): 148–85; Saul N. Scher, "Anthology Drama: TV's Inconsistent Art Form," *Television Quarterly* (Winter 1976–77): 29–34; Tim Brooks and Earle Marsh, *The Complete Directory to Prime Time Network TV Shows: 1946–Present* (New York: Ballantine Books, 1985); Louis Solomon, *The TV Doctors* (New York: Scholastic Book Services, 1974); and David Bianculli, "The Roots of the Problem," *Channels* (October 1986): 42.

10. "Next in Line for Moral Scrutiny: Cable Programming," *Broadcasting* (August 31, 1981): 40.

11. Eli A. Rubinstein, "The TV Violence Report: What Next?" *Journal of Communication* (Winter 1974): 34–40.

12. "TV on Trial," *Broadcasting* (November 5, 1984): 35.

13. "The New Right's TV Hit List," *Newsweek* (June 15, 1981): 101.

14. "Simon Introduces TV Violence Bill," *Broadcasting* (March 30, 1987): 148.

15. "More Violence than Ever Says Gerbner's Latest," *Broadcasting* (February 28, 1977): 20; and "Schneider Attacks Gerbner's Report on TV Violence," *Broadcasting* (May 2, 1977): 57.

16. Max Gunther, "All That TV Violence: Why Do We Love/Hate It?" *TV Guide* (November 6, 1976): 6–10.

17. Studies on violence include: Seymour Feshbach and Robert D. Singer, *Television and Aggression: An Experimental Field Study* (San Francisco: Jossey-Bass, 1970); Melvin S. Heller and Samuel Polsky, *Studies in Violence and Television* (New York: American Broadcasting Company, 1976); Murray Feingold and G. Timothy Johnson, "Television Violence Reactions from Physicians, Advertisers, and the Network," *The New England Journal of Medicine* (February 24, 1977): special article; David P. Phillips and John Eittensley, "When Violence is Rewarded or Punished: The Impact of Mass Media Stories on Homicide," *Journal of Communication* (Summer 1984): 101–116; and George Gerbner, "Science or Ritual Dance? A Revisionist View of Television Violence Effects Research," *Journal of Communication* (Summer 1984): 164–173.

18. Other material dealing with violence includes "A Blizzard of Paper on Violence," *Broadcasting* (May 16, 1977): 22–24; "PTA Ends Hearings on TV Violence but Issue Lingers," *Broadcasting* (February 28, 1977): 21; Harriet Steinberg, "Must Night Fall So Hard?" *Television Quarterly* (Winter 1976–77): 61–64; and "Top Prime-Time Advertisers Reject AMA's TV Violence Rx: Baretta Takes Gloves Off," *Variety* (March 4, 1977): 1.

19. For additional information on situation comedies, see Kaufman, "10 Best Programs," 148–85; David Handler, "Which Old Favorites Are Still Choice and Which Aren't," *TV Guide* (April 4, 1987): 18–23; Larry Wilde, "The Genesis of Comedy," *Television Quarterly* (Winter 1976–1977): 70; Alan Alda, "My Favorite Episodes," *TV Guide* (February 12, 1983): 6–21; and Richard Adler, ed. *All in the Family: A Critical Appraisal* (New York: Praeger, 1979).

20. For additional information on variety shows, see Arthur Schulman and Roger Youman *How Sweet It Was: Television, a Pictorial Commentary* (New York: Bonanza Books, 1966); Ted Sennett, *Your Show of Shows* (New York: Collier Books, 1977); and "ABC, Parton Try to Add Spice of Life to Defunct Variety Format," *Variety* (November 19, 1987): 16.

21. For additional information on specials, see "Stay Tuned for Ten Great TV Events," *Saturday Evening Post* (September 1976): 39; and "Silverman Has a Surprise for the Competition," *Broadcasting* (December 5, 1977): 24–25.

22. For additional information on movies, see Richard Zacks, "Picture Window," *Channels* (May 1986): 40–41; Neil M. Malamuth and Victoria Billings, "Why Pornography? Models of Functions and Effects," *Journal of Communication* (Summer 1984): 117–129; "Coloring the Black and White Past," *Broadcasting* (May 24, 1986): 92; Tom Allen, "TV: Father of the Film," *America* (October 30, 1976):286–288; Saul Kaufman, "Films: TV Versions of Moving Pictures," *New Republic* (August 30, 1975): 20; "TV Movies Exploiting the Drama of Real Life," *The Wall Street Journal* (July 30, 1985): 22; Kenneth Turan, "Two Porcupines About to Embrace," *TV Guide* (July 14, 1984): 37–40.

23. For additional information of talk shows, see Murray B. Levin, *Talk Radio and the American Dream* (Lexington, Ma: Lexington Books, 1987); Dwight Whitney, "Is the Talk Show an Endangered Species," *TV Guide* (July 30, 1977): 2–6; Michael J. Arlen, "Hosts and Guests: Hospitality on Talk Shows," *New Yorker* (January 3, 1977): 62–65; and Glenn Esterly, "Talk Show Hosts: Who Are the Best?" *TV Guide* (December 8, 1979): 26–32.

24. For additional information on game shows, see Arleen Francis, "I Was There from First to Last: What's My Line," *Saturday Evening Post* (September 1976): 43; Dick Russell, "The Return of the TV Outcast," *TV Guide* (December 1, 1979): 21–24; and "What a Deal," *Newsweek* (February 9, 1987): 62–68.

25. For additional information on soap operas, see Ien Ang, *Watching Dallas: Soap Opera and Melodramatic Imagination* (New York: Methuen, 1985); Charles Derry, "Television Soap Opera: Incest, Bigamy, and Fatal Disease," *Journal of the University Film and Video Association* (Winter 1983): 4–16; Ellen Seiter, "Men, Sex, and Money in Recent Family Melodramas," *Journal of the University Film and Video Association* (Winter 1983): 17–27; Bradley S. Greenberg and Dare D'Allessio, "Quantity and Quality of Sex in the Soaps," *Journal of Broadcasting and Electronic Media* (Summer 1985): 309–321; Rhoda Estep and Patrick T. Macdonald, "Crime in the Afternoon: Murder and Robbery in Soap Operas," *Journal of Broadcasting and Electronic Media* (Summer 1985): 323–331; and Michael James Intintoli, *Taking Soaps Seriously: The World of Guiding Light* (New York: Praeger, 1984).

26. Elizabeth M. Perse, "Soap Opera Viewing Patterns of College Students and Cultivation," *Journal of Broadcasting and Electronic Media* (Spring 1986): 175–193; and Lenore Silvian, "Spinoffs from Soapland," *Television Quarterly* (Winter 1976–77): 37–41.

27. George W. Woolery, *Children's Television: The First Thirty-Five Years, 1946–1981. Part I: Animated Cartoon Series* (Metuchen, New Jersey: Scarecrow Press, 1983); George C. Woolery, *Children's Television: The First Thirty-Five Years, 1946–1981. Part II: Live, Film and Tape Series* (Metuchen, New Jersey: Scarecrow Press, 1985); and "TV Crowds Children Out of Daily Schedule," *USA Today* (July 16, 1984): 11A.

28. George Comstock, et al. *Television and Social Behavior: A Technical Report to the Surgeon General's Scientific Advisory Committee on Television and Social Behavior* (Washington, D.C.: Government Printing Office, 1972).

29. Some of the studies concerning children and TV are: John D. Abel and Maureen E. Benison, "Perception on TV Program Violence by Children and Mothers," *Journal of Broadcasting* (Spring 1976): 335; Gerald S. Lesser, *Children and Television: Lessons from Sesame Street* (New York: Random House, 1974); *Learning While They Laugh: Studies of Five Children's Programs on the CBS Television Network* (New York: Columbia Broadcasting System, 1977); Mark M. Miller and Byron Reeves, "Dramatic Content and Children's Sex-Role Stereotypes," *Journal of Broadcasting* (Winter 1976): 35–50; and Lynne Schafer Gross and R. Patricia Walsh, "Factors Affecting Parental Control Over Children's Television Viewing: A Pilot Study," *Journal of Broadcasting* (Summer 1980): 411–419.

30. "Kidvid Program Standards Get Brush-Off from Mark Fowler," *Variety* (December 28, 1983): 2.

31. Jan Cherubin, "Toys Are Programs, Too," *Channels* (May/June 1984): 31–33.

32. "Kidvid: A National Disgrace," *Newsweek* (October 17, 1983): 81–83.

33. "Court Tosses Out FCC's '84 Decision Deregulating Kidvid Commercialization," *Variety* (June 29, 1987): 1; and Jeffrey Colvin, "Children Are Getting Hard to Find," *Fortune* (May 2, 1983): 125.

34. Other material dealing with children's television includes: John P. Murray and Gavriel Salomon, eds., *The Future of Children's Television: Results of the Markel Foundation/Boys Town Conference* (Boys Town, Nebraska: Boys Town Center, 1984); Aimee Door, *Television and Children: A Special Medium for a Special Audience* (Beverly Hills, Ca.: Sage, 1986); James D. Culley, William Lazer, and Charles K. Atkin, "The Experts Look at Children's Television," *Journal of Broadcasting* (Winter 1976): 3–22; Hilde Himmelweit, *Television and the Child* (New York: Television Information Office, 1961); Robert M. Liebert, *The Early Window* (New York: Pergamon Press, 1973); William Melody, *Children's Television: The Economics of Exploitation* (New Haven, Connecticut: Yale University Press, 1975); Fred Rogers, *Mr. Rogers Talks About* . . . (Bronx, N.Y.: Platt and Munk, 1974); Marie Winn, *The Plug-In Drug* (New York: Viking, 1975); Walter Karp, "Where the Do-Gooders Went Wrong," *Channels* (March/April 1984): 41–51; and Bruno Bettelheim, "A Child's Garden of Fantasy," *Channels* (September/October 1985): 54–56.

10

1. "Independent News Agencies Making a Name for Themselves," *Broadcasting* (August 31, 1987): 98–100.

2. "The Race for Network News Laurels," *Broadcasting* (August 31, 1987): 50.

3. "Local Stations: Survival of the Fittest," *Broadcasting* (August 31, 1987): 41–48.

4. Kim Standish, "Satellite News Gathering: The Next Round," *RTNDA Communicator* (September 1986): 37; and Phillip Keirstead, "A Sea of New Goodies for Television News Production," *Television Broadcast* (June 1986): 30.

5. "CBS News Buys Electronic Newsroom System," *Broadcasting* (January 21, 1985): 92.

6. Eric Levin, "How the Networks Decide What's News," *TV Guide* (July 2, 1977): 6–11; and Dennis McDougal, "Broadcast News, L.A.," *Los Angeles Times Magazine* (January 24, 1988): 7–17.

7. Barbara Matusow, *The Evening Stars: The Making of the Network News Anchor* (New York: Ballantine Books, 1984).

8. For more information about controversies and news, see: Dan Nimmo and James E. Combs, *Nightly Horrors: Crisis Coverage by Television Network News* (Knoxville, Tennessee: University of Tennessee Press, 1985); Susanna Barber, *News Cameras in the Courtroom: A Free Press-Fair Trial Debate* (Norwood, New Jersey: Ablex, 1986); "How Much SNG is Too Much?" *Broadcasting* (April 21, 1986): 76; "Rethinking TV News in the Age of Limits," *Newsweek* (March 16, 1987): 79–80; and "TV Journalists Debate Their 'Art'," *Broadcasting* (June 23, 1986): 53.

9. For more information about documentary programs, see A. William Bluem, *Documentary in American Television* (New York: Hastings House, 1965); "Westmoreland Lawyer Calls CBS Vietnam Documentary 'Powerful Work of Fiction'," *Variety* (October 12, 1984): 1; Edwin Diamond, "The Myth of the Dying Documentary," *TV Guide* (April 25, 1981): 8–13; Mike Wolverton, *Reality on Reels: How to Make Documentaries for Video/Radio/ Film* (Houston, Texas: Gulf Publishing Company, 1983); Axel Madsen, *Sixty Minutes: The Power and the Politics of America's Most Popular TV News Show* (New York: Dodd, Mead, and Company, 1984); and Charles Montgomery Hammond, Jr., *The Image Decade: Television Documentary 1965–1975* (New York: Hastings House, 1981).

10. For more information on editorials, see L. S. Feuer, "Why Not a Commentary on Sevareid?" *New Republic* (August 15, 1975): 874–76; and Karl E. Meyer, "Uncommon Commentary of ABC's Howard K. Smith," *Saturday Review* (December 11, 1976). 12.

11. William Taaffee, "TV to Sports: The Bucks Stop Here," *Sports Illustrated* (February 27, 1986): 25.

12. "Emerging from Slump, TV Sports Regains Favor Among Advertisers," *The Wall Street Journal* (August 26, 1987): 23.

13. Stanley Frank, "What TV Has Done to Sports," *TV Guide* (February 4, 1967): 4–8; and Don Kowet, "For Better or for Worse," *TV Guide* (July 1, 1978): 2–6.

14. Melissa Ludtke, "Big Scorers in the Ad Game," *Sports Illustrated* (November 7, 1977): 50.

15. "Struck by Sabres, WKBW-TV Blames It on Pay Cable Rule," *Broadcasting* (March 24, 1976): 55.

16. "ABC Readies Olympic Game Plan," *Los Angeles Times* (July 23, 1984): VI, 1.

17. For more information on magazine shows, see "Westinghouse Puts Magazine Show on All Its Stations in Access Periods," *Broadcasting* (January 24, 1977): 34; "Group W's Fledgling 'P.M.' Did Well Enough in October to Inspire a New Sales Push," *Variety* (December 6, 1978): 1; Tom Shales, "Networks Love TV Newsmagazines," *Los Angeles Times* (May 31, 1985): VI, 1; "TV's Local Magazines," *Newsweek* (December 31, 1979): 50–51; Nancy R. Casplar, "Local Television: the Limits of Prime-Time Access," *Journal of Communication* (Spring 1983): 124–131; and Michelle R. Serra and Richard A. Kallan, "A Sexual Egalitarianism in TV: An Analysis of 'P.M. Magazine'," *Journalism Quarterly* (Autumn 1983): 56–63.

18. For additional information on educational programming, see "How Television Tries to Close the Health Information Gap," *Today's Health* (January 1976): 30–33; "TV and Political Knowledge for Elementary School Pupils," *Intellect* (January 1976): 284–85; and *Voices and Values* (New York: Television Information Office, 1984): 37–44.

19. For more information on religious programming, see George H. Hill, *Airwaves to the Soul: The Influence and Growth of Religious Broadcasting in America* (Saratoga, CA.: R&E Publishers, 1983); Peter G. Horsfield, *Religious Television: The American Experience* (New York: Longman, 1984); "Getting Time on the Tube," *Christianity Today* (May 7, 1976): 27–28; "If You Can't Beat 'Em. . .," *Newsweek* (February 9, 1981): 101; "TV Preachers to Testify," *Newsweek* (August 31, 1987): 64; Roderick Townley, "It's Pitch and Pray and Wish the Scandal Away," *TV Guide* (August 15, 1987): 5–9; and Frank Razelle, *Televangelism: The Marketing of Popular Religion* (Carbondale, Illinois: Southern Illinois University Press, 1987).

20. For more information on public affairs, see Frank Wolf, *Television Programming for News and Public Affairs: A Quantitative Analysis of Networks and Stations* (New York: Praeger, 1972); Roger Simon, "Those Interview Shows—They're Tougher Now, but Are They Better?" *TV Guide* (March 14, 1987): 4–7; "Public Affairs: A Chance to Serve," *Broadcasting* (November 16, 1987): 117; and Michael Bruce MacKuen and Steven Lane Coombs, *More than News: Media Power in Public Affairs* (Beverly Hills, CA: Sage, 1981).

21. For more information on political broadcasting, see Patricia Sellers, "The Selling of the President in '88," *Fortune* (December 21, 1987): 131–136; Gerald R. Ford, "10 Ways to Improve TV's Campaign Coverage," *TV Guide*

(February 8, 1984): 6–10; "TV Networks to Quit Exit Poll Projections," *Variety* (January 18, 1985): 10; Ronald Brownstein, "You'll Have a Different View of the Campaign," *TV Guide* (October 31, 1987): 5–9; Kathleen Hall Jamieson, *Packaging the Presidency: A History and Criticism of Presidential Campaign Advertising* (New York: Oxford University Press, 1984); Austin Kanney, *Channels of Power: The Impact of Television on American Politics* (New York: Basic Books, 1983); John Tebbel and Sarah Miles Watts, *The Press and the Presidency: From George Washington to Ronald Reagan* (New York: Oxford University Press, 1985); Edie N. Goldenberg and Michael W. Traugott, *Campaigning for Congress* (Washington, D.C.: CQ Press, 1984); and Gladys E. Lang and Kurt Lang, *Politics and Television Re-Viewed* (Beverly Hills, CA: Sage, 1984).

22. For more information on special interest programming, see "Home Is Where the Mart Is," *Channels* (Field Guide, 1987): 77–78; Joanmarie Kalter, "Now You Can See How to Call a Goose, Pan for Gold, or Fix Your Horse's Teeth," *TV Guide* (November 21, 1987): 38–40; and "TV in the Senate: One Year Later," *Broadcasting* (June 1, 1987): 36.

11

1. These are some of the formats listed in *Broadcasting/ Cablecasting Yearbook, 1987* (Washington, D.C.: Broadcasting Publications, Inc., 1987): F–68 F–97.

2. These are some of the types of material listed in *Syndicated Radio Programming Directory* (Washington, D.C.: National Association of Broadcasters, 1986), 10–39.

3. "Radio's Syndication Market Shows Robust Signs," *Broadcasting* (July 27, 1987): 64–74.

4. "Radio Programing Tunes in the Satellite Sound," *Broadcasting* (May 26, 1986): 38.

5. "Radio's New Golden Age," *Newsweek* (August 3, 1987): 41.

6. "Funding, Unbundling Top Public Radio Conference," *Broadcasting* (May 4, 1987): 37.

7. *Independent Thinking* (Washington, D.C.: Frazier, Gross, and Kadlec, Inc., 1986): iii.

8. "Theatricals Find Strength in Ad Hoc Numbers," *Broadcasting* (September 30, 1985): 93.

9. "WOR-TV Bags 'Cosby,'" *Broadcasting* (November 10, 1986): 53.

10. "The World of TV Programming, 1984," *Broadcasting* (October 22, 1984): 54–82.

11. "First-run Programming Fueling Syndication Market," *Broadcasting* (January 19, 1987): 106.

12. "Stations Try to Stay Jump Ahead with Checkerboarding," *Broadcasting* (September 22, 1986): 66.

13. "Syndicators Look to Access," *Broadcasting* (October 12, 1987): 70.

14. Steve Behrens, "Will Temptation Undo the Tie That Binds?" *Channels* (May 1987): 41–43.

15. *Broadcasting/Cablecasting Yearbook, 1987,* A–21–A–23.

16. "TV Networks, Producers Battle Over Fees," *Los Angeles Times* (March 17, 1986): IV, 1.

17. The following is based on the budget worksheets of KABC, Los Angeles, California, and Channel 9, KHJ-TV, Los Angeles, California.

18. *The Public Television Station Program Cooperative* (Alexandria, VA: Public Broadcasting System, 1986), 1–3.

19. "Cable Subscribes to Original Programming," *Broadcasting* (October 12, 1987): 50.

20. "NAPTE Takes a Look at Cable," *Broadcasting* (January 21, 1985): 69

21. "Television's Shifting Balance of Power," *Broadcasting* (October 12, 1987): 40–42.

12

1. Four sources from the FCC that give general information about its organization are: *Federal Communications Commission: Broadcasting and Cable Television* (Washington, D.C.: Federal Communications Commission, 1986); *Federal Communications Commission: Spectrum Management* (Washington, D.C.: Federal Communications Commission, 1987); *Federal Communications Commission: Private Radio Services* (Washington, D.C.: Federal Communications Commission, 1986); *The FCC In Brief* (Washington, D.C.: Federal Communications Commission, 1985).

2. "Worsening Jam in CB Radio Swamps FCC," *Broadcasting* (January 19, 1976): 30.

3. Another Deluge of LPTV Filings Inundates FCC," *Broadcasting* (February 23, 1981): 29; and "FCC Gets First MMDS Deluge," *Broadcasting* (October 31, 1983): 55.

4. The Saga of Call Letters: From KAAA to WZZZ," *Broadcasting* (August 6, 1984): 67.

5. *Assignment and Transfer of Control of Broadcast Licenses Statutory and Regulatory Requirements* (Washington, D.C.: Pierson, Ball & Dowd, 1987).

6. Public Law No. 416, June 19, 1934, 73rd Congress (Washington, D.C.: Government Printing Office, 1934), Section 326.

7. Ibid., Section 307(a).

8. "Lots of Lottery Language," *Broadcasting* (November 22, 1982): 42.

9. "Deregulation Bill Passes Senate," *Broadcasting* (April 5, 1982): 36.

10. "Comparative Renewal," *Broadcasting* (August 24, 1987): 27–33.

11. "Supreme Court Upholds Postcard Renewal," *Broadcasting* (June 25, 1984): 38.

12. Charles E. Clift, III, "Station License Revocations and Denials of Renewal, 1970–1978," *Journal of Broadcasting* (Fall 1980): 411–21.

13. John R. Bittner, *Broadcasting and Telecommunications* (Englewood Cliffs, N.J.: Prentice-Hall, Inc., 1985): 347–50.

14. "FCC Will Dump Trafficking Reg for B'Casters," *Variety* (November 19, 1982): 1.

15. Interview with Dan Gingold, Executive Producer, KNXT-TV, Los Angeles, California, January 11, 1979.

16. *A Guide to the Federal Trade Commission* (Washington, D.C.: The Federal Trade Commission, 1987), 3.

17. The entire booklet listed in footnote 16 is good for an overview of the FTC.

18. James Miller, "The President's Advocate: OTP and Broadcast Issues," *Journal of Broadcasting* (Summer 1982): 625–639.

19. Norman Black, "The Deregulation Revolution," *Channels* (September/October 1984): 52–59.

20. Sydney W. Head and Christopher H. Sterling, *Broadcasting in America* (Boston: Houghton Mifflin Company, 1982), 405–7.

21. Erik Barnouw, *A Tower In Babel: A History of Broadcasting in the United States to 1933* (New York: Oxford University Press, 1966), 120.

22. *The Television Code* (New York: National Association of Broadcasters, 1976); and *The Radio Code* (New York: National Association of Broadcasters, 1976).

23. "Setback for NAB's Ad Code," *Variety* (March 5, 1982): 1.

24. A more complete list of broadcasting associations, their officers, and addresses can be found in *Broadcasting/Cable Yearbook, 1987* (Washington, D.C.: Broadcasting Publications, 1987), I–50–I–62.

25. A more complete list of broadcasting awards can be found in *Broadcasting/Cable Yearbook,* F–35–F–38.

26. This information was gathered from interviews with station and network personnel—primarily James Harden, general manager for radio station KNAC, Long Beach, California, 1976; Jay Strong, program director, KCBS, Los Angeles, California, 1987; and Walt Baker, vice president of programming, KHJ-TV, Los Angeles, California, 1987.

27. Insightful books and articles dealing with the pros and cons of government regulation are Edwin G. Krasnow and Lawrence D. Longley, *The Politics of Broadcast Regulation* (New York: St. Martin's Press, 1972); *The FCC and Broadcasting* (Washington, D.C.: Federal Communications Commission, 1985): Marvin R. Bensman, *Broadcast Regulation: Selected Cases and Decisions* (Lanhan, MD: University Press of America, 1985); John R. Bittner, *Broadcast Law and Regulation* (Englewood Cliffs, NJ: Prentice-Hall, 1982); and Barton T. Carter and others, *The First Amendment and the Fourth Estate: Regulation of Electronic Mass Media* (Mineola, NY: Foundation Press, 1985).

13

1. Three good sources for general information on broadcast regulation are Marvin R. Bensman, *Broadcast Regulation: Selected Cases and Decisions* (Lanham, MD: University Press of America, 1985); John R. Bittner, *Broadcast Law and Regulation* (Englewood Cliffs, NJ: Prentice-Hall, 1982); and Stuart N. Brotman, *Communications Policymaking at the Federal Communications Commission* (Washington, D.C.: The Annenberg Washington Program, 1987).

2. "From Fighting Bob to the Fairness Doctrine," *Broadcasting* (January 5, 1976): 46.

3. Erik Barnouw, *The Golden Web* (New York: Oxford University Press, 1968), 227–36.

4. Don R. Pember, *Mass Media in America* (Chicago: Science Research Associates, 1974), 283.

5. Giraud Chester, Garnet R. Garrison, and Edgar E. Willis, *Television and Radio* (New York: Appleton-Century-Crofts, 1971), 133–36.

6. Pember, *Mass Media in America*, 283.

7. "Looking Back to WLBT(TV)," *Broadcasting* (April 16, 1984): 43.

8. "The Checkered History of License Renewal," *Broadcasting* (October 16, 1978): 30.

9. Sydney W. Head and Christopher H. Sterling, *Broadcasting in America* (Boston: Houghton Mifflin Co., 1982): 503–5.

10. John H. Pennybacker, "Comparative Renewal Hearings: Another Dialogue Between Commission and Court," *Journal of Broadcasting* (Fall 1980): 532–47.

11. "RKO Unfit Licensee, Says FCC Judge," *Broadcasting* (August 17, 1987): 35–36.

12. "The Next Best Thing to Renewal Legislation," *Broadcasting* (January 10, 1977): 20.

13. "Crossownership Rules for TV/Cable Should Be Repealed, Says Marsh Media," *Broadcasting* (December 22, 1980): 64; and "CapCities-ABC Gets Breather from the FCC," *Variety* (April 16, 1987): 1.

14. "Where Things Stand," *Broadcasting* (October 5, 1987): 10.

15. General sources dealing with electronic media and the First Amendment are: T. Carter Barton and others, *The First Amendment and the Fifth Estate: Regulation of Electronic Mass Media* (Mineola, NY: Foundation Press, 1986); Maurice R. Cullen, Jr., *Mass Media and the First Amendment: An Introduction to the Issues, Problems and*

Practices (Dubuque, Iowa: William C. Brown Company, 1981); and Lucas A. Powe, Jr., *American Broadcasting and the First Amendment* (Berkeley, CA: University of California Press, 1987).

16. Head, 471–72.

17. "New York Court Finds Shield Law Protection Not Absolute," *Broadcasting* (July 13, 1987): 51.

18. "Family Viewing's Inauspicious Demise," *Broadcasting* (January 7, 1985): 204.

19. "Briefs Filed in Preferred Case," *Broadcasting* (March 3, 1986): 40.

20. "Journalists Win First Amendment Victory on Exit Polls," *Broadcasting* (December 23, 1985): 35.

21. Wayne Overbeck and Rick D. Pullen, *Major Principles of Media Law* (New York: Holt, Rinehart, and Winston, 1985).

22. "FCC Launches Attack on Indecency," *Broadcasting* (April 20, 1987): 35.

23. "High Court Upholds Ruling Striking Down Utah Indecency Statute," *Broadcasting* (March 30, 1987): 143.

24. "Supreme Court Hears Porn Video Case," *Electronic Media* (March 10, 1986): 6.

25. For an excellent overview, see the special issue "Defamation and the First Amendment: New Perspectives," *The William and Mary Law Review* (1984).

26. "Good News for Broadcasters and Libel Suits," *Broadcasting* (October 15, 1984): 70.

27. "As TV News Reporting Gets More Aggressive, It Draws More Suits," *The Wall Street Journal* (January 21, 1983): 1; "$19.2 Mil Victory for Newton in NBC Libel Suit," *Variety* (December 18, 1986): 1; and "Herbert Libel Win Against CBS Overturned," *Broadcasting* (January 20, 1986): 70.

28. "The General's Retreat," *Newsweek* (March 4, 1985), 59–60; and Don Kowet, *A Matter of Honor: General William C. Westmoreland Versus CBS* (New York: Macmillan, 1984).

29. "Media Wins Big Libel Victory," *Variety* (April 22, 1986): 1; and "Supreme Court Hands Press Another Victory," *Broadcasting* (June 30, 1986): 51.

30. Susanna Barber, *News Cameras in the Courtroom: A Free Press-Fair Trial Debate* (Norwood, New Jersey: Ablex, 1987).

31. "Supreme Court Says Constitution Does Not Ban Cameras in Courts," *Broadcasting* (February 2, 1981): 43.

32. "Bright Outlook for Cameras in Supreme Court," *Broadcasting* (August 4, 1986).

33. "Providing Coverage, Protecting Victims: Rape Trials on Trial," *Broadcasting* (April 30, 1984): 134.

34. More information regarding editorializing cases can be found in "From Fighting Bob to the Fairness Doctrine"; and William L. Rivers, Theodore Peterson, and Jay W. Jensen, *The Mass Media and Modern Society* (San Francisco: Rinehart Press, 1971), 228–29.

35. "Public Broadcasters Granted Right to Editorialize," *Broadcasting* (July 9, 1984): 33.

36. *Public Law No. 416,* Section 315.

37. Ibid.

38. "Appeals Court Agrees with FCC: Broadcasters May Sponsor Debates," *Broadcasting* (May 12, 1984): 69.

39. More information of Section 315 can be found in "From Fighting Bob to the Fairness Doctrine"; "Section 315: Be Prepared," *Broadcasting* (March 29, 1976): 44; David H. Ostroff, "Equal Time: Origins of Section 18 of the Radio Act of 1927," *Journal of Broadcasting* (Summer 1980): 367–380; "Supreme Court Rules Vid Nets Erred in Not Selling Carter Campaign Air Time," *Variety* (July 2, 1981): 1; and "Attack on 315," *Broadcasting* (October 12, 1987).

40. *In the Matter of Editorializing by Broadcast Licensees,* 13 FCC 1246, June 1, 1949.

41. *Public Law No. 416,* Section 315.

42. Material regarding the fairness doctrine can be found in David Schoenbrun, "Is Perfect Fairness Possible?" *Television Quarterly* (Special Election Issue, 1976): 77–79; "From Fighting Bob to the Fairness Doctrine"; Fred W. Friendly, *The Good Guys, The Bad Guys, and the First Amendment: Free Speech vs. Fairness in Broadcasting* (New York: Random House, 1976); Steven J. Simmons, *The Fairness Doctrine and the Media* (Berkeley: The University of California Press, 1978); Louis F. Cooper and Robert E. Emeritz, *Pike and Fischer's Desk Guide to the Fairness Doctrine* (Bethesda, Maryland: Pike and Fischer, 1985); "Fairness Held Unfair," *Broadcasting* (August 10, 1987): 27–39K; "Fairness Doctrine: Gone But Not Forgotten," *Current* (September 22, 1987): 6; and "FCC Finds First Fairness Violation since Fowler," *Broadcasting* (October 29, 1984): 34.

14

1. *Broadcasting/Cablecasting Yearbook, 1987* (Washington, D.C.: Broadcasting Publications, Inc., 1987), D–3.

2. Excellent information about advertising and business practices can be found in Kim B. Rotzoll and James E. Haefner, *Advertising in Contemporary Society: Perspectives Toward Understanding* (Cincinnati: South-Western Publishing, 1986); David Samuel Barr, *Advertising in Cable: A Practical Guide for Advertisers* (Englewood Cliffs, New Jersey: Prentice-Hall, 1985); Charles Warner, *Broadcast and Cable Selling* (Belmont, Calif.: Wadsworth Publishing Company, 1986); and Elizabeth J. Heighton and Don R. Cunningham, *Advertising in the Broadcast and Cable Media* (Belmont, Calif.: Wadsworth Publishing Company, 1984).

3. These are actual rate cards, but the stations did not want to be identified because their rates change so often.

4. Heighton and Cunningham, 32.

5. Thomas Moor, "Culture Shock Rattles the TV Networks," *Fortune* (April 14, 1986): 22–27.

6. *Broadcasting/Cablecasting Yearbook,* G–1–G–7.

7. For more information on commercial production, see Norman D. Cary, *The Television Commercial: Creativity and Craftsmanship* (New York: Decker, 1971); Lincoln Diamant, *The Anatomy of a Television Commercial* (New York: Hastings House, 1971); and "Playing to Your Fears—and Your Funny Bone," *TV Guide* (March 10, 1984): 18–22.

8. Lincoln Diamant, *Television's Classic Commercials* (New York: Hastings House, 1970); and Jim Hall, *Mighty Minutes: An Illustrated History of Television's Best Commercials* (New York: Harmony Books, 1984).

9. Several of the books issued periodically and carried by many libraries are *The Annual Register of Grant Support, The Catalog of Federal Domestic Assistance,* and *The Foundation Directory.*

10. Gregory J. Liptak, "1980—The Year the Tier Became Clear," *TVC* (December 15, 1980): 72–76.

11. "ESPN To Hike Fees to Affiliates in 1985," *Variety* (June 1, 1984): 1.

12. "CBS's Compensation Cure: Expand Commercial Inventory," *Broadcasting* (March 9, 1987): 35.

13. "More Blurbs Headed for TV," *Variety* (January 18, 1985): 1.

14. "Nielson Offers Tempered Approach to Alcohol Ads," *Broadcasting* (February 4, 1985): 31

15. "Harris Survey Shows Viewers Think Contraceptive Ads OK," *Broadcasting* (March 30, 1987): 178.

16. Alice E. Courtney and Thomas W. Whipple, *Sex Stereotyping in Advertising* (Lexington, Mass: Lexington Books, 1983).

17. Edward L. Palmer and Cynthia N. McDowell, "Children's Understanding of Nutritional Information Presented in Breakfast Cereal Commercials," *Journal of Broadcasting* (Summer 1981): 295–302; "NAB Asks FCC to Clarify Its TV Dereg Order," *Broadcasting* (November 12, 1984): 58; and "Interactive Toys Hit Animation Programming," *Broadcasting* (February 9, 1987): 110.

15

1. Several accounts of early ratings can be found in Frederick Lumley, *Measurement in Radio* (Columbus, Ohio: Ohio State University Press, 1934); Paul F. Lazarfield and Frank N. Stanton, *Radio Research* (New York: Duell, Sloan, and Pearce, 1942); Mark James Banks, "History of Broadcast Audience Research in the United States, 1920–1980" (Ph.D. dissertation, Knoxville: University of Tennessee, 1981); and Hugh M. Beville, Jr., *Audience Ratings: Radio, Television and Cable* (Hillsdale, NJ: Lawrence Erlbaum Associates, 1985).

2. *The Nielsen Ratings in Perspective* (Northbrook, IL: Nielsen Media Research, 1980 (with 1986 updates); "Peoplemeter Progress," *Broadcasting* (September 21, 1987): 38; and "Nielsen on Nielsen," *Broadcasting* (July 6, 1981): 57.

3. Shirley Biagi, "The Measure of the Meters," *Emmy* (August 1987): 90–94.

4. *Arbitron Ratings Today* (New York: Arbitron Ratings Company, n.d.); *Arbitron Ratings Radio* (New York: Arbitron Ratings Company, 1987); and "ScanAmerica Launch Delayed," *Broadcasting* (June 22, 1987): 45.

5. Rich Barbieri, "Perfecting the Body Count," *Channels* (June 1987): 15.

6. *Broadcasting/Cablecasting Yearbook* (Washington, D.C.: Broadcasting Publications, Inc., 1987), I–32–I–36.

7. *Turning the Numbers into Sales Strategies* (New York: Arbitron Ratings Company, 1987), 3–5.

8. Pros and cons of ratings (mostly cons) can be found in many cleverly written articles or sections of books, including Dick Adler, "The Nielsen Ratings— and How I Penetrated Their Secret Network," *New York Times* (October 1, 1974): 74B; "Roundtable: Research and Ratings," *Emmy* (Spring 1981): 13–16; Bob Shanks, *The Cool Fire* (New York: W. W. Norton and Company, 1976), 244–58; Michael Wheeler, "Life and Death in the Little Black Box," *Television Quarterly* (May-July 1976): 5–14; three articles in Barry G. Cole, *Television* (New York: The Free Press, 1970), 385–408; and "Television in the Peoplemeter Age," *Broadcasting* (September 7, 1987): 35–39.

9. Details can be found in Committee on Interstate and Foreign Commerce, House of Representatives, *Broadcast Ratings: The Methodology, Accuracy, and Use of Ratings in Broadcasting* (Washington, D.C.: Government Printing Office, 1963–1965).

10. Elizabeth J. Heighton and Don R. Cunningham, *Advertising in the Broadcast and Cable Media* (Belmont, CA.: Wadsworth Publishing Company, 1984), 199–201.

11. Details can be found in CONTAM, *Television Ratings Revisited: A Further Look at Television Audiences* (New York: Television Information Office, 1971).

12. Details can be found in Committee on Interstate and Foreign Commerce, House of Representatives, *Broadcast Ratings: A Progress Report on Industry and Programs Involving Broadcast Ratings* (Washington, D.C.: Government Printing Office, 1966).

13. Details can be found in *ARMS— What It Shows, How It Has Changed Radio Measurement* (Washington, D.C.: National Association of Broadcasters, 1966).

14. Details can be found in *CRAM— Cumulative Radio Audience Method (New York: National Broadcasting Company, 1966).*

16

1. Listings of these companies can be found in *Broadcasting/ Cablecasting Yearbook, 1987* (Washington, D.C.: Broadcasting Publications, Inc., 1987), sections F, G, H, and I.

2. Bob Shanks, *The Cool Fire* (New York: W. W. Norton, 1976), 113–114.

3. "Screen Actors Guild Okays First Phase of AFTRA Merger by a Wide Margin," *Variety* (August 18, 1981): 1; and "Largest Voter Turnout in SAG History Rejects Merger with Screen Extras Guild," *Variety* (March 20, 1984): 1.

4. For a more complete list of unions see *Broadcasting/Cablecasting Yearbook, 1987,* I–62–I–63.

5. *Sixteenth Report: Broadcast Programs in American Colleges and Universities, 1986–1987* (Washington, D.C.: Broadcast Education Association, 1986): iii; and U.S. Department of Labor, *Supplement to Employment and Earnings* (Washington, D.C.: Bureau of Labor Statistics, 1986): 119

6. For more information see Lynne S. Gross, *The Internship Experience* (Prospect Heights, IL: Waveland Press, 1987); and Ronald H. Claxton and B. A. Powell, *The Guide to Mass Media Internships* (Boulder, Colorado: University of Colorado Press, 1980).

7. Many books exist to give advice on how to get a job in the electronic media industries. Some of them are: Paul Allman, *Exploring Careers in Video* (New York: Rosen Publishing Group, 1985); David W. Berlyn, *Exploring Careers in Cable/TV* (New York: Rosen Publishing Group, 1985); Dan Blum, *Making It in Radio: Your Future in the Modern Medium* (Hartford, CT: Continental Media Co., 1983); Jan Bone, *Opportunities in Telecommunications* (Lincolnwood, IL: National Textbook Company, 1984); E. L. Beckinger and Jules B. Singer, *Exploring Careers in Advertising* (New York: Rosen Publishing Group, 1985); Jon S. Denny, *Careers in Cable TV* (New York: Harper and Row, 1984); Tom Logan, *How to Act and Eat at the Same Time: The Business of Landing a Professional Acting Job* (Washington, D.C.: Communications Press, 1984); Donn Pearlman, *Breaking in Broadcasting: Getting a Good Job in Radio or TV Out Front or Behind the Scenes* (Chicago: Bonus Books, 1986); Maxine Reed, *Career Opportunities in Television, Cable, and Video* (New York: Facts on File, 1986); Rod Vahl, *Exploring Careers in Broadcast Journalism* (New York: Richards Rosen Press, 1983); John C.Zacharis, *Exploring Careers in Communications and Telecommunications* (New York: Rosen Publishing Group, 1985); and Caroline A. Zimmerman, *How to Break Into the Media Professions* (New York: Doubleday, 1981).

17

1. *International Television Association: The Organization for Professional Video Communicators* (Irving, Texas: ITVA, n.d.).

2. Speech by Barry Cole, Professor of the Annenberg School of Communications, University of Pennsylvania, at the Summer Faculty Workshop, Washington, D.C., June 15, 1987.

3. "RCC," *Broadcasting* (October 4, 1982): 39–47.

4. "Lottery Makes Interesting Cellular Bedfellows," *Broadcasting* (October 8, 1984): 48.

5. "Car Phones: From Toy to Valued Tool," *Los Angeles Times* (July 7, 1987): I–1.

6. Daniel Seligman, "Life Will Be Different When We're All On-Line," *Fortune* (February 4, 1985): 68–71.

7. Interview with Ann Ward, Engineering Librarian, General Dynamics, Pomona, Calif., July 21, 1987.

8. Daniel Seligman, "Life Will Be Different," 68.

9. *Hayes Smartcom II* (Norcross, Georgia: Hayes Microcomputer Products, 1984): 9.2–9.18.

10. *1986 DIALOG Database Catalog* (Palo Alto, Calif.: DIALOG Information Services, Inc., 1986).

11. "Time, AT&T in on New Videotex Service," *Broadcasting* (July 17, 1985): 68.

12. "Is ISDN Worth Waiting For?" *Business Communications Review* (March-April 1986): 13–17.

13. "CDs and Their Laser Kin: The Beam Brightens," *Channels* (Field Guide, 1987): 93; and "Expert Systems: A New Era in Videodiscs," *E-ITV* (August 1987): 23–25.

14. *Facilities for Journalists* (Los Angeles, Calif.: Games of the XXIIIrd Olympiad, 1984): 27–33.

15. "The 'Write' Stuff," *Online Today* (April 1987): 22–23.

16. "Technology in the Workplace," *The Wall Street Journal* (June 12, 1987): 9D.

17. "CBS News Buys Electronic Newsroom System," *Broadcasting* (January 21, 1985): 92.

18. *Where to Go When You Want to Tell Everyone Where to Go* (San Diego, Calif: Architectural Signs and Graphics, n.d.).

19. "Not Really Combat But It Sure Seems Like It," *Los Angeles Times* (July 7, 1987): IV, 1.

20. Roger Pawley, "Video Training System 'Cracks Down' On Corporate Overhead," *E-ITV* (June 1987): 30–31.

21. Neil Hickey, "Goodbye Boob Tube, Hello Smart Set," *TV Guide* (July 10, 1982): 28–31.

22. Peter W. Bernstein, "Atari and the Video Game Explosion," *Fortune* (July 27, 1981): 40–46.

23. Richard C. Morgan, "Soy Group Gets World TV Report," *The Atlanta Constitution* (August 14, 1979): 21.

24. "Holiday Inn, Comsat Venture Would Create Communications Network," *Broadcasting* (November 5, 1984): 75.

25. "Joint Venture Installs Earth Stations for Shopping Mall Teleconferencing," *Aviation Week and Space Technology* (April 21, 1986): 57–58.

26. "Firms Are Cool to Meetings by Television," *The Wall Street Journal* (July 26, 1983): 2.

27. "Videoconference Use Expands to Meet Rising Business Needs," *Aviation Week and Space Technology* (July 22, 1985): 157–164; and "Teleconferencing: Nobody's Baby Now," *E-ITV* (November 1987); 26–29.

28. "ITFS at the Crossroads Again," *E-ITV* (March 1985): 66–71.

29. Nancy Nelson, "Around the Nation with ITFS," *E-ITV* (June 1985): 28–29.

30. "PBS Gets Go-Ahead on 82 of 102 ITFS Channels," *E-ITV* (February 1984): 5.

31. *PBS: Facts About Us* (Alexandria, Virginia: Public Broadcasting Service, 1987): 3.

32. Jeanette Ross, "The Politics of ITFS: Questions of Control and Construction," *E-ITV* (June 1985): 26–27.

33. R. T. Rock, J. S. Duva, and J. E. Murray, *Training by Television: A Study of Learning and Retention* (Port Washington, New York: Special Services Center, n.d.).

34. "Broadcast News, Inc.," *Newsweek* (January 4, 1988): 34–35.

35. John Barwich and Stewart Kranz, *Profiles in Video* (White Plains, New York: Knowledge Industry Publications, 1975).

36. Judith M. Brush and Douglas P. Brush, *Private Television Communications: Into the Eighties* (Berkeley Heights, New Jersey: International Television Association, 1981).

18

1. Sydney W. Head, *World Broadcasting Systems* (Belmont, Ca.: Wadsworth Publishing Company, 1985), 14–15.

2. Joe Roezin, "The Status of International Television," *Video Systems* (October 1977): 26.

3. David Hodes, "HDTV: Still Searching for the Standard," *View* (September 1, 1986): 52–53.

4. "The Privatization of Europe," *Broadcasting* (May 31, 1986): 61–68.

5. Sydney Head, *World Broadcasting Systems,* 225.

6. "Global TV," *Electronic Media* (April 21, 1986): G1–G17.

7. "The Path Once Taken," *Channels* (Field Guide, 1988): 20–21.

8. "Canada Wants to Raise CBC's Native Program Levels to 90 percent," *Broadcasting* (March 9, 1987): 43.

9. Shawn Tully, "U.S.-Style TV Turns on Europe," *Fortune* (April 13, 1987): 95–98.

10. "April in Cannes: MIP Programing Blooms Profusely," *Broadcasting* (April 20, 1981): 77.

11. Sydney Head, *World Broadcasting Systems,* 197.

12. Interview with Ichio Morre, Training Director, Tokyo Broadcasting System, June 5, 1986.

13. "Britain's TEN Pay-See Web Closing Down," *Variety* (June 4, 1985): 1.

14. Interview with Kate Harrison, Project and Research Officer for the Public Interest Advocacy Centre, Sydney, Australia, June 17, 1987.

15. "From Ahmadabad to Makapura," *TV Guide* (June 19, 1976): 10–12.

16. "European States Face Problem of Controlling Their Neighbors' TV," *The Wall Street Journal* (March 22, 1982): 1.

17. "Soviets Open Video Shops," *Electronic Media* (April 21, 1986): G12–G13.

18. "Videotext—In France It's the Rage," *Los Angeles Times* (September 12, 1986): 1 and 15.

19. "Formation of the BBC," *BBC Fact Sheet 1* (May 1982): 1–2; and R. W. Burns, *British Television: The Formative Years* (London: Peter Peregrinos, 1986).

20. "The Constitution of the BBC," *BBC Fact Sheet 9* (May 1981): 1–2.

21. "License Fees," *BBC Fact Sheet 11* (May 1982): 1–2.

22. William E. McCavitt, *Broadcasting Around the World* (Blue Ridge Summit, Pa.: TAB Books, 1981): 236.

23. *It's Your BBC* (London: British Broadcasting Corporation, n.d.): 5–6.

24. "The Growth of Television," *BBC Fact Sheet 32* (May 1981): 1–2.

25. "Britain's BBC-TV," *Broadcasting* (November 3, 1986): 43–53.

26. *This is Independent Broadcasting* (London: Independent Broadcasting Authority, n.d.), 3–14.

27. *Advertising on Independent Broadcasting* (London: Independent Broadcasting Authority, n.d.), 3–10.

28. Alan Coren, "It's Sausages and Kidneys vs. Morning TV," *TV Guide* (June 4, 1983): 18–19.

29. Chris Auty, "The New English Channel," *American Film* (December 1982): 17.

30. *A Guide to the BBC's Programmes and Services for Education* (London: British Broadcasting Corporation, n.d.); and *Learning with Television and Radio* (London: Independent Broadcasting Authority, n.d.).

31. "House of Commons Defeats Bill Allowing Ads on BBC," *Variety* (January 24, 1985), 14.

32. "Technical Boom Around Corner for U. K. Video Biz," *Variety* (January 15, 1982): 24.

33. Cheryl Rhodes, "The Growth of Videotext in Britain and France," *DataCast* (January/February 1982): 23–35.

34. William McCavitt, *Broadcasting Around the World*, 52–75; "TV: Medium With a Message," *U.S. News and World Report* (February 1, 1982): 34; and Ellen Propper Mickiewitz, *Media and the Russian Public* (New York: Preager, 1981).

35. "Europe Braces for Free-Market TV," *Fortune* (February 20, 1984): 74–82.

36. "Canada Redefines B'Cast Regs," *Variety* (February 1, 1983): 1.

37. *History and Development of the ABC* (Sydney: Australian Broadcasting Corporation, 1983); and Allan G. Brown, *Commercial Media in Australia: Economics, Ownership, Technology, and Regulation* (St. Lucia, Australia: University of Queensland Press, 1986).

38. *Tokyo Broadcasting System* (Tokyo: Tokyo Broadcasting System, n.d.).

39. Zhenzhi Guo, "A Chronicle of Private Radio in Shanghai," *Journal of Broadcasting and Electronic Media* (Fall 1986): 379–392; Max Wilk, " 'I Love Lucy' Would be a Great Leap Forward," *TV Guide* (October 7, 1978): 32–35; and "CBS TV Package Sold to China," *Variety* (July 20, 1984): 1.

40. Sydney Head, *World Broadcasting Systems,* 27–28.

41. Bruce Henstell, "Rede Globe," *Emmy* (January/February 1984): 40–42.

42. Alan Brender, "After a 25–Year Wait . . ." *TV Guide* (August 14, 1976): 24–26.

43. "Saudi TV Pace-Setter of Broadcasting in Arab World," *Variety* (April 16, 1982): 16.

44. Interview with Abdul Aziz Abbas, Controller of TV Programmes, Radio-TV Malaysia, January 6, 1987.

45. Sydney Head, *World Broadcasting Systems,* 32.

46. *Facts About VOA* (Washington, D.C.: United States Information Agency, 1987): 2.

47. *BBC Broadcasts to the World* (London: British Broadcasting Corporation, n.d.).

48. *The Voice of America: A Brief History and Current Operations* (Washington, D.C.: United States Information Agency, 1986).

49. Donald R. Browne, "The International Newsroom: A Study of Practices at the Voice of America, BBC, and Deutsche Welle," *Journal of Broadcasting* (Summer 1983): 205–231.

50. *Facts About VOA*, 2; and *BBC Broadcasts to the World*.

51. "IFRB Backs U.S. Claims of Soviet Jamming," *Broadcasting* (September 22, 1986): 84.

52. "U.S. Broadcasts to Cuba Miss Targeted Start-Up," *Los Angeles Times* (January 30, 1985): VI, 1.

53. "USIA's High Tech Propaganda War," *Broadcasting* (February 6, 1984): 51.

54. *Armed Forces Radio and Television Service (AFRTS)* (Washington, D.C.: Department of Defense, n.d.); and Craig R. Stephen, "The American Forces Network, Europe: A Case Study in Military Broadcasting," *Journal of Broadcasting and Electronic Media* (Winter 1986): 33–46.

55. Speech by David Leach, Managing Director, Orion Telecommunications, Ltd. at the Annenberg Summer Workshop, Washington, D.C., June 17, 1987.

56. "U.S. Prepared for ITU Conference on Wider AM Band," *Broadcasting* (February 10, 1986): 68.

57. "Space WARC Reaches Consensus," *Broadcasting* (September 16, 1985): 40.

58. *Twenty Years Via Satellite* (Washington, D.C.: COMSAT, 1986).

59. *INTELSAT Report 1985–1986* (Washington, D.C.: INTELSAT, 1986).

60. "International Communications Satellite Competition Examined," *Broadcasting* (June 29, 1987): 44.

61. *Twenty Years Via Satellite.*

62. *COMSAT Guide to the INTELSAT, MARISAT, and COMSTAR Satellite Systems* (Washington, D.C.: COMSAT, n.d.).

63. "Colino, Others Plead Guilty to Defrauding Intelsat," *Broadcasting* (July 20, 1987): 23.

A

AAAA American Association of Advertising Agencies

ABC American Broadcasting Company.

above-the-line Creative costs of a particular program or series for such items as talent and producer.

ACT Action for Children's Television.

actuality An event that really occurs; often used to describe an interview in which someone talks about what actually happened.

advertising agency An organization that decides on and implements an advertising strategy for a customer.

affiliate A station or system that receives programming from a broadcast or cable network.

AFM American Federation of Musicians.

AFRTS Armed Forces Radio and Television Service.

AFTRA American Federation of Television and Radio Artists.

allocation table A list compiled by the FCC that indicates where FM or TV stations can be located.

alternator A device for converting mechanical energy into electrical energy in the form of alternating current (AC).

AM Amplitude modulation: changing the height of a transmitting radio wave according to the sound being broadcast.

amplifier A circuit, tube, transistor, or other apparatus that draws power from a source other than the input signal and then produces as an output an enlarged reproduction of the essential features of the input.

analog A device or circuit in which the output varies as a continuous function of the input.

antenna A wire or set of wires or rods used both to send and to receive radio waves.

antitrust act An act of Congress designed to oppose business deals made to control or centralize industries.

AP Associated Press.

APR American Public Radio.

ASCAP American Society of Composers, Authors, and Publishers.

ascertainment A process stations used to undertake to keep their licenses; it involves interviewing community leaders to learn what they believe are the major problems in the community.

AT&T American Telephone and Telegraph Company.

audience flow Maintaining a specific audience for a period of time, such as a whole evening, by airing one type of program, such as situation comedies.

audimeter An electronic device that used to be attached to TV sets by the A. C. Nielsen Company in order to determine, for audience measurement purposes, when a set was turned on and to what station it was tuned.

audion A three-electrode vacuum tube invented by Lee De Forest that was instrumental in amplifying voice so that it could be sent over wireless.

average measures Audience measurement counts of the average number of households listening to or watching a station over a preselected period of time.

B

balance sheet A statement of the assets, liabilities, and net worth of a company at a particular time, such as the close of a fiscal year.

GLOSSARY

473

bandwidth The number of continuous frequencies within given limits that are allowable for transmission of a given signal; for example, a TV station has a much broader bandwidth than an AM radio.

barter To give goods or services in return for other goods or services.

basic cable Channels, often supported by advertising, for which the subscriber does not pay a large extra fee.

BBC British Broadcasting Corporation.

beat An area in which a news reporter is expected to gather news.

below-the-line Costs that are constant, regardless of the particular program or series, such as those for camera operators, engineers, and videotape.

bicycling A method by which the same program is shown on different stations or cable systems at different times. The program is not sent over wires, microwave, or satellite, but rather mailed, flown, or driven from one place to another.

bidirectional Picking up on two sides; used to refer to microphones with two live sides.

blacklisting A phenomenon of the 1950s when many people in the entertainment business were accused of leaning toward communism and, as a result, could not find work.

blackout Not allowing a particular program, usually sports, to be shown in a particular area because the event is not sold out.

blanket license The right to use a large catalogue of musical selections by paying one set fee.

block programming Airing one form of program, such as dramas, for a block of time, such as an evening.

BMI Broadcast Music, Inc.

booster A carrier frequency amplifier that strengthens a signal at one fixed point so that it can be retransmitted to another fixed point.

BRC Broadcast Rating Council.

broadcast standards The department at a network or station that decides the general standards of acceptability of program content.

broker A person or company that acts like a real estate agent for people who wish to buy and sell radio and TV stations and cable TV systems.

C

call letters A series of government-assigned letters that identify a transmitting station.

canned Preproduced programs, commercials, or program elements that arrive at the station already on tape or film.

carbon mike A microphone that operated on the variations of resistance of carbon contacts, a nonmetallic conductive material.

cardioid A heart-shaped pickup pattern for a microphone.

carrier wave A high-frequency wave that can be sent through the air and is modulated by a lower-frequency wave containing information.

cartridge A self-contained case of magnetic tape wound in a continuous loop; also the portion of a turntable pickup arm that receives vibration.

cassette A two-reeled, self-contained case for magnetic tape.

cathode-ray tube A tube in which an electronic beam can be focused to a small cross section on a luminescent screen and can then be varied in position and density to produce what appears to be a moving picture.

CATV Community Antenna Television; an old name for cable TV.

CBS Columbia Broadcasting Company.

CCD A charge-coupled device; a type of camera that uses a silicon chip to convert the image to electronics

cease and desist A formal order stating that a person or company must stop a certain practice immediately.

cellular radiotelephones Mobile telephones primarily used in automobiles.

chain broadcasting An early term for network broadcasting.

chain regulation FCC ruling of 1941 that gives the affiliates the right to reject a network program.

character generator A system used to type words onto a TV screen electronically.

circulation The number of homes that tune in to a particular station over a set period of time, usually a week.

citizen's band A radio communication service intended for short-distance personal and business communication.

clear and present danger News information that, if publicized, would threaten the security of the nation

close-captioning Words that appear on the screen for the deaf; the converter to make the words appear must be purchased.

closed-circuit Television signals that are transmitted via a self-contained wire system rather than broadcast through the air.

close-up An object or any part of it seen at close range.

coaxial cable A transmission line in which one conductor completely surrounds the other, making a cable that is not susceptible to external fields from other sources.

coincidental telephone technique
An audience-measurement method whereby people are called on the telephone and asked what they are watching on TV or listening to on radio at that moment.

color coding system The system by which red, blue, and green information is encoded on a TV signal.

commission A percentage of money given to a salesperson based on the amount of the sale.

compact disc A small high-fidelity record produced digitally using laser technology.

comparative license renewal A procedure by which a group may challenge the license of incumbent station owners.

compulsory license fee A copyright fee that is a percentage of earnings rather than a fee for each copyrighted work used.

computer graphics Designs and figures created electronically instead of with paints or other art supplies.

COMSAT Communications Satellite Corporation.

construction permit (CP) A document issued by the FCC that allows the recipient to begin building a radio or television station.

CONTAM Committee on Nationwide Television Audience Measurement.

control track The portion of videotape that contains sync information, such as horizontal and vertical picture alignment.

converter box A device for allowing all the cable TV channels to appear on the TV screen.

co-op advertising Joint participation in the content and cost of a commercial by two entities, usually a national company that manufactures the product and a local company that sells it.

copyright tribunal The government organization that decides how copyright fees should be assessed and distributed.

cosmic rays Rays in the vicinity of 10 megahertz that have high-penetrating power produced by transmutations of atoms in outer space.

counterprogramming Scheduling programs to reduce the audience size of a competitor's programs.

coverage The number of homes a particular radio or TV station has the potential to reach if all conditions are perfect.

CPB Corporation for Public Broadcasting.

CPM Cost per thousand; the price that an advertiser pays for each thousand households a commercial reaches.

crane A mechanism on which a camera can be mounted so that the camera can move from close to the floor to a fairly high distance above the floor.

cross-ownership A situation in which one company owns various media in one market.

crystal A thick slab or plate of quartz ground to a thickness that causes it to vibrate at a specific frequency when energy is applied.

CTW Children's Television Workshop.

cue tone A subaudio signal on a tape that enables a specially equipped tape recorder to start or stop the tape at an appropriate point.

cume The total number of households that tune in to a particular station at different times.

cycles per second The number of times a second that an alternating wave goes from zero to a negative peak to zero to a positive peak and back to zero.

D

data banks Computers filled with information that can be accessed by the customers.

daytimers AM radio stations authorized to operate only during daylight hours.

DAT Digital audio tape.

DBS Direct Broadcast Satellite; a process of transmission and reception whereby signals sent to a satellite can be received directly by TV sets in homes that have small satellite receiving dishes on their roofs.

decoder A device that unscrambles TV signals that have been purposely made unintelligible.

defamation of character Utterings either written or oral that attack the reputation of a person.

deintermixture An attempt to strengthen UHF stations by making some markets all UHF.

demodulate To operate on a previously modulated wave in such a way that it will have the same characteristics as the original wave.

demographics Information pertaining to vital statistics of a population, such as age, sex, marital status, and geographic location.

deregulation The removal of laws and rules that spell out government policies.

detector A part of a radio receiver that demodulates the wave or separates the carrier wave from the information it carries.

DGA Directors Guild of America.

diary A booklet used for audience measurement in which people write down the programs they watch or listen to.

digital A device or circuit in which the output varies in discrete "on-off" steps.

directional antenna An antenna that sends out radio waves more effectively in some directions than in others.

director The person who calls for the various shots and generally oversees production once it is in the TV studio or on remote location.

disc A phonograph record; also a device resembling a phonograph record that is used to play video information.

dissolve Moving gradually from one picture to another in such a way that the two pictures overlap for a brief time.

distant signal importation The bringing in of stations from other parts of the country by a cable system.

domestic syndication A process whereby a network can participate in distribution of United States programming produced by another entity but aired on the network.

downconverter A device for changing high frequencies to lower-frequency signals.

downlink A facility that can receive signals from a satellite.

downstream A term used to designate signals that go from a central source to a home.

duopoly A rule stating that a network organization cannot operate two or more networks that cover the same territory at the same time.

E

EBS Emergency Broadcasting System.

EFP Electronic field production; a term applied to portable remote productions and equipment.

electrode A conducting element that emits or collects electrons or ions or controls their movement by electric field.

electromagnetic spectrum A continuous frequency range of wave energies including radio waves and light waves.

electron gun A structure that produces and controls an electron beam in a TV camera.

electronic publishing Providing material such as that traditionally supplied by newspapers through a network that usually includes phone lines and computers.

electronic scanning A method that analyzes the density of areas to be copied and translates this into a moving arrangement of electrons that can later reproduce the densities in the form of a picture.

ENG Electronic news gathering; specifically, portable video equipment first used in obtaining on-the-spot news stories.

equalizer An attachment for a turntable that can eliminate high frequencies and, therefore, scratchy noises

equal time A rule stemming from Section 315 of the Communications Act stating that TV and radio stations should give the same treatment and opportunity to all political candidates for a specific office.

F

fairness A policy that evolved from FCC decisions, court cases, and congressional actions stating that radio and TV stations had to present all sides of the controversial issues they discussed.

family hour A policy that stated all programs aired between 7:30 and 9:00 P.M. should be suitable for children as well as adults.

FBC Fox Broadcasting Company.

FCC Federal Communications Commission.

feevee A name for pay-TV that developed as a counterterm to conventional network-station free TV.

fiber optics Glass strands through which information can be sent.

film chain Equipment used for placing slides or motion picture film on TV, usually consisting of at least one slide projector, at least one film projector, a multiplexer, and a TV camera.

filter A device that eliminates certain colors and allows others to pass.

financial interest A process whereby a network helps finance programming supplied to it by a production company in return for some sort of profit from the programming.

fixed buy A selling practice in which the advertiser decides the exact time that its commercials will air.

flat-screen TV A form of TV reception being developed that generally does not use a picture tube and can hang on the wall like a picture.

floor manager The person in charge of studio operations during production, cuing the talent and communicating with the director.

FM Frequency modulation; placing a sound wave upon a carrier wave in such a way that the number of recurrences is varied.

font A style of lettering such as gothic or italics.

footprint The section of the earth that a satellite's signal covers.

format The type of programming a radio station selects, usually described in terms of the music it plays.

frame One complete scanning cycle of a TV camera or receiver.

franchise A special right granted by a government or corporation to operate a facility such as cable TV.

FRC Federal Radio Commission.

free-lancer A person who is not under contract for regular work, but sells his or her services, such as acting or writing, to organizations on a per-project basis.

freeze Immobilization or cessation of an activity, such as a stop in the assigning of radio stations.

frequency The number of recurrences of a periodic phenomenon, such as a carrier wave during a set time period, such as a second.

frequency discounts Reduced fees charged to an advertiser depending on the number of ads he or she buys.

FTC Federal Trade Commission.

futures The practice of selling network shows several years before they are actually available for syndication.

G

gamma rays Electromagnetic radiations with wavelengths in the vicinity of 10 megahertz.

GE General Electric Company.

grid A pipe structure for hanging lights, usually attached to a TV studio ceiling.

ground station A receiving station for information transmitted by a satellite.

guild An organization similar to a union, usually organized for above-the-line personnel.

H

ham A person who operates a radio station as a hobby rather than a business.

head A small electromagnet used to read, record, or erase information on magnetic tape.

headend The part of the cable TV facility that receives signals from various sources.

helical A method of videotape recording whereby video information is placed on the tape at a slant.

hertz A frequency unit of one cycle per second; abbreviated Hz.

high-definition TV TV that scans above one thousand lines a frame.

hub Part of a cable TV system that receives signals from the headend and sends them by cable to subscribers in its area.

HUT Homes using TV; the percentage of homes that have a TV set tuned to any station.

hyping The practice of airing special programs or holding special contests in order to increase audience size during a rating period.

I

IATSE International Alliance of Theatrical and Stage Employees and Moving Picture Machine Operators of the United States and Canada.

IBEW International Brotherhood of Electrical Workers.

iconoscope The earliest form of TV camera tube in which a beam of electrons scanned a photoemissive mosaic.

IEEE Institute of Electrical and Electronic Engineers.

image orthicon A TV camera tube used for over twenty years in which a photoemitting surface focused an image on a glass target that was scanned by an electronic beam.

independent A station that is not aligned with one of the major networks.

infrared rays Waves just beyond the red end of visible light with wavelengths longer than light, but shorter than radio waves.

input The current, voltage, or power coming into a component of an electronic system, usually by means of a connector.

INS International News Service.

instantaneous measure The number of households tuned to a station at a particular moment.

instant replay Playing back videotape material immediately after it is recorded, usually by means of a videodisc.

INTELSAT International Telecommunications Satellite Organization.

interactive cable Two-way capability that allows interaction between the subscriber and sources provided by the cable.

intercom A two-way communication system without a central switchboard that allows all to hear and often allows all to talk.

IPS Inches per second.

iris The adjustable opening in a camera lens that admits light.

ISDN Integrated Services Digital Network; technology that allows two-way processing, storing, and transporting of any information simultaneously in voice, data, graphics, or video.

ITFS Instructional Television Fixed Service.

ITU International Telecommunication Union.

J

jack panel A board with many sockets to which the wires of a circuit are connected at one end and a male plug can be inserted in the other.

K

key A switching device for turning a circuit off or on that usually consists of concealed spring contacts and an exposed handle or button.

kinescope An early form of poor-quality TV program reproduction that basically involved making a film of a TV screen.

L

laser Acronym for light amplification by simulated emission of radiation; a device for transforming light of various frequencies into a very narrow, intense beam.

laserdisc Uses a laser beam to read information embedded in a plastic, recordlike device.

lens The part of a camera that projects an image on the television pickup tube.

libel To say or print something unfavorable and false about a person.

license A document the FCC issues to stations authorizing them to operate.

light waves The portion of the electromagnetic spectrum that is visible to the human eye.

line of sight A straight line between a broadcast antenna and a receiver.

local origination Programs produced about the local community, particularly as it refers to programming created by cable TV systems.

log A sheet that lists a breakdown of a station's program schedule.

long waves Radio waves at low frequencies.

lottery The involvement of chance, prize, and consideration(money) for a game or contest; also a random selection method.

LPTV Low-power TV; television stations that broadcast to a very limited area because they do not transmit with much power.

M

magazine concept Placing ads of various companies within a program rather than having the entire program sponsored by one company.

market An area, often consisting of one city, that a particular station serves.

master antenna A system's main antenna that sometimes collects signals for an apartment complex so they can be sent on by wire.

matching funds Money provided when an equal amount of money has been raised by other means.

MCTV Multichannel TV; another name for MMDS.

mechanical scanning An early form of scanning by which a rotating device, such as a disc, broke up a scene into a rapid succession of narrow lines for conversion into electrical impulses.

medium waves Radio waves at medium frequencies.

microwave Radio waves, 1,000 megahertz and up, that can travel fairly long distances.

minicam A small, lightweight television camera generally used in remote locations to cover news events.

miniseries Drama series presented over several nights.

MMDS Multichannel Multipoint Distribution Service; an over-the-air service of several channels that operates at higher frequencies than broadcast TV.

modem A device that allows computer-generated data to be sent over phone lines.

modulate To vary the amplitude or frequency of one wave by placing another on it.

monaural One channel of sound coming from one direction.

monitor To listen to something without interrupting it; also a device for seeing a TV picture directly from the video output.

MSA Metropolitan Survey Area.

multiple system owner A company that owns several cable TV operations.

multiplex To transmit two or more messages at the same time on a single channel.

music licensing Paying for the use of music by giving a set fee to a company that then distributes the revenue according to the popularity of the music.

must-carry Stations in the local area that cable TV systems have been required to put on their channels.

N

NAB National Association of Broadcasters.

NABET National Association of Broadcast Employees and Technicians.

NABTS North American Broadcast Teletext Standard.

NAEB National Association of Educational Broadcasters.

NARB National Advertising Review Board.

narrowcasting Airing program material that appeals to a small segment of the population.

NATPE National Association of Television Program Executives.

NBC National Broadcasting Company.

needle-drop fee Money paid for the right to use a specific piece of music with a specific program.

NET National Educational Television.

network feed A program or other information coming from a network to a station, sometimes at a time other than when the station plans to air it.

NPR National Public Radio.

NTI Nielsen Television Index.

NTSC National Television System Committee.

O

offline A form of videotape editing done with tapes dubbed from the masters, not the masters themselves.

OFS Operation Fixed Service; a data service located in the 2,500 megahertz band planned for corporation communication.

omnidirectional Not favoring any one direction; used to refer to a microphone that picks up sound on all sides.

online Videotape editing done with the master tapes.

open reel A tape recorder for which tape is threaded manually from one reel to another.

oscilloscope A mechanism that makes visible instantaneous values of scanned information and may or may not produce a permanent record.

output The current, voltage, or power coming out of a component of an electronic system, usually by means of a connector.

overbuild The process by which a company installs a second cable system in an area where another company already has a cable system.

overnights Ratings that are available the day after they are taken.

owned A station having financial control and programming supplied by a network.

P

PAL Phase Alternate Line; a system that scans at 625 lines instead of 525, the American standard.

pay-cable A system for which subscribers pay to see commercialfree programming over a cable TV channel.

pay per view Charging the customer for a particular program watched.

pay-TV A method by which people pay in order to receive television programming free of commercials.

PBS Public Broadcasting Service.

pedestal A mechanism that holds a camera and allows it to be raised or lowered, usually by hydraulic or pneumatic means.

per-inquiry A selling practice whereby an advertiser pays a station an amount based on the number of people who respond to the ad.

petition to deny A process by which citizen groups express the desire for a license to be taken away from a station.

PGA Producers Guild of America.

phosphor A layer of material on the inner face of the TV tube that fluoresces when bombarded by electrons.

photosensitive surface An area capable of emitting electrons when it is hit by light.

pilot A film or tape of a single program of a proposed series that is prepared in order to obtain acceptance and commercial support.

play list A radio station's listing of the musical selections it has aired over a set period of time.

plumbicon An improved vidicon tube with a lead-oxide target that cuts down on image retention or lag.

potentiometer (pot) An electromechanical device that varies resistance in a circuit and, therefore, increases and decreases loudness.

preamplifier An amplifier that raises a low-level source so that its signal may be further processed.

preemption Temporarily removing a program from the air in order to broadcast a special event or another program.

pressure groups Organizations that try to change television organization or programming.

prime time The time of day when most people are tuned to broadcasting, generally driving time for radio and evening hours for TV.

prime-time access An FCC ruling declaring that stations should program their own material rather than network fare during one hour of prime time.

prism A transparent glass with shape and properties that allow it to refract white light into its "rainbow" of colors.

producer The person in charge of a particular program or series that oversees such things as the budget, personnel, and facilities.

profit and loss statement A bookkeeping form that shows the amount of income and expenses of a company and whether the company has made or lost money.

projected TV TV picture seen on a large screen.

promise versus performance A procedure whereby broadcasters promised what they would do during a license period and then were judged on the fulfillment of their promises when their license came up for renewal.

public access Programming conceived and produced by members of the public for cable TV channels.

Q

quadraphonic Sound reproduction using four channels through four separate speakers to blend separate sounds.

R

RADAR Radio All-Dimension Audience Research.

radio frequency The portion of the electromagnetic spectrum from about 30 kilohertz to 300,000 megahertz.

radio waves The waves of the radio frequency band of the electromagnetic spectrum.

random sampling A method of selection whereby each unit has the same chance of being selected as any other unit.

rate card A listing of the prices a station charges for advertisements at different times of the day and under different circumstances.

rating The percentage of households that are watching or listening to a particular program.

rate protection Guaranteeing a customer a certain advertising fee, even if the rate card increases.

RCA Radio Corporation of America.

recall An audience-measurement method by which people are asked what they have watched or heard in the past.

receiver The part of a communication system that converts electric waves into visible or audible form.

residuals Payments made to those involved in a production when the program is rerun.

resolution The degree to which detail can be distinguished.

retransmit To feed the output of a transmitter onto another receiver or transmitter.

run-of-schedule A selling practice that allows a station, rather than the customer, to decide exactly when the commercial will run.

S

SAG Screen Actors Guild.

sample A part that is representative of a whole.

satellite A human-made object that orbits the earth and can pick up and transmit radio signals.

saticon A high-resolution television camera tube.

scan To examine the density of an area point by point and convert that information into an electronic code that can later recreate the density.

scarcity theory Reasoning that broadcasters should be regulated because there are not enough station frequencies for everyone to have one.

season The period of time from the start of one block of programs to the start of another block of new and rescheduled programs.

Section 315 The portion of the Communications Act that states that political candidates running for the same office must be given equal treatment.

SEG Screen Extras Guild.

SESAC Society of European Stage Artists and Composers.

shader A person who adjusts remote controls for cameras in order to keep color and other electronic elements consistent.

share The percentage of households watching a particular program in relation to all programs available at that time.

short waves Radio waves at high frequencies.

simulcast To broadcast over two facilities at once, such as a concert that is broadcast over a TV station and a radio station at the same time.

siphoning A process whereby pay-TV systems might drain programming from networks by paying a higher price for it initially.

skycam A device from which a camera can be suspended above a site.

slander False statements harmful to a person's character or reputation.

slant track Another name for helical recording.

SMATV Satellite Master Antenna TV; a system that incorporates signals received by a satellite dish with broadcast signals received by master antenna, usually for an apartment complex.

SMPTE Society of Motion Picture and Television Engineers.

SNG Satellite news gathering; news gathered with trucks equipped with satellite uplinks.

software Program material.

special effects generator A piece of equipment often incorporated within the switcher that enables pictures to form wipes, stars, squares, cutouts, and other patterns and effects.

spectacular An early term for what is now known as a special.

spot A commercial inserted in or between programs.

stage manager Another name for floor manager.

static Noise caused by weather and electrical charges in the atmosphere.

station representative A company that sells time for a number of stations.

steady-cam A harness device that attaches around the waist to keep the camera steady.

stereophonic Sound reproduction using two channels through two separate speakers to give a feeling of reality.

storyboard A chart that contains step-by-step pictorialization of a commercial or program.

story line A basic idea of a plot in summary form.

strike To tear down and clean up a set.

stringer A person who gathers news information and is paid only for the material used.

strip Airing programs on a daily basis at the same time each day.

stylus A device that picks up a signal from a phonograph record.

subaudio Sounds that have a lower frequency than can be heard by the human ear.

subscription TV Scrambled programs broadcast over the air that can be descrambled when a subscriber pays a fee for the service.

superstation A broadcast station that is put on satellite and shown by cable systems.

sweeps Audience measurement reports that encompass the entire country.

switcher A piece of equipment used to select the TV picture going out over the air or going to a video tape recorder.

sync The precise matching of electron beams.

synchronous satellite A satellite that travels in orbit at such a rate that it appears to hang stationary above the earth.

syndicate To sell a radio or TV program outside the network structure to a number of different stations.

syndicated exclusivity A cable TV rule that stated if a local station was carrying a particular program, the cable system could not show that same program on a service imported from a distant location.

T

take To change the on-air television picture quickly.

target The portion of a TV camera tube that is scanned by the electron beam.

technical director The person who operates the switcher during production and oversees the technical crew.

telecommunications An umbrella term that covers broadcasting, electronic media, telephone, and computer technologies.

teleconferencing Renting satellite time to transmit video information from place to place.

teletext Words, numbers, and graphics placed on the scan retrace of a broadcast signal.

teletype A form of telegraph in which the striking of keys on a keyboard produces electrical impulses that cause corresponding keys on an instrument at a distant point to type.

testimonial A type of commercial in which a person states the value that a certain product has had for him or her.

tiering Charging cable subscribers different rates for different services.

time The broadcast space that a commercial occupies.

time-period measurement An audience-measurement calculation indicating the percentage of households that tune in a station during a certain period, such as a half-hour.

"toll" station A name for the type of programming WEAF initiated in 1922 that allowed anyone to broadcast a public message by paying a fee, similar to the way that one pays a toll to communicate a private message by telephone.

top 40 A radio station format that involves repeatedly playing the most popular forty songs.

traffic The department of a TV or radio station that handles the log and schedules commercials.

translator A low-powered TV transmitter usually used to send a signal into an area with poor reception.

transmitter A piece of equipment that generates and amplifies a carrier wave and modulates it with information that can be radiated into space.

transponder The part of a satellite that carries a particular program.

transverse quadraplex A type of video tape recorder that uses two-inch tape and places the signal on the tape vertically.

tripod A three-legged structure on which a camera can be placed.

TSA Total Survey Area.

tuner The portion of a receiver that can select frequencies.

turntable A device for spinning and playing records.

turret A round plate in front of a camera holding several lenses, each of which can be rotated into a "taking" position.

U

UHF Ultrahigh frequency; the area in the spectrum between 300 and 3,000 megahertz.

ultraviolet rays Electromagnetic radiations with wavelengths beyond the visible violet end of the spectrum in the vicinity of 10 megahertz.

underwrite To finance an undertaking such as a TV show.

unidirectional Picking up on one side; used to refer to microphones with only one live side.

unit manager The person responsible for the facilities for a particular show or shows.

universe A total number from which a sample is selected.

uplink A facility that can send a signal to a satellite for further distribution.

upstream A term to designate signals that go from a home to the central source of a cable TV company.

V

vacuum tube An electron tube evacuated of air to the extent that its electrical characteristics are unaffected by the remaining air.

VCR Videocassette recorder.

vertical blanking interval The retrace of the electron beam from the bottom to the top of the TV screen.

VHF Very high frequency; the area in the spectrum between 30 and 300 megahertz.

VHS Video home system.

videocassette A magnetic tape in a closed container that can either record or play back video programming when it is inserted into a videocassette tape recorder.

videodisc A device shaped somewhat like a phonograph record that contains video and audio information and can display this information on a TV screen connected to a videodisc player.

videotape A plastic iron oxide-coated tape of various widths for recording video and audio signals, as well as other technical code information.

vidicon A camera tube in which an electron beam scans the surface of a photoconductor.

viewfinder A monitor on a TV camera that allows the camera operator to see the picture that the camera is taking.

VOA Voice of America.

VTR Video tape recorder.

VU meter Volume unit meter; a device that measures the power level of an audio wave and, therefore, indicates the volume of the sound.

W

wavelength The distance between points of a corresponding phase in electromagnetic waves.

WGA Writers Guild of America.

window The period between the time a movie is shown in a theater and the time it is released for showing on other media such as pay-TV or network TV.

wireless Any apparatus that transmits messages by means of radio frequencies rather than devices connected to each other by wires.

wire recording An early form of magnetic recording that used wire rather than tape.

wire service An organization that supplies news to stations by means of teletype.

X

X rays Penetrating radiation similar to light with wavelengths in the vicinity of 10 megahertz.

zoom lens A lens with a variable focal length, which allows a TV camera to frame more or less of a scene without the camera's being moved.

Following are selected sources of information regarding telecommunications. For more complete references, see the notes for each chapter.

A

"After 50 Years, the Feeling is Still Mutual." *Broadcasting* (September 10, 1984): 43–50.

"After 10 Years of Satellites, the Sky's No Limit." *Broadcasting* (April 9, 1984): 43–68.

Alten, Stanley R. *Audio in Media.* Belmont, Calif.: Wadsworth, 1981.

Ang, Ien. *Watching Dallas: Soap Opera and Melodramatic Imagination.* New York: Methuen, 1985.

Arlen, Michael J. *Living Room War.* New York: Viking, 1969.

Avery, Robert K., and Robert Pepper. "An Institutional History of Public Broadcasting." *Journal of Communication* (Summer 1980): 126–38.

B

Baldwin, Thomas E., and D. Stevens McVoy. *Cable Communication.* Englewood Cliffs, N.J.: Prentice-Hall, 1983.

Barber, Susanna. *News Cameras in the Courtroom: A Free Press-Fair Trial Debate.* Norwood, New Jersey: Ablex, 1987.

Barnouw, Erik. *The Golden Web: A History of Broadcasting in the United States, 1933–1953.* New York: Oxford University Press, 1968.

———. *The Image Empire: A History of Broadcasting in the United States from 1953.* New York: Oxford University Press, 1970.

———. *A Tower in Babel: A History of Broadcasting in the United States to 1933.* New York: Oxford University Press, 1966.

———. *Tube of Plenty: The Development of American Television.* New York: Oxford University Press, 1975.

Barr, David Samuel. *Advertising in Cable: A Practical Guide for Advertisers.* Englewood Cliffs, New Jersey: Prentice-Hall, 1985.

Beech, Linda. *TV Favorites.* New York: Scholastic Book Services, 1971.

Bensman, Marvin R. *Broadcast Regulation: Selected Cases and Decisions.* Lanham, MD: University Press of America, 1985.

Bernstein, Peter W. "The Race to Feed Cable TV's Maw." *Fortune* (May 4, 1981): 308–18.

Bilby, Kenneth. *The General: David Sarnoff and the Rise of the Communications Industry.* New York: Harper and Row, 1986.

Bittner, John R. *Broadcasting and Telecommunication.* Englewood Cliffs, N.J.: Prentice-Hall, 1985.

Black, Norman. "The Deregulation Revolution." *Channels* (September-October 1984): 52–59.

Broadcasting/Cable Yearbook, 1987. Washington, D.C.: Broadcasting Publications, Inc., 1987.

Brooks, Tim, and Earle Marsh. *The Complete Directory to Prime Time Network Shows: 1946–Present.* New York: Ballantine Books, 1985.

Browne, Donald R. "The International Newsroom: A Study of Practices at the Voice of America, BBC, and Deutsche Welle." *Journal of Broadcasting* (Summer 1983): 205–31.

Burrows, Tom, and Don Wood. *Television Production: Disciplines and Techniques.* Dubuque, Iowa: William C. Brown Publishers, 1986.

Busterna, John C. "Division of Ownership as a Criterion in FCC Licensing since 1956." *Journal of Broadcasting* (Winter 1976): 101–10.

Buxton, Frank, and Bill Owen. *The Big Broadcast, 1920–1950.* New York: Viking, 1972.

C

"Cable TV's Costly Trip to the Big Cities." *Fortune* (April 18, 1983): 82–87.

Carlson, Margaret B. "Where MGM, the NCAA, and Jerry Falwell Fight for Cash." *Fortune* (January 23, 1984): 169–71.

Carnegie Commission on Public Television. *Public Television: A Program for Action.* New York: Harper and Row, 1967.

BIBLIOGRAPHY

Carnegie Commission on the Future of Public Broadcasting. *A Public Trust.* New York: Carnegie Commission, 1979.

Comstock, George, et al. *Television and Social Behavior: A Technical Report to the Surgeon General's Scientific Advisory Committee on Televisions and Social Behavior.* Washington, D.C.: Government Printing Office, 1972.

Cook, Bruce. "High Tech: The New Videocassettes." *Emmy* (Summer 1980): 40–44.

Cooper, Robert B., Jr. "Home Reception Using Backyard Satellite TV Receivers." *Radio-Electronics* (January 1980): 57.

Correll, Charles J., and Freeman F. Gosden. *All About Amos 'n Andy.* New York: Rand McNally, 1929.

Culley, James D., William Lazer, and Charles K. Atkin. "The Experts Look at Children's Television." *Journal of Broadcasting* (Winter 1976): 3–22.

D

"Defamation and the First Amendment." *William and Mary Law Review* (1984): special issue.

DeForest, Lee. *Father of Radio: Autobiography of Lee De Forest.* Chicago: Wilcox and Follett, 1950.

Diamant, Lincoln. *Television's Classic Commercials.* New York: Hastings House, 1971.

Dunlap, Orrin E., Jr. *Radio's 100 Men of Science.* New York: Harper and Brothers, 1944.

E

Eastman, Susan Tyler, Sydney W. Head, and Lewis Klein. *Broadcast/ Cable Programming.* Belmont, Calif.: Wadsworth, 1985.

Estep, Rhoda, and Patrick T. Macdonald. "Crime in the Afternoon: Murder and Robbery in Soap Operas." *Journal of Broadcasting and Electronic Media* (Summer 1985): 323–31.

F

Feshbach, Seymour, and Robert D. Singer. *Television and Aggression: An Experimental Field Study.* San Francisco: Jossey-Bass, 1970.

Fessenden, Helen M. *Fessenden: Builder of Tomorrows.* New York: Coward-McCann, 1940.

"The First Amendment and the Fifth Estate." *Broadcasting* (January 5, 1976): 45–101.

"The First 60 Years of NBC." *Broadcasting* (June 9, 1986): 49–64.

Friendly, Fred. *Due to Circumstances Beyond Our Control.* New York: Random House, 1967.

"From Fighting Bob to the Fairness Doctrine." *Broadcasting* (January 5, 1976): 46.

G

Garvey, Daniel E. "Introducing Color Television: The Audience and Programming Problems." *Journal of Broadcasting* (Fall 1980): 515–26.

Gerbner, George. "Science or Ritual Dance? A Revisionist View of Television Violence Effects Research." *Journal of Communication* (Summer 1984): 164–73.

Glick, Edwin L. "The Life and Death of the Liberty Broadcasting System." *Journal of Broadcasting* (Spring 1979): 117–36.

Gross, Lynne Schafer. *The New Television Technologies.* Dubuque, Iowa: William C. Brown Publishers, 1986.

H

Harmon, Jim. *The Great Radio Comedians.* Garden City, N.J.: Doubleday, 1970.

Hasling, John. *Fundamentals of Radio Broadcasting.* New York: McGraw-Hill, 1980.

Hazlett, Thomas W. "The Policy of Exclusive Franchising in Cable Television." *Journal of Broadcasting and Electronic Media* (Winter 1987): 1–20.

Head, Sydney. *World Broadcasting Systems.* Belmont, CA: Wadsworth Publishing Company, 1985.

Head, Sydney, and Christopher H. Sterling. *Broadcasting in America.* Boston: Houghton Mifflin, 1982.

Heighton, Elizabeth J., and Don. R. Cunningham. *Advertising in the Broadcast and Cable Media.* Belmont, Calif.: Wadsworth, 1984.

Hofer, Stephen F. "Philo Farnsworth: Television Pioneer." *Journal of Broadcasting* (Spring 1979): 153–66.

Horwitz, Robert Britt. "For Whom the Bell Tolls: Causes and Consequences of the AT&T Divestiture." *Critical Studies in Mass Communication* (June 1986): 119–53.

I

"Independents: State of the Art, 1985." *Broadcasting* (January 7, 1985): 69–90.

K

Kalter, Suzy. *The Complete Book of M*A*S*H.* New York: Abrams, 1984.

Kerr, Gregory. "Lights, Camera. . .Congress!" *Emmy* (May/June 1987): 41–47.

L

Lazarfield, Paul F., and Frank N. Stanton. *Radio Research.* New York: Duell, Sloan, and Pearce, 1942.

Leigh, Frederic A. "Effects of FCC Class D Rules Changes." *Feedback* (Fall 1984): 18–21.

LeRoy, David J. "Who Watches Public Television?" *Journal of Communication* (Summer 1980): 189–97.

Lessing, Edwin Lawrence. *Man of High Fidelity: Edwin Howard Armstrong.* New York: Lippincott, 1956.

Levin, Murray B. *Talk Radio and the American Dream.* Lexington, MA: Lexington Books, 1987.

Link, Tom. "Saving What We've Seen." *Emmy* (Winter 1980): 39–41.

"LPTV." *Broadcasting* (February 23, 1981): 39–66.

M

McCavitt, William E. *Broadcasting Around the World.* Blue Ridge Summit, Pa.: TAB Books, 1981.

Macy, John W. *To Irrigate a Wasteland: The Struggle to Shape a Public Television System in the United States.* Berkley: University of California Press, 1974.

Marconi, Degna. *My Father Marconi.* New York: McGraw-Hill, 1962.

Meadow, Charles T., and Albert S. Tedesco. *Telecommunications for Management.* New York: McGraw-Hill, 1985.

Metz, Robert. *CBS: Reflections in a Bloodshot Eye.* Chicago: Playboy Press, 1975.

Mickiewitz, Ellen Propper. *Media and the Russian Public.* New York: Preager, 1981.

Minow, Newton, John Bartlow Martin, and Lee M. Mitchell. *Presidential Television.* New York: Basic Books, 1973.

Mullally, Donald P. "Radio: The Other Public Medium." *Journal of Communication* (Summer 1980): 189–97.

N

Norback, Craig T., and Peter G. Norback, eds. *TV Guide Almanac.* New York: Ballantine Books, 1980.

P

Pepper, Robert M. *The Formation of the Public Broadcasting Service.* New York: Arno Press, 1979.

Perse, Elizabeth M. "Soap Opera Viewing Patterns of College Students and Cultivation." *Journal of Broadcasting and Electronic Media* (Spring 1986): 175–93.

Peyser, Tony. "Pat Weaver: Visionary or Dilettante?" *Emmy* (Fall 1979): 32–34.

Philips, Mary Alice Mayer. *CATV: A History of Community Antenna Television.* Evanston, Ill.: Northwestern University Press, 1972.

Podmore, Christopher, and Denise Faguy. "The Challenge of Optical Fibres." *Telecommunications Policy* (December 1986): 341–51.

Prentice, Stan. *Television: From Analog to Digital.* Blue Ridge Summit, PA: 1985.

Public Attitudes Toward Television and Other Media in a Time of Change. New York: Television Information Office, 1987.

Public Law No. 416, June 19, 1934, 73rd Congress. Washington, D.C.: Government Printing Office, 1934.

Q

Quinlan, Sterling. *Inside ABC: American Broadcasting Company's Rise to Power.* New York: Hastings House, 1979.

R

Reeves, Michael G. and Tom W. Hoffer. "The Safe, Cheap, and Known: A Content Analysis of the First (1974) PBS Program Cooperative." *Journal of Broadcasting* (Fall 1976): 549–66.

Robertson, James, and Gerald G. Yokom. "Educational Radio: The Fifty-Year-Old Adolescent." *Educational Broadcasting Review* (April 1973): 107–15.

"Rountable: Research and Ratings." *Emmy* (Spring 1981): 13–16.

Rubenstein, Eli A., George A. Comstock, and John P. Murry, eds. *Television and Social Behavior. Vol. 4. Television in Day-To-Day Life.* Washington, D.C.: U.S. Government Printing Office, 1972.

Rubin, Rebecca B., and Alan M. Rubin. *Communications Research: Strategies and Sources.* Belmont, Calif.: Wadsworth Publishing Company, 1986.

S

Sarnoff, David. *Looking Ahead: The Papers of David Sarnoff.* New York: McGraw-Hill, 1968.

Scher, Saul N. "Anthology Drama: TV's Inconsistent Art Form." *Television Quarterly* (Winter 1976–77): 29–34.

Schoenbrun, David. "Is Perfect Fairness Possible?" *Television Quarterly* (Special Election Issue 1976): 77–79.

Seligman, Daniel. "Life Will Be Different When We're All On-Line." *Fortune* (February 4, 1985): 68–71.

Settle, Irving. *A Pictorial History of Radio.* New York: Grossett and Dunlap, 1967.

Severin, Werner J. "Commercial vs. Non-Commercial Radio During Broadcasting's Early Days." *Journal of Broadcasting* (Summer 1981): 295–302.

Shanks, Bob. *The Cool Fire.* New York: Norton, 1976.

Shulman, Arthur, and Roger Youman. *The Television Years.* New York: Popular Library, 1973.

Singleton, Lox A. *Telecommunications in the Information Age.* Cambridge, MA: Ballinger Publishing Company, 1986.

Sklar, Robert. *Prime-Time America.* Oxford: Oxford University Press, 1980.

Stamberg, Susan. *Every Night at Five: Susan Stamberg's "All Things Considered."* New York: Pantheon, 1982.

"STV: Scratching Out Its Place in the New-Video Universe." *Broadcasting* (April 7, 1980): 46–62.

V

"Videoconference Use Expands to Meet Rising Business Needs." *Aviation Week and Space Technology* (July 22, 1985): 157–64.

W

Warner, Charles. *Broadcast and Cable Selling.* Belmont, CA: Wadsworth Publishing Company, 1984.

White, Paul. *News on the Air.* New York: Harcourt Brace, 1947.

Whitehouse, George E. *Understanding the New Technologies of the Mass Media.* Englewood Cliffs, New Jersey: Prentice-Hall, 1986.

Y

Yoakam, Richard D., and Charles F. Cremer. *ENG: Television News and the New Technology.* New York: Random House, 1985.

Z

Zacharis, John C. *Exploring Careers in Communications and Telecommunications.* New York: Rosen Publishing Group, 1985.

Zacks, Richard. "Picture Window." *Channels* (May 1986): 40–41.

Zettl, Herbert. *Television Production Handbook.* Belmont, Calif.: Wadsworth Publishing Company, 1984.

INDEX

_A_

above-the-line, 297
Academy of Television Arts and
    Sciences, 318, 319
accounting, 401
ACE Awards, 319
Action for Children's Television
    (ACT), 254–55, 321, 366
"actual malice," 334
Adult Learning Service, 126
Advanced Compatible Television
    (ACTV), 220
advertising, 347–55. _See also_
    commercials
    agencies, 50, 81, 86–87, 353–54,
      404
    associations, 318
    on the BBC, 436, 440
    and blacklisting, 80
    and broadcast standards, 320
    on cable TV, 142, 144, 147, 301
    in Canada, 443
    on Channel 4, 439
    of cigarettes, 95, 315, 365
    and codes, 317
    complaints against, 363–66
    during the depression, 44, 48–50
    on drama, 238
    on early radio, 41
    on early TV, 81
    effects of, 20–21
    in Europe, 442
    and the FCC, 328
    on foreign systems, 432
    during the freeze, 75
    FTC involvement in, 313–14, 323
    Hoover's opinion of, 40
    and LPTV, 169
    on Mutual, 43
    and news, 56
    political, 279
    postwar, 61
    profit of, 20–21
    and ratings, 374, 386, 388
    rise of, 39–40
    and Section 315, 339
    self-regulation, 317
    and sports, 269
    on teletext, 170, 440
    during World War II, 60
Advertising Research Foundation, 318
Advisory Opinions, 314
affiliate relations representatives, 401
affiliates
    cable TV, 301
    and the FCC, 312
    and Paley, 41
    public television, 300
    radio, 286–87

and spots, 352
    television, 293–95
"Africans, The," 123–24
AGB, 377
agents, 406–7
Agnew, Spiro, 96, 321
AIDS, 365
Alabama Educational Television
    Commission, 329
Alda, Alan, 242
Aldrich Family, 46
Alexanderson, Ernst F. W., 33, 38, 70
Ali, Muhammad, 270
Ali-Frazier fight, 139
all-channel receiver bill, 88
Allen, Gracie, 46
Allen, Steve, 81, 94
All-Industry Music License
    Committee, 235–36
"All in the Family," 97, 243
Allison, Fran, 76, 253
All Radio Methodology Study
    (ARMS), 390
"All Things Considered," 110
Alpha Epsilon Rho, 318
alternating current (AC), 29
alternator, 33
AM, 212–18
    allocations of, 78, 310
    bandwidth, 215–16
    early stations, 44
    and the ITU, 447
    music on, 237
    stereo, 65, 215
American Advertising Federation, 318
American Association of Advertising
    Agencies, 318
"American Bandstand," 85, 237
American Bar Association (ABA), 335
American Bell Telephone Company,
    172
American Broadcasting Company
    (ABC)
    after-school specials, 255
    and Betacam, 202
    bought by Capital Cities, 99
    cable channels, 145, 150
    and color TV, 84
    and documentaries, 89
    and ESPN, 151
    and fiber optics, 224
    and films, 85, 94, 247
    founding of, 42
    and libel, 334
    and LPTV, 168
    and news, 260–61
    radio networks, 64–65, 263, 322
    and sports, 269
    and violence, 241
American Express, 143. _See also_
    Warner-Amex